Probability and Its Applications

Published in association with the Applied Probability Trust

Probability and Its Applications

The *Probability and Its Applications* series publishes research monographs, with the expository quality to make them useful and accessible to advanced students, in probability and stochastic processes, with a particular focus on:

- Foundations of probability including stochastic analysis and Markov and other stochastic processes
- Applications of probability in analysis
- Point processes, random sets, and other spatial models
- Branching processes and other models of population growth
- Genetics and other stochastic models in biology
- Information theory and signal processing
- Communication networks
- Stochastic models in operations research

For further volumes:
www.springer.com/series/1560

Pierre Collet • Servet Martínez • Jaime San Martín

Quasi-Stationary Distributions

Markov Chains, Diffusions and Dynamical Systems

Springer

Pierre Collet
Centre de Physique Théorique, UMR7644
CNRS, École Polytechnique
Palaiseau, France

Servet Martínez
CMM-DIM, Faculty of Mathematical
and Physical Sciences
University of Chile
Santiago, Chile

Jaime San Martín
CMM-DIM, Faculty of Mathematical
and Physical Sciences
University of Chile
Santiago, Chile

Series Editors:
Søren Asmussen
Department of Mathematical Sciences
Aarhus University
Aarhus, Denmark

Joe Gani
Centre for Mathematics and its Applications
Mathematical Sciences Institute
Australian National University
Canberra, Australia

Peter Jagers
Mathematical Statistics
University of Gothenburg
Göteborg, Sweden

Thomas G. Kurtz
Department of Mathematics
University of Wisconsin-Madison
Madison, USA

ISSN 1431-7028 Probability and Its Applications
ISBN 978-3-642-33130-5 ISBN 978-3-642-33131-2 (eBook)
DOI 10.1007/978-3-642-33131-2
Springer Heidelberg New York Dordrecht London

Library of Congress Control Number: 2012951442

Mathematics Subject Classification: 60J70, 60JXX, 60J10, 60J60, 60J65, 92Dxx, 37DXX, 35PXX

Printed on acid-free paper

Springer is part of Springer Science+Business Media (www.springer.com)

This book is dedicated to all surviving small communities and the endangered species.

The authors P.C., S.M. and J.S.M. are indebted to Sonia Lund, María Angélica Salazar and Valeska Amador for their patience, support and encouragements along all these years.

Preface

This book is devoted to the study of the main concepts on survival for killed Markov processes and dynamical systems. The first computations on the number of survivals after a long time in branching process was done by Kolmogorov (1938). Later on Yaglom (1947) showed that the limit behavior of sub-critical branching processes conditioned to survival was given by a proper distribution. This pioneer work triggered an important activity in this field.

Quasi-stationary distributions (QSD) capture the long term behavior of a process that will be surely killed when this process is conditioned to survive. A basic and useful property is that the time of killing is exponentially distributed when starting from a QSD, which implies that the rate of survival of a process must be at most exponential in order that QSD can exist.

The study of QSD on finite state irreducible Markov chains started with the pioneering work Darroch and Seneta (1965). For these processes many of the fundamental ideas in the topic can be easily developed. In particular, the Perron–Frobenius theory for finite positive matrices gives all the required information. Thus, the exponential rate of survival is the Perron–Frobenius eigenvalue. The QSD is the Perron–Frobenius normalized eigenmeasure, and the chain of trajectories that are never killed (Q-process) is governed by an h-process where h is the Perron–Frobenius eigenfunction. This relation between a QSD and the kernel of the Q-process is also encountered in many other processes. In the finite case, the fact that the eigenvalues are isolated simplifies the study. Since the dominant eigenvalue is simple, the QSD attracts all the conditioned measures, and is the quasi-limiting distribution, or the Yaglom limit. When the transition kernel is symmetric, the spectral decomposition contains all the data that allow one to understand the survival phenomenon of the chain.

For general countable state Markov chains, this study started in Seneta and Vere-Jones (1996). We give a detailed proof of the characterization of QSDs as a finite mass eigenmeasure due to Pollett, and we give a more general result on the existence of QSDs, which states that if there is no entrance from infinity then exponential survival is a necessary and sufficient condition to have a QSD. We follow the original proof done in Ferrari et al. (1995a). We study also in details the symmetric chains and monotone chains. We also give conditions in order that the exponential decay rate be equal to the Kingman decay parameter of the transition probability.

An important literature has been devoted to time-continuous birth-and-death chains in the last 25 years, many of these works take their basis in the spectral representation given in McGregor (1957a, 1957b). We revisit some of these results. An important problem appears because under exponential survival, the set of QSDs can be a continuum or a singleton. We give the criterion, due to Van Doorn (1991), to identify which case occurs. This is done in terms of the parameters of the chain. When there is a continuum of QSDs one can identify the Yaglom limit, it corresponds to the minimal QSD This result is due to Good (1968). On the other hand, the spectrum can be continuous, and the fact that, in general, there is no spectral gap raises a delicate technical problem. But a useful point is that the bottom of the spectrum is the exponential rate of survival. For birth-and-death chains, we derive the classification of the associated Q-process (as transient, recurrent or positive recurrent) by studying the exponential asymptotic rate function of the survival probability when the chain is restricted to be above some barrier. For some particular birth-and-death chains, we identify explicitly the QSD and other significant parameters and properties.

We present some results for one dimensional diffusions on the half-line killed at the origin. We study the problem of QSD for diffusions that have nice behavior at 0 and at ∞. In this framework, we also study the Q-process and its recurrence classification in relation with spectral properties of the generator of the initial diffusion. We also consider one dimensional diffusions that present singularities at 0 and/or at ∞, and we study for them the problems of existence and properties of QSD. This generalization is motivated by some models in mathematical ecology. These chapters on one dimensional diffusions find their source of inspiration in the work of Mandl (1961). The main tools come from the spectral theory of the Sturm–Liouville operators. The study of QSD in population dynamics is a very active topic nowadays, we include some of the results in the chapters on birth-and-death chains and diffusions.

We also consider the case of dynamical systems (with discrete time evolution) with a trap in the phase space. A QSD in this context deals with trajectories which do not fall into the trap. There are many analogies with the stochastic case concerning the questions, the results and their proofs. The QSDs for dynamical systems have been first studied in the context of expanding systems in Pianigiani and Yorke (1979). We discuss in particular Gibbs QSDs for symbolic systems and absolutely continuous QSDs for repellers. For both cases, we use a similar technique of proof based on quasi-compact operators analogous to those used in the thermodynamic formalism.

One of the main objects in our study is the associated Q-process, that is, the process having trajectories that survive forever, as well as its relations with the QSD and the spectral properties. For a branching process, the Q-process was introduced by Spitzer (unpublished) and in Lamperti and Ney (1968). In this book, we study the Q-process and describe it as an h-process with respect to the eigenfunction h having the same eigenvalue as the minimal QSD. This is done for Markov chains, one-dimensional diffusions and expanding dynamical systems, giving a unified approach. Thus, it appears that the Gibbs invariant measure for topological Markov chains described by Bowen–Ruelle–Sinai can be interpreted as the stationary measure of a Q-process associated to a process that is killed in a region having the Pianigiani–Yorke as its QSD.

The problem of existence of a QSD is in a sense more difficult than the problem of proving the existence of an invariant measure since one has to determine also the rate of decay of the probability of survival. This also implies that the equation for the QSD is nonlinear.

Several techniques have been used in the literature to prove existence and some of them will be explained in details in the present book. Without trying to be exhaustive, one can mention the abstract techniques relying on the Tychonov fixed point theorem (see Tychonov 1935), Krein's lemma (Oikhberg and Troitsky 2005), and the Krein–Rutman Theorem (Dunford and Schwartz 1958). Birkhoff's Theorem on Hilbert's projective metric can also be useful (see Eveson and Nussbaum 1995). Many results are based on spectral techniques, and in particular on various extensions of the Perron–Frobenius Theorem like in the theory of the Ruelle–Perron–Frobenius operator Ruelle (1978), Bowen (1975), the spectral theory of quasi-compact positive operators (see, for example, Ionescu Tulcea and Marinescu 1950 and Nussbaum 1970), and the uniform ergodic theorem (Yosida and Kakutani 1941).

The literature on QSDs is quite large and covers many fundamental and applied domains. This book does not pretend to cover them all. We have not included the developments of QSDs in all directions, for instance, we do not treat the case of branching processes where several seminal results can be found in the book Athreya and Ney (1972) and where there is a lot of active research nowadays. We have selected some of the topics which either have called the attention of specialists, or where the authors of this book have made some contributions.

This book is not intended to be a complete exposition of the theory of QSDs. It mostly reflects the (present) interest of the authors in some parts of this vast field.

In Chap. 2 dealing with general aspects of QSDs, we have mainly gathered some of our own research. We also point out that only Sects. 2.1, 2.2, 2.3 are needed to read independently the other chapters.

Paris, France
Santiago, Chile

Pierre Collet
Servet Martínez
Jaime San Martín

Acknowledgements

The authors acknowledge their institutions for the facilities given to write this book: the Center of Theoretical Physics at the Ecole Polytechnique, the Center of Mathematical Modeling and the Department of Engineering Mathematics of the Universidad de Chile, the first two being units of the CNRS. We have had the support of various institutions and programs: Fondap, Conicyt, Fondecyt, Presidential Chair, Nucleus Millennium, Basal program, Franco–Chilean cooperation, Guggenheim fellowship, Chair MMB, MathAmsud program.

Besides the common work among the authors, the origins of the book can be found at the end of the 1980s in collaborations with Pablo Ferrari and Pierre Picco, at University Sao Paulo and CPT Marseille, where we first learned on QSD and collaborated with Harry Kesten from Cornell University. Later on the collaboration with Bernard Schmitt and Antonio Galves from University of Bourgogne and University of Sao Paulo triggered our interest in the case of dynamical systems. We are deeply indebted to all of them and also to the collaboration with Patrick Cattiaux, Andrew Hart, Amaury Lambert, Véronique Maume-Deschamps, Maria Eulalia Vares and Bernard Ycart. Finally, we acknowledge Sylvie Méléard from CMAP at Ecole Polytechnique for our collaboration on QSD in population dynamics.

Contents

Chapter 1
Introduction

1.1 Quasi-Stationary Distributions

In the framework of this theory, there is a Markov process evolving in a domain where there is a set of forbidden states that constitutes a trap. The process is said to be killed when it hits the trap and it is assumed that this happens almost surely. We investigate the behavior of the process before being killed, more precisely we study what happens when one conditions the process to survive for a long time.

Let us fix some notation. The state space is $\mathcal{X}$ endowed with a measurable structure. We consider a Markov process $Y = (Y_t : t \geq 0)$ taking values in $\mathcal{X}$ and $(\mathbb{P}_x : x \in \mathcal{X})$ is the family of distributions, $\mathbb{P}_x$, with the initial condition $x \in \mathcal{X}$. As said, there is a measurable set of forbidden states which we denote by $\partial\mathcal{X}$ (sometimes we will denote this set by $\mathcal{X}^{\text{tr}}$ because it does not necessarily coincides with the boundary of $\mathcal{X}$). Its complement $\mathcal{X}^{\mathbf{a}} := \mathcal{X} \setminus \partial\mathcal{X}$ is the set of allowed states. Let $T = T_{\partial\mathcal{X}}$ be the hitting time of $\partial\mathcal{X}$. We will assume that there is sure killing at $\partial\mathcal{X}$, so $\mathbb{P}_x(T < \infty) = 1$ for all $x \in \mathcal{X}^{\mathbf{a}}$. We will be mainly concerned with the trajectory before extinction $(Y_t : t < T)$, so without loss of generality, we can assume $Y_t = Y_T$ when $t \geq T$, that is, the set of forbidden states is absorbing, or a trap. A sketch of various trajectories is shown in Fig. 1.1.

Among the main concepts in our study are the quasi-stationary distributions (QSDs, for short), these are distributions that are invariant under time evolution when the process is conditioned to survive. More precisely, a probability measure ν on $\mathcal{X}^{\mathbf{a}}$ is called a QSD (for the process killed at $\partial\mathcal{X}$) if for every measurable set B contained in $\mathcal{X}^{\mathbf{a}}$,

$$\mathbb{P}_\nu(Y_t \in B | T > t) = \nu(B), \quad t \geq 0.$$

Here, as usual, $\mathbb{P}_\nu = \int_{\mathcal{X}^{\mathbf{a}}} \mathbb{P}_x \, d\nu(x)$.

The existence of a QSD for killed processes, its description, the convergence to it of conditioned processes, its role in the process conditioned to never be killed, and the behavior of the killing time constitute the core of this work. On the other hand, since only few results exist for general processes, we prefer to introduce these problems in concrete contexts, as discrete Markov chains and dynamical systems.

P. Collet et al., *Quasi-Stationary Distributions*, Probability and Its Applications,
DOI 10.1007/978-3-642-33131-2_1,

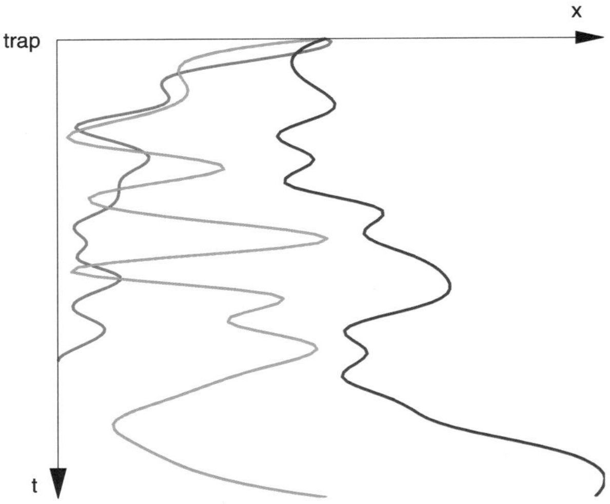

Fig. 1.1 Trajectories killed at the trap $x = 0$

1.2 Markov Chains

The study of QSDs on Markov chains have been extensively developed since the pioneering work (Yaglom, 1947) on branching process, and the classification of killed processes introduced in Vere-Jones (1962). The description of QSDs for finite state Markov chains was done in Darroch and Seneta (1965).

Below we describe some of the main problems of the theory on long time survival and QSDs on Markov chains. To avoid technical difficulties in this section, we prefer to introduce the ideas for discrete time Markov chains. But, in the next chapters, except the one devoted to QSDs of dynamical systems, we will only consider continuous time because this is the case of the major part of the bibliography devoted to QSDs of Markov processes. Some of the questions we set will be answered in the following chapters where we will also give precise references.

So, let us assume $\mathcal{X}$ is countable and $Y = (Y_t : t \in \mathbb{Z}_+ = \{0, 1, \ldots\})$ is a discrete time Markov chain taking values in $\mathcal{X}$. We assume that $(Y_t : t < T)$ is irreducible on $\mathcal{X}^{\mathbf{a}}$ and that the process is surely killed, $\mathbb{P}_x(T < \infty) = 1$ for $x \in \mathcal{X}^{\mathbf{a}}$. In this context, a QSD is a probability measure ν on $\mathcal{X}^{\mathbf{a}}$ that satisfies

$$\mathbb{P}_\nu(Y_t = x | T > t) = \nu(x) \quad \forall x \in \mathcal{X}^{\mathbf{a}}, t \in \mathbb{Z}_+.$$

From the irreducibility condition, one obtains that every QSD ν charges all points in $\mathcal{X}^{\mathbf{a}}$. On the other hand, since Y is absorbed at time T, the condition of a QSD takes the form

$$\mathbb{P}_\nu(Y_t = x) = \nu(x)\mathbb{P}_\nu(T > t) \quad \forall x \in \mathcal{X}^{\mathbf{a}}, t \in \mathbb{Z}_+.$$

Let ν be a QSD. We proceed to prove that when starting from ν, the killing time T at $\partial\mathcal{X}$ is geometrically distributed, that is,

$$\exists \alpha = \alpha(\nu) \in [0, 1] \quad \text{such that} \quad \mathbb{P}_\nu(T > t) = \alpha^t \quad \forall t \in \mathbb{Z}_+.$$

Since ν is a QSD, the Markov property gives the following equalities for all $t, s \geq 0$:

$$\begin{aligned}\mathbb{P}_\nu(T > t+s) &= \sum_{x\in\mathcal{X}^{\mathbf{a}}} \mathbb{P}_\nu(T > t+s, Y_s = x)\\ &= \sum_{x\in\mathcal{X}^{\mathbf{a}}} \mathbb{P}_\nu(T > t+s | Y_s = x)\mathbb{P}_\nu(Y_s = x)\\ &= \sum_{x\in\mathcal{X}^{\mathbf{a}}} \mathbb{P}_x(T > t)\nu(Y_s = x)\mathbb{P}_\nu(T > s) = \mathbb{P}_\nu(T > s)\mathbb{P}_\nu(T > t).\end{aligned}$$

Then, the geometrical distribution of T starting from ν follows. We avoid trivial situations, and so $\alpha(\nu) \in (0, 1)$. We denote by $\theta(\nu) := -\log\alpha(\nu) \in (0, \infty)$ the exponential rate of survival of ν, it verifies $\mathbb{P}_\nu(T > t) = e^{-\theta(\nu)t}$ for all $t \in \mathbb{Z}_+$. Furthermore, when ν is a QSD, we get

$$\forall \theta < \theta(\nu): \quad \mathbb{E}_\nu\left(e^{\theta T}\right) < \infty.$$

Hence, a necessary condition for the existence of a QSD is that some exponential moments be finite:

$$\forall x \in \mathcal{X}^{\mathbf{a}} \exists \theta > 0: \quad \mathbb{E}_x\left(e^{\theta T}\right) < \infty.$$

A solidarity argument, based upon irreducibility, shows that the quantity

$$\theta^* = \sup\left\{\theta : \mathbb{E}_x\left(e^{\theta T}\right) < \infty\right\}$$

does not depend on $x \in \mathcal{X}^{\mathbf{a}}$, we call it the exponential rate of survival of the process Y (killed at $\mathcal{X}^{\mathbf{a}}$). The above discussion shows that $\theta^* > 0$ is a necessary condition for the existence of a QSD. When this condition holds, we say that the process Y is exponentially killed.

Let $P = (p(x, y) : x, y \in \mathcal{X})$ be the transition matrix of Y, $P_{\mathbf{a}} = (p(x, y) : x, y \in \mathcal{X}^{\mathbf{a}})$ be the transition kernel restricted to the allowed states, and denote by ${P_{\mathbf{a}}}^{(t)} = (p^{(t)}(x, y) : x, y \in \mathcal{X}^{\mathbf{a}})$ the tth power of $P_{\mathbf{a}}$ for $t \in \mathbb{Z}_+$.

Let ν be a probability distribution on $\mathcal{X}^{\mathbf{a}}$, we denote by $\nu' = (\nu(x) : x \in \mathcal{X}^{\mathbf{a}})$ the associated row probability vector indexed by $\mathcal{X}^{\mathbf{a}}$. We have $\mathbb{P}_\nu(Y_t = y) = \nu' {P_{\mathbf{a}}}^{(t)}(y)$. In the case ν is a QSD, there exists an $\alpha \in (0, 1)$ such that $\mathbb{P}_\nu(T > t) = \alpha^t$ for all $t \in \mathbb{Z}_+$. So, if ν is a QSD, ν' must verify

$$\forall t \in \mathbb{Z}_+: \quad \nu' {P_{\mathbf{a}}}^{(t)} = \alpha^t \nu'.$$

This is equivalent to $\nu' P_{\mathbf{a}} = \alpha\nu'$ and so ν' must be a left-eigenvector of $P_{\mathbf{a}}$ with the eigenvalue $\alpha \in (0, 1)$. Observe that the eigenvalue is related to ν' by the equality $\alpha = \nu' P_{\mathbf{a}}\mathbf{1} = \mathbb{P}_\nu(T > 1)$, where $\mathbf{1}$ is the unit vector (all its components are 1). Note that the equation of a QSD is nonlinear in ν because α also depends on ν.

Let $\mathcal{X}^{\mathbf{a}}$ be finite; the sure killing condition is equivalent to $P_{\mathbf{a}}\mathbf{1} \leq \mathbf{1}$ and at some $x_0 \in \mathcal{X}^{\mathbf{a}}$ there is a loss of mass, that is, $\sum_{y\in\mathcal{X}^{\mathbf{a}}} p(x_0, y) < 1$. In this case, the existence and uniqueness of a QSD follow straightforwardly from the Perron–Frobenius theory. Indeed, since $P_{\mathbf{a}}$ is a positive, irreducible, strictly substochastic matrix, this theory ensures the existence of a simple eigenvalue $\alpha \in (0,1)$ of $P_{\mathbf{a}}$ that has left- and right-eigenvectors ν and ψ, respectively, that can be chosen to be strictly positive. Moreover, the associated eigenvectors of any other eigenvalue cannot be positive. We can impose the following two normalizing conditions:

$$\nu'\psi = 1, \qquad \nu'\mathbf{1} = \mathbf{1},$$

where $\nu'\psi = \sum_{x\in\mathcal{X}^{\mathbf{a}}} \psi(x)\nu(x)$. In particular, ν is a probability vector. Let us see that ν is a QSD. From $\nu'\mathbf{1} = 1$, we get

$$\mathbb{P}_\nu(T > t) = \nu' P^{(t)}\mathbf{1} = \alpha^t,$$

which shows that ν is a QSD, and it is clearly the unique one. Also $\theta = -\log\alpha$ is the exponential rate of survival.

The Perron–Frobenius Theorem also gives

$$\exists\beta \in [0,\alpha) \quad \text{such that} \quad {P_{\mathbf{a}}}^{(t)} = \alpha^t \psi\nu' + O\big(\beta^t\big),$$

or equivalently, $p^{(t)}(x,y) = \alpha^t\psi(x)\nu(y) + O(\beta^t)$. So, in the finite case, one has

$$\forall x \in \mathcal{X}^{\mathbf{a}}\colon\quad \mathbb{P}_x(T > t) = \sum_{y\in I^{\mathbf{a}}} p^{(t)}(x,y) = \alpha^t\psi(x) + O\big(\beta^t\big).$$

When a probability measure ν defined on $\mathcal{X}^{\mathbf{a}}$ satisfies the property

$$\forall y \in \mathcal{X}^{\mathbf{a}} \exists \nu(y) := \lim_{t\to\infty} \mathbb{P}_x(Y_t = y | T > t),$$

it is called a quasi-limiting distribution, or a Yaglom limit. As defined, it is independent of $x \in \mathcal{X}^{\mathbf{a}}$.

Notice that

$$\forall x \in \mathcal{X}^{\mathbf{a}}, \qquad \alpha = -\lim_{t\to\infty} \frac{1}{t}\mathbb{P}_x(T > t),$$

so the Perron–Frobenius eigenvalue is the exponential rate of survival of the process starting from any initial distribution.

When the chain evolves in an infinite countable set, the following problems on QSDs can be set:

Problem 1.1 The problem of existence of a QSD. From the discussion, a necessary condition for the existence of a QSD is that the process is exponentially killed. Under this condition, we will give the proof that a sufficient condition for the existence of a QSD is that the process does not come from infinity in finite time (that is, for all fixed positive time, it is not possible to be killed before this time when the process starts arbitrarily far away). As shown for birth-and-death chains, this condition on non-coming from infinity is clearly unnecessary for the existence of a QSD.

Problem 1.2 The problem of uniqueness of a QSD. As we shall see for exponentially killed time-continuous birth-and-death chains, there are two possibilities: (i) if the chain comes from infinity in finite time, there is a unique QSD, and (ii), in the opposite case, there is a continuum of QSDs, in fact, there is one QSD ν_θ for each $\theta \in (0, \theta^*]$, where θ^* is the exponential rate of survival of the process. The problem of the interpretation of the extremal QSD ν_{θ^*} appears.

Problem 1.3 The problem of the domains of attraction. If ρ is a probability distribution on $\mathcal{X}^{\mathbf{a}}$ and we consider the conditional evolution $\mathbb{P}_\rho(Y_t \in \bullet | T > t)$, we would like to know if it converges; and when the answer is positive and there are several QSDs, we would like to know to which of the QSDs it converges.

For birth-and-death chains, we will give the proof that this limit exists for all Dirac measures $\rho = \delta_x$ and it coincides with the extremal QSD, that is,

$$\forall y \in \mathcal{X}^{\mathbf{a}}: \quad \nu_{\theta^*}(y) = \lim_{t\to\infty} \mathbb{P}_x(Y_t = y | T > t).$$

When there is a unique QSD (for instance, for birth-and-death chains that are exponentially absorbed and coming from infinity in finite time), a problem is when all distributions ρ belong to the domain of attraction of the QSD.

For initial distributions different from the Dirac measures, a complete answer to the question of domains of attraction is very hard. In fact, it is answered in a satisfactory way only (in the context of diffusions) for the Brownian motion with strictly negative constant drift.

Let us study the process of trajectories that survive forever. From the Markov property, we get

$$\mathbb{P}_x(Y_1 = x_1, \ldots, Y_k = i_k, T > t) = \mathbb{P}_x(Y_1 = i_1, \ldots, Y_k = i_k)\mathbb{P}_{x_k}(T > t - k).$$

In the finite case, we can go further. In fact, the computations already done give the following ratio limit result on the survival probabilities:

$$\forall x, y \in \mathcal{X}^{\mathbf{a}}: \quad \lim_{t\to\infty} \frac{\mathbb{P}_x(T > t - k)}{\mathbb{P}_y(T > t)} = \alpha^k \frac{\psi(x)}{\psi(y)}.$$

Hence,

$$\lim_{t\to\infty} \mathbb{P}_{x_0}(Y_1 = x_1, \ldots, Y_k = x_k | T > t) = \prod_{l=1}^{k} \left(\alpha \frac{\psi(x_l)}{\psi(x_{l-1})} p(x_{l-1}, x_l) \right).$$

Therefore, in the finite case, the process of trajectories that survive forever, $Z = (Z_t : t \geq 0)$, whose law is given by

$$\mathbb{P}_x(Z_1 = x_1, \ldots, Z_k = x_k) := \lim_{t\to\infty} \mathbb{P}_x(Y_1 = x_1, \ldots, Y_k = x_k | T > t), \qquad x \in \mathcal{X}^{\mathbf{a}},$$

is a Markov chain taking values on $\mathcal{X}^{\mathbf{a}}$, with transition probability

$$\tilde{p}(x,y) := \mathbb{P}_x(Z_1 = y) = \alpha \frac{\psi(y)}{\psi(x)} p(x,y).$$

Observe that $\widetilde{P} = (\tilde{p}(x,y) : x, y \in \mathcal{X}^{\mathbf{a}})$ is a stochastic matrix because

$$\sum_{y \in \mathcal{X}^{\mathbf{a}}} \tilde{p}(x,y) = \alpha \sum_{y \in \mathcal{X}^{\mathbf{a}}} p(x,y) \frac{\psi(y)}{\psi(x)} = 1.$$

In this finite case, and under irreducibility, the chain Z is obviously positive recurrent, and from the equalities

$$(\nu\psi)' P(y) = \sum_{x \in I^{\mathbf{a}}} \nu(x)\psi(x)\tilde{p}(x,y) = \alpha\psi(y) \sum_{x \in I^{\mathbf{a}}} \nu(x) p(x,y) = \nu(y)\psi(y),$$

$\nu\psi = (\nu(x)\psi(x) : x \in I^{\mathbf{a}})$ is the stationary distribution for Z. To avoid confusion, we denote by $\widetilde{\mathbb{P}}$ the law of the process Z given by the transition matrix $\widetilde{P}$.

For a general Markov chain, the following problem is put:

Problem 1.4 Existence and classification of the process Z of trajectories that survive forever. For birth-and-death chains, we show the existence of the process Z. To do it, we extend to birth-and-death chains the result showing that the ratio between the components of the right-eigenvector correspond to the limit ratios of the survival probabilities. In relation with the classification of the process Z (following Vere-Jones classification), we supply a criterion based on the exponential rate of survival which ensures when the process Z is positive recurrent, null-recurrent or transient.

We have discussed the convergence in distribution of the initial piece of trajectory,

$$\lim_{t\to\infty} \mathbb{P}_\rho(Y_0 = x_0, \ldots, Y_k = x_k | T > t) = \widetilde{\mathbb{P}}_\rho(Z_0 = x_0, \ldots, Z_k = x_k).$$

In the finite case, the quasi-limiting behavior allows getting the convergence of the final piece of trajectory. It is given by

$$\lim_{t\to\infty} \mathbb{P}_\rho(Y_t = x_0, \ldots, Y_{t-l} = x_l | T > t) = \mathbb{P}_\nu(Y_t = x_0, \ldots, Y_{t-l} = x_l | T > l),$$

where ν is the quasi-limiting distribution.

Note that this limit distribution on the last piece of the trajectory requires the existence of the quasi-limiting distribution and the limit ratio of the survival probabilities. Hence, by previous discussion, this limit behavior can be also retrieved for the birth-and-death chains that are exponentially killed.

1.3 Diffusions

In this section, we give a simple example of a QSD for a diffusion. We refer to Pinsky (1985) for a more general case and to Chaps. 6 and 7 for other diffusions in dimension one.

In this section, we denote by $\mathcal{X}$ a compact, connected domain of $\mathbb{R}^n$ with a regular boundary, and the process $Y_.$ will be the n-dimensional Brownian motion killed on $\partial\mathcal{X}$. We will prove the existence of a QSD for this process. When necessary, we will use the notation $Y_.^x$ to denote the process starting at x.

Theorem 1.1 *Under the above hypothesis, the Brownian motion killed on $\partial\mathcal{X}$ has a QSD which is absolutely continuous with respect to the Lebesgue measure.*

Proof From the spectral theory of elliptic partial differential operators (see, for example, Davies 1989), it is well known that the Dirichlet Laplacian in $\mathcal{X}$ has a largest eigenvalue $-\lambda < 0$ with an associated eigenfunction u which is nonnegative and bounded, namely

$$\Delta u = -\lambda u,$$

$u \geq 0$ and $u_{|\partial\mathcal{X}} = 0$.

From Ito's formula (see, for example, Gikhman and Skorokhod 1996) we have for any function f which is twice continuously differentiable and with compact support contained in the interior of $\mathcal{X}$ that

$$e^{\lambda t/2} f\left(Y_{t\wedge T}^x\right) = f(x) + \frac{1}{2}\int_0^{t\wedge T} \left(\Delta f\left(Y_{s\wedge T}^x\right) + \lambda f\left(Y_{s\wedge T}^x\right)\right) ds + \mathcal{M}_t$$

where $(\mathcal{M}_.)$ is a martingale. Therefore,

$$\begin{aligned} e^{\lambda t/2}\mathbb{E}_x\left(f(Y_{t\wedge T})\right) &= f(x) + \frac{1}{2}\mathbb{E}\left(\int_0^{t\wedge T} \left(\Delta f\left(Y_{s\wedge T}^x\right) + \lambda f\left(Y_{s\wedge T}^x\right)\right) ds\right) \\ &= f(x) + \frac{1}{2}\mathbb{E}\left(\int_0^{t\wedge T} \left(\Delta f\left(Y_{s\wedge T}^x\right) + \lambda f\left(Y_{s\wedge T}^x\right)\right) ds\right). \end{aligned}$$

Using integration by parts, we get

$$\mathbb{E}\left(\int_{\mathcal{X}} u(x)\left(\Delta f\left(Y_{s\wedge T}^x\right) + \lambda f\left(Y_{s\wedge T}^x\right)\right) dx\right) = 0.$$

Therefore, by Fubini's Theorem, we obtain

$$e^{\lambda t/2}\int_{\mathcal{X}} u(x)\mathbb{E}_x\left(f(Y_{t\wedge T})\right) dx = \int_{\mathcal{X}} u(x) f(x)\, dx.$$

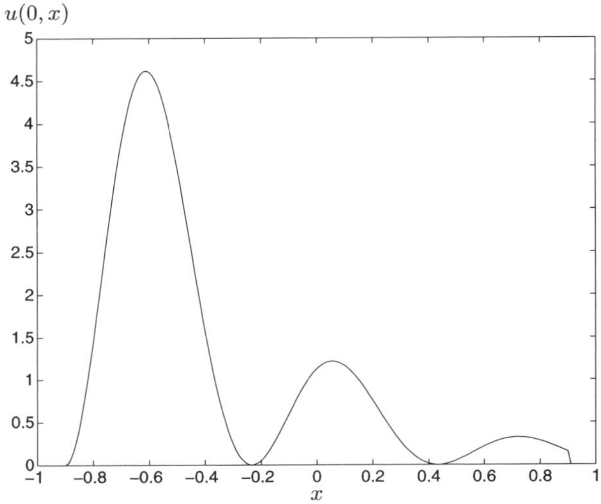

Fig. 1.2 The density of $u(0,x)$ of the initial distribution

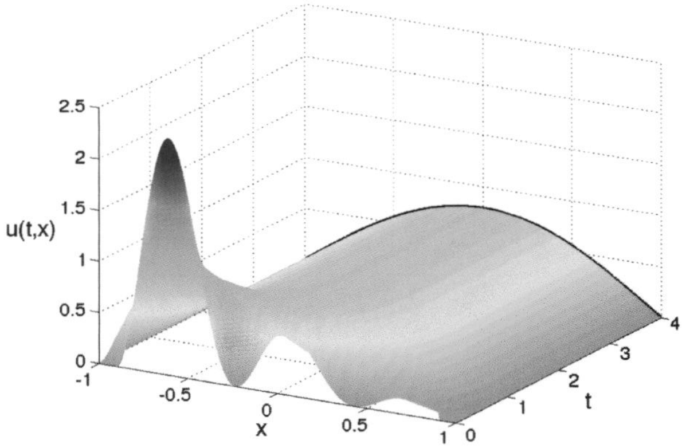

Fig. 1.3 Time evolution $u(t,x)$ of the density

Let ν denote the positive measure with density u ($d\nu = u\,dx$), the above formula can be rewritten

$$\mathbb{E}_\nu\big(f(Y_{t\wedge T})\big) = e^{-\lambda t/2} \int_{\mathcal{X}} f\,d\nu.$$

Since this formula holds for any function f which is twice continuously differentiable and with compact support contained in the interior of $\mathcal{X}$, and since u vanishes on the boundary, we conclude that ν is a QSD. Note also that, since u is bounded and hence integrable, we can normalize ν to a probability measure. □

One can give other proofs of this theorem avoiding the use of the spectral theory of elliptic operators, using, for example, the spectral theory. Results for more general diffusions in dimension one will be proved in Chaps. 6 and 7.

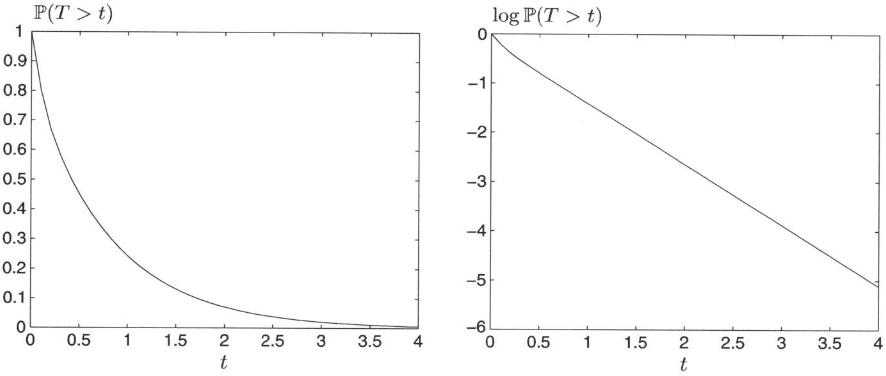

Fig. 1.4 *(Left)* $\mathbb{P}(T > t)$ as a function of t. *(Right)* $\log \mathbb{P}(T > t)$ as a function of t

Spectral theory allows, in particular, one to prove the existence of a Yaglom limit and to show that the rate of convergence is exponential (we will see more on this in later chapters). Here we illustrate these properties using a simulation for the one dimensional Brownian motion on the interval $[-1, 1]$. We start with the initial condition which is absolutely continuous with respect to the Lebesgue measure with density drawn in Fig. 1.2.

In Fig. 1.3, we show the time evolution of the density, normalized to have integral one. We clearly see convergence to the density of the Yaglom limit ($\pi \cos(\pi x/2)/4$ in this case) which is drawn in black at the end of the surface.

One can also look at the time evolution of the survival probability $\mathbb{P}(T > t)$. This is shown in Fig. 1.4 on the left. On the right, we plotted the logarithm of this quantity as a function of time, clearly showing convergence to a straight line, namely that $\mathbb{P}(T > t)$ decays asymptotically exponentially.

1.4 Dynamical Systems

A classical system (in physics, chemistry, biology, etc.) is described by the set $\mathcal{X}$ of all its possible states, often called the phase space of the system. At a given time, all the properties of the system can be recovered from the knowledge of the instantaneous state $x \in \mathcal{X}$. The system is observed using the so-called observables which are real-valued functions on $\mathcal{X}$. Most often the space of states $\mathcal{X}$ is a metric space (so we can speak of nearby states). In many physical situations, there are even more structures on $\mathcal{X}$: $\mathbb{R}^d$, Riemannian manifolds, Banach or Hilbert spaces, etc.

As a simple example one can consider a mechanical system with one degree of freedom. The state of the system at a given time is given by two real numbers: the position q and momentum p. The state space is therefore $\mathbb{R}^2$. A continuous material (solid, fluid, etc.) is characterized by the field of local velocities, pressure, density, temperature, etc. In that case, one often uses phase spaces which are Banach spaces.

As time goes on, the instantaneous state changes (unless the system is in a situation of equilibrium). The time evolution is a rule giving the change of the state with time. It comes in several flavors and descriptions summarized below:

(i) Discrete time evolution. This is a map f from the state space $\mathcal{X}$ into itself producing a new state from an old one after one unit of time. If x_0 is the state of the system at time zero, the state at time one is $x_1 = f(x_0)$, and more generally, the state at time n is given recursively by $x_n = f(x_{n-1})$. This is often written $x_n = f^n(x_0)$ with $f^n = f \circ f \circ \cdots \circ f$ (n times). The sequence $(f^n(x_0))$ is called the trajectory or the orbit of the initial condition x_0.

(ii) Continuous-time semi-flow. This is a family $(\varphi_t : t \in \mathbb{R}^+)$ of maps of $\mathcal{X}$ satisfying

$$\varphi_0 = \mathrm{Id}, \qquad \varphi_s \circ \varphi_t = \varphi_{s+t}.$$

The set $\{\varphi_t(x_0) : t \in \mathbb{R}^+\}$ is called the trajectory (orbit) of the initial condition x_0. Note that if we fix a time step $\tau > 0$, and observe only the state at times $n\tau$ ($n \in \mathbb{N}$), we obtain a discrete time dynamical system given by the map $f = \varphi_\tau$. A flow is a family $(\varphi_t : t \in \mathbb{R})$ of maps of $\mathcal{X}$ satisfying the above two conditions for any t and s in $\mathbb{R}$. Note in particular that φ_{-t} is the inverse of φ_t.

(iii) A differential equation on a manifold associated to a vector field $\mathbf{F}$

$$\frac{d\mathbf{x}}{dt} = \mathbf{F}(\mathbf{x}).$$

This is, for example, the case of a mechanical system in the Hamiltonian formalism. Under regularity conditions on $\mathbf{F}$, the integration of this equation leads to a flow.

(iv) There are other more complicated situations like nonautonomous systems (in particular, stochastically forced systems), systems with memory, systems with delay, etc., but we will not consider them below.

A dynamical system is a set of states $\mathcal{X}$ equipped with a time evolution. If there is more structure on $\mathcal{X}$, one can put more structure on the time evolution itself. For example, in the case of discrete time, the map f may be measurable, continuous, differentiable, etc.

Assume now that the phase space $\mathcal{X}$ is equipped with a Borel structure, and let ν_0 be a Borel probability measure on $\mathcal{X}$. To any dynamical system on $\mathcal{X}$, we can now associate a stochastic process. We explain how this is done for the case of a discrete time evolution, the case of continuous time is similar. Let f be a measurable map of the phase space $\mathcal{X}$ as above, and let ν_0 be a probability measure on $\mathcal{X}$. Let $\mathcal{Y}$ be a measurable space, and let Y be a measurable map from $\mathcal{X}$ to $\mathcal{Y}$. We can consider Y as a $\mathcal{Y}$-valued random variable by the usual construction. Namely, for any measurable $B \in \mathcal{Y}$, we define

$$\mathbb{P}(Y \in B) = \nu_0\big(Y^{-1}(B)\big).$$

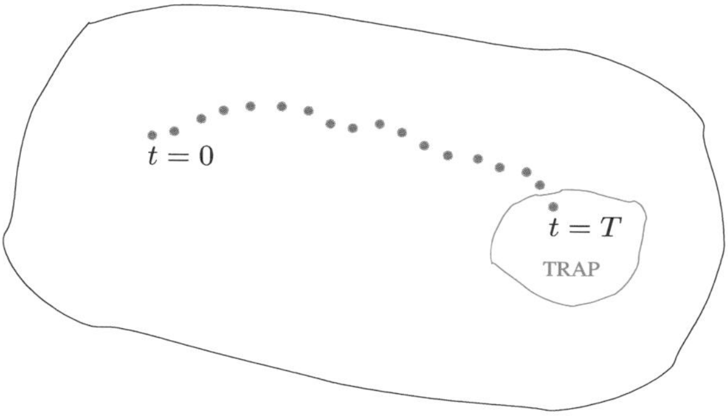

Fig. 1.5 Time evolution with a trap

More generally, we can define a discrete time stochastic process on $\mathscr{Y}$ by the sequence $(Y_n : n \in \mathbb{Z}_+)$ of random variables defined by

$$Y_n = Y \circ f^n.$$

In these processes defined by dynamical systems, the time evolution is deterministic, but the initial condition is chosen at random. This randomness is propagated by the time evolution. There is no loss of generality if we assume that $\mathcal{X} = \mathscr{Y}$ and Y is the identity mapping.

A measure ν on $\mathcal{X}$ is called an invariant measure for the map f if for any measurable set $B \in \mathcal{B}(\mathcal{X})$ we have

$$\nu\big(f^{-1}(B)\big) = \nu(B).$$

We recall that f^{-1} denotes here the set-theoretic inverse, or in other words, $f^{-1}(B)$ is the set of preimages of points in B

$$f^{-1}(B) = \big\{x : f(x) \in B\big\}.$$

The definition is the same for a semi-flow.

As explained above, to the map f and the measure ν, we can associate stochastic processes. If the measure ν is invariant, it is easy to verify that these processes are stationary.

We now consider the particular setting corresponding to quasi-stationary measures. We assume that a measurable subset $\mathcal{X}^{\text{tr}}$ of the phase space $\mathcal{X}$ is given, which is a hole or a trap. Namely, when an orbit reaches $\mathcal{X}^{\text{tr}}$, the time evolution stops (although the dynamical system may be well defined on $\mathcal{X}^{\text{tr}}$). See Fig. 1.5 for a sketch of this evolution.

Let $T = T_{\mathcal{X}^{\text{tr}}}$ be the first time an initial condition x reaches the trap $\mathcal{X}^{\text{tr}}$, namely, $f^j(x) \notin \mathcal{X}^{\text{tr}}$ for $0 \leq j < T(x)$ and $f^{T(x)}(x) \in \mathcal{X}^{\text{tr}}$. It is convenient to define $T = 0$ on $\mathcal{X}^{\text{tr}}$.

In this situation, it is natural to define the sequence $(\mathcal{X}_n)$ of sets of initial conditions which do not reach $\mathcal{X}^{\text{tr}}$ before time n, namely $\mathcal{X}_n = \{T > n\}$. The sets $\mathcal{X}_n$ can also be defined recursively by $\mathcal{X}_0 = \mathcal{X} \backslash \mathcal{X}^{\text{tr}}$, and

$$\mathcal{X}_{n+1} = \mathcal{X}_n \cap \big\{x : f^{n+1}(x) \notin \mathcal{X}^{\text{tr}}\big\}.$$

Note that we also have

$$\mathcal{X}_n = \bigcap_{j=0}^{n} f^{-j}(\mathcal{X}_0).$$

We recall that there are several natural questions associated to this situation.

Problem 1.5 Given an initial distribution ν_0 on $\mathcal{X}$, what is the probability that a trajectory has survived up to time $n > 0$? (For example, if ν_0 is the Dirac measure on one point.)

In other words, what is the behavior of $\nu_0(\mathcal{X}_n)$. Often one can say something only for large n.

Problem 1.6 Assume a trajectory initially distributed with ν_0 has survived up to time $n > 0$. What is the distribution at that time (of the surviving trajectory)?

In other words, can we say something about

$$\frac{\nu_0(f^{-n}(B) \cap \mathcal{X}_n)}{\nu_0(\mathcal{X}_n)},$$

B a measurable subset of $\mathcal{X}$.

We recall that the Yaglom limit of the measure ν_0 is defined as the limit of the above measure when n tends to infinity (if the limit exists). It may depend on the initial distribution ν_0.

A related object is a quasi-stationary measure (QSD).

We recall that ν (a probability measure on $\mathcal{X}$) is quasi-stationary if for any $n \geq 0$

$$\frac{\nu(f^{-n}(B) \cap \mathcal{X}_n)}{\nu(\mathcal{X}_n)} = \nu(B)$$

for any measurable set B. If there is no trap, this is a stationary measure.

Problem 1.7 Are there trajectories which never reach the trap A? ($T = \infty$).

If so, how are they distributed?

How is this related to Problem 1.6?

To get some intuition about these questions in the case of dynamical systems, we now discuss a simple example where explicit computations can be performed.

The phase space $\mathcal{X}$ is the unit interval $[0, 1]$. Consider the map f given by $f(x) = 3x \pmod 1$. In Fig. 1.6 we show the graph of f, and in Fig. 1.7 the geometrical construction of one iteration.

It is easy to verify that the Lebesgue measure Leb is invariant, namely, for any Borel set B

$$\text{Leb}\big(f^{-1}(B)\big) = \text{Leb}(B). \tag{1.1}$$

Indeed, it is enough to verify (1.1) for a finite union of disjoint sub-intervals of $[0, 1]$, and in fact for each interval separately. By a simple computation, one obtains

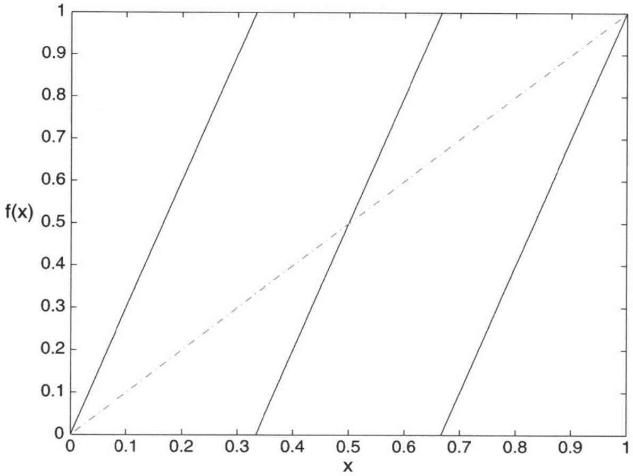

Fig. 1.6 The map $f(x) = 3x \pmod 1$

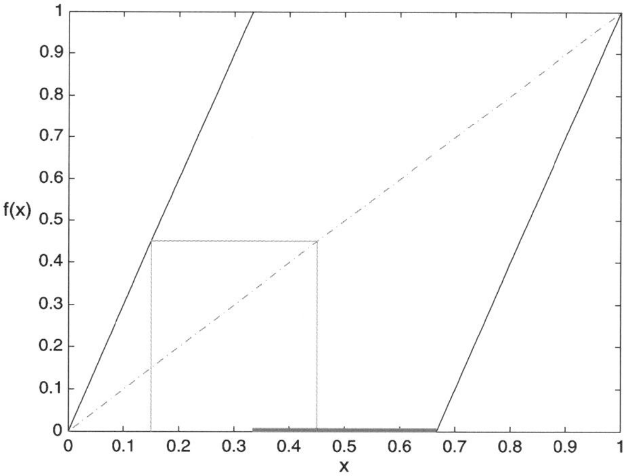

Fig. 1.7 The map $f(x) = 3x \pmod 1$ with the trap $]1/3, 2/3[$ (*in red*), geometrical construction of an iterate (*in green*) falling into the trap after one step

immediately for the interval $I = (a, b) \subset [0, 1]$

$$f^{-1}(I) = (a/3, b/3) \cup \big((a+1)/3, (b+1)/3\big) \cup \big((a+2)/3, (b+2)/3\big),$$

and therefore,

$$\mathrm{Leb}\big(f^{-1}(I)\big) = 3(b/3 - a/3) = b - a = \mathrm{Leb}(I).$$

As explained above, given an initial distribution, we have a discrete time stochastic process.

The probability space is $[0, 1]$ and the process is defined recursively by $Y_{n+1} = f(Y_n)$.

We use the interval $]1/3, 2/3[$ as a trap.

The normalized Lebesgue measure $\nu = 3\text{Leb}/2$ on $\mathcal{X}_0 = [0, 1/3] \cup [1/3, 1]$ is quasi-stationary.

The proof is an easy computation from the definition in the discrete time case. As we will see later (see Lemma 8.2), it is enough to check that

$$\frac{\nu(f^{-1}(B) \cap \mathcal{X}_1)}{\nu(\mathcal{X}_1)} = \nu(B),$$

for any Borel set $B \subseteq \mathcal{X}_0$. Moreover, by completion, it is enough to check this identity for a finite union of disjoint intervals, and this follows if we can check it for any interval $(a, b) \subset \mathcal{X}_0$. A simple computation (as above for the invariant measure) leads to

$$\mathcal{X}_1 = [0, 1/9] \cup [2/9, 1/3] \cap [2/3, 7/9] \cap [8/9, 1].$$

Therefore,

$$\text{Leb}(\mathcal{X}_1) = \frac{4}{9} \quad \text{and} \quad \nu(\mathcal{X}_1) = \frac{2}{3}.$$

As above, for $I = (a, b)$, we have

$$f^{-1}(I) = (a/3, b/3) \cup \big((a+1)/3, (b+1)/3\big) \cup \big((a+2)/3, (b+2)/3\big),$$

and therefore, if $I \subset \mathcal{X}_0$,

$$f^{-1}(I) \cap \mathcal{X}_1 = (a/3, b/3) \cup \big((a+2)/3, (b+2)/3\big).$$

This implies

$$\text{Leb}\big(f^{-1}(I) \cap \mathcal{X}_1\big) = \frac{2}{3}(b-a).$$

Therefore,

$$\frac{\nu(f^{-1}(I) \cap \mathcal{X}_1)}{\nu(\mathcal{X}_1)} = \frac{3\text{Leb}(f^{-1}(I) \cap \mathcal{X}_1)/2}{2/3} = \frac{3}{2}(b-a) = \nu(I).$$

It is easy to verify inductively that

$$\nu(\mathcal{X}_n) = \left(\frac{2}{3}\right)^n.$$

Namely, we have the answer to Problem 1.5: the probability of surviving up to time n when the initial condition is distributed according to the QSD ν.

The set of initial conditions which never die is

$$K = \bigcap_n \mathcal{X}_n$$

which is the Cantor set. K is of zero Lebesgue measure. Moreover, during the recursive construction, if we start with the Lebesgue measure (or the QSD ν), $\mathcal{X}_n$ is a union of triadic intervals which have the same length and hence the same weight. We get at the end the Cantor measure which is singular with respect to the Lebesgue measure (or the QSD ν). So there is little connection with the nice QSD. Intuitively, a QSD (or a Yaglom limit) describes the distribution of trajectories which have survived for a large time but most of them are on the verge of falling in the trap.

Problem 1.7 deals with trajectories that will never see the trap, this is very different. In the case of dynamical systems, these trajectories concentrate on a very small set which is invariant and disjoint from the trap.

Problem 1.6 (distribution of survivors at large time) and Problem 1.7 (eternal life) have very different answers.

Chapter 2
Quasi-Stationary Distributions: General Results

In this chapter, we introduce the main concepts in a general context of killed processes. Thus, in Sect. 2.2, we give the definition of quasi-stationary distributions (QSDs). In Theorem 2.2 of Sect. 2.3, we show that starting from a QSD the killing time is exponentially distributed, and in Theorem 2.6 of Sect. 2.4, we show that the killing time and the state of killing are independent random variables. In Theorem 2.11 of Sect. 2.7, we give a theorem of existence of a QSD in a topological setting without any assumption on compactness or spectral properties.

2.1 Notation

We will consider a Markov process $Y = (Y_t : t \geq 0)$ taking values in a state space $\mathcal{X}$ which is endowed with a σ-field $\mathcal{B}(\mathcal{X})$. Let us fix some notation:

- $\mathcal{M}(\mathcal{X})$ is the set of real measurable functions defined on $\mathcal{X}$; $\mathcal{M}_+(\mathcal{X})$ (respectively, $\mathcal{M}_b(\mathcal{X})$) is the set of positive (respectively, bounded) elements in $\mathcal{M}(\mathcal{X})$;
- $\mathcal{P}(\mathcal{X})$ is the set of probability measures on $(\mathcal{X}, \mathcal{B}(\mathcal{X}))$.

When $\mathcal{X}$ has a topological structure, $\mathcal{B}(\mathcal{X})$ denotes its Borel σ-field. In this case, we denote by $\mathcal{C}(\mathcal{X})$ the set of continuous functions and by $\mathcal{C}_b(\mathcal{X})$ its bounded elements.

Let v be a measure on $\mathcal{X}$ and $f \in \mathcal{M}(\mathcal{X})$. If $\int f\, dv$ is well defined then we set $v(f) = \int f\, dv$. This is the case if $f \in L^1(v)$, or when $f \in \mathcal{M}_+(\mathcal{X})$, or when $v \in \mathcal{P}(\mathcal{X})$ and $f \in \mathcal{M}_b(\mathcal{X})$.

Let us fix $\mathcal{X}$ a topological space endowed with its Borel σ-field $\mathcal{B}(\mathcal{X})$. Let Y be a Markov process taking values in $\mathcal{X}$ satisfying the conditions we fix below.

Let Ω be the set of right continuous trajectories on $\mathcal{X}$, indexed by points in $[0, \infty)$. So $\omega = (\omega_s : s \in [0, \infty)) \in \Omega$ means that $\omega_s \in \mathcal{X}$ and $\omega_s = \lim_{h \to 0^+} \omega_{s+h}$ for all $s \in [0, \infty)$. Let $(\Omega, \mathcal{F})$ be a measurable space where $\mathcal{F}$ contains the σ-field generated by all the projections $\mathrm{pr}_s : \Omega \to \mathcal{X}, \omega \to \mathrm{pr}_s(\omega) = \omega_s, s \geq 0$. For $t \geq 0$ let $\Theta_t : \Omega \to \Omega$ be the operator shifting the trajectories by t, so $\Theta_t(\omega) = (\omega_s : s \geq t)$.

P. Collet et al., *Quasi-Stationary Distributions*, Probability and Its Applications,
DOI 10.1007/978-3-642-33131-2_2, © Springer-Verlag Berlin Heidelberg 2013

The process $Y = (Y_t : t \geq 0)$ takes values on $\mathcal{X}$ and its trajectories belong to Ω. We denote by $\mathbb{F} = (\mathcal{F}_t : t \geq 0)$ a filtration of σ-fields such that Y is adapted to $\mathbb{F}$, so for all $t \geq 0$ $\mathcal{F}_t \supseteq \sigma(Y_s : s \leq t)$ the σ-field generated by the process up to time t. Let $(\mathbb{P}_x : x \in \mathcal{X})$ be a family of probability measures defined on $(\Omega, \mathcal{F})$. We assume that Y is a Markov process with respect to the family $(\mathbb{P}_x : x \in \mathcal{X})$, that is,

- $\mathbb{P}_x(Y_0 = x) = 1$ for all $x \in \mathcal{X}$;
- For all $A \in \mathcal{F}$ the function $x \to \mathbb{P}_x(A)$ is $\mathcal{B}(\mathcal{X})$-measurable;
- $\mathbb{P}_x(\Theta_t(Y) \in A | \mathcal{F}_t) = \mathbb{P}_{Y_t}(A)$ $\mathbb{P}_x$- a.s., $\forall x \in \mathcal{X}$ and $\forall A \in \mathcal{F}$.

We assume that $\mathcal{X}$ is a Polish space (metric, separable and complete) and that the Markov process Y is standard in the sense of Definition 9.2 in Blumenthal and Getoor (1968). In particular, Y is strong Markov. We denote by

$$T_B = \inf\{t > 0 : Y_t \in B\}, \qquad B \in \mathcal{B}(\mathcal{X}),$$

the hitting time of B, where as usual we put $\inf \emptyset = \infty$. Since the Markov process Y is standard, T_B is a stopping time. When $B = \{x_0\}$ is a singleton, we put T_{x_0} instead of $T_{\{x_0\}}$.

2.2 Killed Process and QSD

In this theory, there is a set of forbidden states for the process, denoted by $\partial\mathcal{X}$. We assume that $\partial\mathcal{X} \in \mathcal{B}(\mathcal{X})$ and $\emptyset \neq \partial\mathcal{X} \neq \mathcal{X}$, this last condition is to avoid trivial situations. Its complement $\mathcal{X}^{\mathbf{a}} := \mathcal{X} \setminus \partial\mathcal{X}$ is called the set of allowed states. Let

$$T = T_{\partial\mathcal{X}},$$

be the hitting time of $\partial\mathcal{X}$, it is called the killing time (or extinction time in population models).

All the quantities in this study will only depend on $Y^T = (Y_{t\wedge T} : t \geq 0)$, the process stopped at $\partial\mathcal{X}$, and so there is no loss of generality in assuming $Y_t = Y_T$ for $t \geq T$. Then $Y = Y^T$ and T is an absorption time. We note that except when we look for the region where the process is absorbed, we will be mainly concerned with the trajectory before killing $(Y_t : t < T)$.

In our framework, we will assume that there is sure killing at $\partial\mathcal{X}$,

$$\forall x \in \mathcal{X}^{\mathbf{a}}: \quad \mathbb{P}_x(T < \infty) = 1. \tag{2.1}$$

In particular, this means that there is no explosion. Condition (2.1) is equivalent to

$$\forall \rho \in \mathcal{P}(\mathcal{X}^{\mathbf{a}}): \quad \mathbb{P}_\rho(T < \infty) = 1,$$

where as usual we put $\mathbb{P}_\rho = \int_{\mathcal{X}^{\mathbf{a}}} \mathbb{P}_x \, d\rho(x)$ for any $\rho \in \mathcal{P}(\mathcal{X}^{\mathbf{a}})$. Relation (2.1) obviously implies

$$\forall x \in \mathcal{X}^{\mathbf{a}}: \quad \mathbb{P}_x(T < \infty) > 0, \tag{2.2}$$

then,

$$\forall \rho \in \mathcal{P}(\mathcal{X}^{\mathbf{a}}) \; \exists t(\rho) \; \forall t \geq t(\rho): \quad \mathbb{P}_\rho(T > t) < 1. \tag{2.3}$$

Condition (2.2) (and hence the stronger condition (2.1)) implies that there cannot exist stationary distributions supported by $\mathcal{X}^{\mathbf{a}}$ because if ρ is one of them, from (2.3) we should have $\rho(Y_t \in \mathcal{X}^{\mathbf{a}}) < 1$ for $t \ge t(\rho)$, which contradicts the stationarity because $\rho(Y_0 \in \mathcal{X}^{\mathbf{a}}) = 1$.

Let us define the quasi-stationary distributions.

Definition 2.1 $\nu \in \mathcal{P}(\mathcal{X}^{\mathbf{a}})$ is said to be a QSD (for the process killed at $\partial\mathcal{X}$) if

$$\forall B \in \mathcal{B}(\mathcal{X}^{\mathbf{a}})\ \forall t \ge 0: \quad \mathbb{P}_\nu(Y_t \in B | T > t) = \nu(B).$$

So, ν is a QSD if

$$\forall B \in \mathcal{B}(\mathcal{X}^{\mathbf{a}})\ \forall t \ge 0: \quad \mathbb{P}_\nu(Y_t \in B, T > t) = \nu(B)\mathbb{P}_\nu(T > t). \tag{2.4}$$

Since Y is stopped at time T, we have $\mathbb{P}_\nu(Y_t \in B, T > t) = \mathbb{P}_\nu(Y_t \in B)$ for all $B \in \mathcal{B}(\mathcal{X}^{\mathbf{a}})$, and the condition of QSD takes the form

$$\forall B \in \mathcal{B}(\mathcal{X}^{\mathbf{a}})\ \forall t \ge 0: \quad \mathbb{P}_\nu(Y_t \in B) = \nu(B)\mathbb{P}_\nu(T > t). \tag{2.5}$$

From the previous observation, condition (2.2) ensures that a QSD cannot be stationary.

2.3 Exponential Killing

Let ν be a QSD. The first remark is that when starting from ν, the killing time at $\partial\mathcal{X}$ is exponentially distributed (see Ferrari et al. 1995b). Let us state it.

Theorem 2.2 *If ν be a QSD, then*

$$\exists \theta(\nu) \ge 0 \quad \textit{such that} \quad \mathbb{P}_\nu(T > t) = e^{-\theta(\nu)t} \quad \forall t \ge 0,$$

that is, starting from ν, T is exponentially distributed with parameter $\theta(\nu)$.

Proof Since ν is a QSD, from relation (2.4) we get that all $g \in \mathcal{M}_+(\mathcal{X}^{\mathbf{a}})$ and all $g \in \mathcal{M}_b(\mathcal{X}^{\mathbf{a}})$ verify

$$\forall t \ge 0: \quad \mathbb{E}_\nu\big(g(Y_t)\mathbb{1}_{T>t}\big) = \nu(g)\mathbb{P}_\nu(T > t). \tag{2.6}$$

By taking the measurable function $g(x) = \mathbb{P}_x(T > s)$, $x \in \mathcal{X}^{\mathbf{a}}$, and since $\int_{\mathcal{X}^{\mathbf{a}}} g\, d\nu = \mathbb{P}_\nu(T > s)$, we get

$$\forall t \ge 0: \quad \mathbb{E}_\nu\big(\mathbb{P}_{Y_t}(T > s)\mathbb{1}_{T>t}\big) = \mathbb{P}_\nu(T > s)\mathbb{P}_\nu(T > t).$$

From the Markov property, it is verified that $\forall t, s \ge 0$

$$\begin{aligned}\mathbb{P}_\nu(T > t + s) &= \mathbb{E}_\nu(\mathbb{1}_{T>t+s}) = \mathbb{E}_\nu\big(\mathbb{1}_{T>t}\mathbb{E}(\mathbb{1}_{T>t+s}|\mathcal{F}_t)\big)\\ &= \mathbb{E}_\nu\big(\mathbb{1}_{T>t}\mathbb{E}_{Y_t}(\mathbb{1}_{T>s})\big) = \mathbb{P}_\nu(T > s)\mathbb{P}_\nu(T > t).\end{aligned}$$

So, there exists some $\theta \ge 0$ such that $\mathbb{P}_\nu(T > t) = e^{-\theta t}$ $\forall t \ge 0$. Then, the result follows. □

Remark 2.3 For discrete time, the argument in the proof of Theorem 2.2 shows that when starting from a QSD ν, the killing time at $\partial\mathcal{X}$ is geometrically distributed, so $\mathbb{P}_\nu(T > n) = \kappa(\nu)^n$ for all $n \geq 0$, where $\kappa(\nu) = \mathbb{P}_\nu(T > 1)$.

The coefficient $\theta(\nu)$ verifies $\theta(\nu) = -\frac{1}{t}\log\mathbb{P}_\nu(T > t)\ \forall t > 0$. So, when the process starts from ν, $\theta(\nu)$ is its exponential rate of survival, which is stationary in time.

Theorem 2.2 gives $\mathbb{P}_\nu(Y_{0+} \in \partial\mathcal{X}) = 0$. Then, from (2.3) we get $0 < \mathbb{P}_\nu(T > t) < 1$ for ν a QSD, and so $\theta(\nu) \in (0, \infty)$.

From (2.4) and Theorem 2.2, we get that $\nu \in \mathcal{P}(\mathcal{X}^{\mathbf{a}})$ is a QSD if and only if it verifies

$$\exists\theta \in (0,\infty)\ \forall B \in \mathcal{B}(\mathcal{X}^{\mathbf{a}})\ \forall t \geq 0:\quad \mathbb{P}_\nu(Y_t \in B, T > t) = \nu(B)e^{-\theta t}. \tag{2.7}$$

In this case, $\theta = \theta(\nu)$.

When ν is a QSD, we get

$$\forall\theta < \theta(\nu):\quad \mathbb{E}_\nu\left(e^{\theta T}\right) < \infty,$$

and we deduce

$$\forall\theta < \theta(\nu),\ \nu\text{-a.s. in } x:\quad \mathbb{E}_x\left(e^{\theta T}\right) < \infty. \tag{2.8}$$

The exponential moment condition can be written in the following way.

Proposition 2.4 *We have the equality*

$$\theta_x^* := \sup\left\{\theta : \mathbb{E}_x\left(e^{\theta T} < \infty\right)\right\} = \liminf_{t\to\infty} -\frac{1}{t}\log\mathbb{P}_x(T > t), \tag{2.9}$$

and a necessary condition for the existence of a QSD is the existence of a positive exponential moment, or equivalently, a positive exponential rate of survival:

$$\exists x \in \mathcal{X}^{\mathbf{a}}:\quad \theta_x^* > 0. \tag{2.10}$$

Proof The fact that (2.10) is necessary for the existence of a QSD follows from (2.8) and (2.9). We now prove the equality (2.9). Note that the result will follow once we show the following relation for a nonnegative random variable W:

$$\sup\left\{\theta : \mathbb{E}\left(e^{\theta W}\right) < \infty\right\} = \liminf_{t\to\infty} -\frac{1}{t}\log\mathbb{P}(W > t). \tag{2.11}$$

The inequality $\leq$ in (2.11) follows from Markov's Inequality applied to all nonnegative θ such that $\mathbb{E}(e^{\theta W}) < \infty$,

$$\mathbb{E}\left(e^{\theta W}\right) \geq e^{\theta t}\mathbb{P}(W > t)\quad \forall t \geq 0.$$

Let us prove the other inequality. First, observe that for every distribution function F, Fubini's Theorem gives

$$\begin{aligned}\int_0^\infty \theta e^{\theta t}\big(1 - F(t)\big)\,dt &= \int_0^\infty e^{\theta t}\,dt\int_{t^+}^\infty dF(u) = \int_{0^+}^\infty\left(\int_0^u \theta e^{\theta t}\,dt\right)dF(u)\\ &= \int_0^\infty \left(e^{\theta u} - 1\right)dF(u) = \int_0^\infty e^{\theta u}\,dF(u) - \big(1 - F(0)\big).\end{aligned}$$

Then

$$\int_0^\infty e^{\theta t}\,dF(t) = \big(1 - F(0)\big) + \int_0^\infty \theta e^{\theta t}\big(1 - F(t)\big)\,dt. \tag{2.12}$$

Note that

$$\liminf_{t\to\infty} -\frac{1}{t}\log \mathbb{P}(W > t) = \sup\{\theta' : \exists C < \infty\ \forall t \geq 0, \mathbb{P}(W > t) \leq Ce^{-\theta' t}\}. \tag{2.13}$$

Now, if $\theta' > 0$ satisfies $\mathbb{P}(W > t) \leq Ce^{-\theta' t}$ $\forall t \geq 0$, from (2.12) we get that any $\theta \in (0, \theta')$ verifies:

$$\begin{aligned}\mathbb{E}\big(e^{\theta W}\big) &= \mathbb{P}(W > 0) + \int_0^\infty \theta e^{\theta t}\mathbb{P}(W > t)\,dt \leq \mathbb{P}(W > 0) + C\theta \int_0^\infty e^{-(\theta'-\theta)t}\,dt \\ &= \mathbb{P}(W > 0) + C\left(\frac{1}{\theta' - \theta}\right) < \infty.\end{aligned} \tag{2.14}$$

The inequality $\geq$ in (2.11) follows from (2.13) and (2.14). □

Remark 2.5 Note that if ν is a QSD then $\mathbb{E}_\nu(e^{\theta(\nu)T}) = \infty$. Then, if $\theta > 0$ satisfies the condition

$$\sup\big(\mathbb{E}_x\big(e^{\theta T}\big) : x \in \mathcal{X}^{\mathbf{a}}\big) < \infty,$$

there cannot exists a QSD ν with $\theta(\nu) = \theta$ because otherwise we should have

$$\infty = \mathbb{E}_\nu\big(e^{\theta(\nu)T}\big) \leq \sup\big(\mathbb{E}_x\big(e^{\theta(\nu)T}\big) : x \in \mathcal{X}^{\mathbf{a}}\big) < \infty.$$

2.4 Independence Between the Exit Time and the Exit State

Let us show that when the process starts from a QSD, the killing time and the state where it is killed are independent random variables (see Martínez 2008).

Theorem 2.6 *Let ν be a QSD (for the process killed at $\partial\mathcal{X}$). Then T and Y_T are $\mathbb{P}_\nu$-independent random variables.*

Proof We already proved that the QSD property implies $\mathbb{P}_\nu(T \in (s, s + ds]) = \theta(\nu)e^{-\theta(\nu)s}\,ds$. Let $E \in \mathcal{B}(\partial\mathcal{X})$ be a measurable set contained in $\partial\mathcal{X}$. The Markov property and the QSD definition $\mathbb{P}_\nu(Y_s \in \bullet | T > s) = \nu(\bullet)$ imply that for all real $a > 0$ and every integer $n \geq 0$,

$$\begin{aligned}\mathbb{P}_\nu\big(T \in \big(na, (n+1)a\big], Y_T \in E\big) &= \mathbb{P}_\nu(T > na, Y_{(n+1)a} \in E) \\ &= \mathbb{P}_\nu(Y_{(n+1)a} \in E | T > na)\mathbb{P}_\nu(T > na) \\ &= \mathbb{P}_\nu(Y_a \in E)e^{-\theta(\nu)na}.\end{aligned}$$

We sum over all $n \geq 0$ to get

$$\mathbb{P}_\nu(Y_T \in E) = \left(\frac{1}{1 - e^{-\theta(\nu)a}}\right)\mathbb{P}_\nu(Y_a \in E).$$

So, the independence relation follows:

$$\begin{aligned}\mathbb{P}_\nu(T \le a, Y_T \in E) &= \mathbb{P}_\nu(Y_a \in E) = \left(1 - e^{-\theta(\nu)a}\right)\mathbb{P}_\nu(Y_T \in E)\\ &= \mathbb{P}_\nu(T \le a)\mathbb{P}_\nu(Y_T \in E).\end{aligned}$$

□

From this relation, we find

$$\frac{d}{da}\mathbb{P}_\nu(Y_a \in E)\bigg|_{a=0} = \theta(\nu)\mathbb{P}_\nu(Y_T \in E). \tag{2.15}$$

Remark 2.7 Let us give the proof of the above property in discrete time. For $n \ge 0$, we have

$$\begin{aligned}\mathbb{P}_\nu(T = n+1, Y_T \in E) &= \mathbb{P}_\nu(T > n, Y_{n+1} \in E)\\ &= \mathbb{P}_\nu(Y_{n+1} \in E | T > n)\mathbb{P}_\nu(T > n)\\ &= \mathbb{P}_\nu(Y_1 \in E)\mathbb{P}_\nu(T > n).\end{aligned}$$

When we sum over all $n \ge 0$ and use $\mathbb{P}_\nu(T > n) = \kappa(\nu)^n$, we obtain

$$\mathbb{P}_\nu(Y_T \in E) = \frac{1}{1 - \kappa(\nu)}\mathbb{P}_\nu(Y_1 \in E).$$

Since $\mathbb{P}(T = n+1) = (1 - \kappa(\nu))\kappa(\nu)^n$, we find the result

$$\mathbb{P}_\nu(T = n+1, Y_T \in E) = \mathbb{P}_\nu(Y_T \in E)\mathbb{P}_\nu(T = n+1) \quad \forall n \ge 0.$$

Example: The independent case Let Y^k, $k = 1, \dots, K$, be K independent Markov processes taking values in $\mathcal{X}_k$. Assume that each Y_k is killed at time T_k when it hits $\partial\mathcal{X}_k$. Let $\partial\mathcal{X} = \bigcup_{k=1}^K \partial\mathcal{X}_k \times \prod_{l \ne k} \mathcal{X}_l$. Then, the product measure $\nu = \bigotimes_{k=1}^K \nu^k$ is a QSD for the product process $Y = (Y^k : k = 1, \dots, K)$ killed at the region $\partial\mathcal{X}$. Indeed, the killing time is $T = \inf\{T^k : k = 1, \dots, K\}$. Moreover, in this case the exponential rate of survival is $\theta = \sum_{k=1}^K \theta_k$. From the equalities $\mathbb{P}_\nu(T^k \in (t, t+dt], T^l > t\ \forall l \ne k) = \theta_k e^{-\theta t}\,dt$ and $\mathbb{P}_\nu(T \in (t, t+dt]) = \theta K e^{-\theta t}\,dt$, we find $\mathbb{P}_\nu(T = T_k) = \mathbb{P}_\nu(Y_T \in \partial\mathcal{X}_k \times \prod_{l \ne k} \mathcal{X}_l) = \theta_k/\theta$.

2.5 Restricting the Exit State

Let $E \subset \partial\mathcal{X}$ be such that $\mathbb{P}(Y_T \in E) < 1$. We will consider the process that hits the boundary at $\partial\mathcal{X} \setminus E$. For this purpose, let $\Omega^* = \Omega \setminus \{Y_T \in E\} = \{Y_T \notin E\}$ with the restricted σ-field $\mathcal{F}^* = \mathcal{F}|_{\Omega^*}$. We endow this space with the family of conditional probability measures

$$\mathbb{P}^*_x(A) = \mathbb{P}_x(A | Y_T \notin E), \quad x \in \mathcal{X}^* := \mathcal{X} \setminus E, A \in \mathcal{F}^*.$$

Consider the process $Y^* = (Y^*_t : t \ge 0)$, with $Y^*_t := Y_{t \wedge T}$ on Ω^*, so it is the restriction of Y^T to Ω^*. The process Y^* takes values in $\mathcal{X}^*$. We define $\partial\mathcal{X}^* = \partial\mathcal{X} \setminus E$

and the killing time $T^* = \inf\{t \geq 0 : Y_t^* \in \partial\mathcal{X}^*\}$. The set of allowable states is $\mathcal{X}^{\mathbf{a}} = \mathcal{X}^* \setminus \partial\mathcal{X}^*$.

Proposition 2.8 *Y^* is a Markov process with respect to the probability measures $(\mathbb{P}_x^* : x \in \mathcal{X}^*)$. The process $Y_{T^*}^*$ takes values in $\partial\mathcal{X}^*$ with the measure $\mathbb{P}_x^*(Y_{T^*}^* \in C) = \mathbb{P}_x(Y_T \in C | Y_T \notin E)$ for $x \in \mathcal{X}^*$, $C \in \mathcal{B}(\partial\mathcal{X} \setminus E)$.*

Moreover, if ν is a QSD for Y killed at T, the probability measure ν^* defined on $\mathcal{X}^{\mathbf{a}}$ by

$$\nu^*(B) = \frac{1}{\mathbb{P}_\nu(Y_T \notin E)} \int_B \mathbb{P}_x(Y_T \notin E)\nu(dx), \quad B \in \mathcal{B}(\mathcal{X}^{\mathbf{a}}), \tag{2.16}$$

is a QSD for the process Y^* killed at T^*.

Proof Let us see that Y^* is a Markov process. Since $\mathbb{P}_x^*(Y_t^* = x \ \forall t \geq 0) = 1$ for $x \in \partial\mathcal{X}^*$, we can assume that the starting point satisfies $x \in \mathcal{X}^{\mathbf{a}}$. For $s \geq 0$ let $\mathcal{F}_s^*$ be the σ-field $\mathcal{F}_s$ restricted to $\Omega^* = \{Y_T \notin E\}$. For $0 \leq s < t$ we have

$$\begin{aligned}
\mathbb{E}_x^*\big(g(Y_t^*)\big|\mathcal{F}_s^*\big) &= \mathbb{E}_x\left(\frac{1}{\mathbb{P}_x(Y_T \notin E)} g(Y_{t\wedge T})\mathbb{1}_{Y_T \notin E}\,\middle|\,\mathcal{F}_s^*\right)\mathbb{1}_{Y_T \notin E} \\
&= \mathbb{E}_{Y_{s\wedge T}}\left(g(Y_{(t-s)\wedge T})\mathbb{1}_{Y_T \notin E}\frac{1}{\mathbb{P}_x(Y_T \notin E)}\right)\mathbb{1}_{Y_T \notin E} = \mathbb{E}_{Y_s^*}^*\big(g(Y_{t-s}^*)\big).
\end{aligned}$$

Then, the Markov property holds.

The only thing left to prove is that the measure ν^* given in (2.16) is a QSD for the process Y^* killed at T^*. We have

$$\begin{aligned}
&\mathbb{P}_{\nu^*}^*\big(Y_t^* \in B, T^* > t\big) \\
&\quad = \int_{\mathcal{X}^{\mathrm{a}}} \mathbb{P}_x^*\big(Y_t^* \in B, T > t\big)\nu^*(dx) \\
&\quad = \int_{\mathcal{X}^{\mathrm{a}}} \frac{1}{\mathbb{P}_x(Y_T \notin E)}\mathbb{P}_x(Y_t \in B, T > t, Y_T \notin E)\frac{\mathbb{P}_x(Y_T \notin E)}{\mathbb{P}_\nu(Y_T \notin E)}\nu(dx) \\
&\quad = \frac{\mathbb{P}_\nu(Y_t \in B, T > t, Y_T \notin E)}{\mathbb{P}_\nu(Y_T \notin E)}.
\end{aligned}$$

Then, from the QSD property of ν and the Markov property satisfied by Y, we find for $B \in \mathcal{B}(\mathcal{X}^{\mathbf{a}})$ that

$$\begin{aligned}
\mathbb{P}_{\nu^*}^*\big(Y_t^* \in B, T^* > t\big) &= \frac{1}{\mathbb{P}_\nu(Y_T \notin E)}\int_B \mathbb{P}_y(Y_T \notin E)\mathbb{P}_\nu(T > t)\nu(dy) \\
&= \frac{\mathbb{P}_\nu(T > t)}{\mathbb{P}_\nu(Y_T \notin E)}\int_B \mathbb{P}_y(Y_T \notin E)\nu(dy).
\end{aligned}$$

We conclude that

$$\mathbb{P}_{\nu^*}^*\big(Y_t^* \in B\big|T^* > t\big) = \frac{\int_B \mathbb{P}_y(Y_T \notin E)\nu(dy)}{\int_{\mathcal{X}^{\mathrm{a}}} \mathbb{P}_z(Y_T \notin E)\nu(dz)} = \nu^*(B).$$

Then ν^* fulfills the QSD requirements. □

2.6 Characterization of QSD by the Semigroup

Let $(P_t : t \geq 0)$ be the semigroup of the process before killing at $\partial\mathcal{X}$ (often this semigroup is only denoted by (P_t)). Then

$$P_t f(x) = \mathbb{E}_x\big(f(Y_t), T > t\big), \quad x \in \mathcal{X}^{\mathbf{a}}, \tag{2.17}$$

and it acts on the set $\mathcal{M}_b(\mathcal{X}^{\mathbf{a}})$ of real measurable bounded functions defined on $\mathcal{X}^{\mathbf{a}}$. The Markov property implies that it verifies the semigroup equation

$$P_t \circ P_s = P_{t+s} = P_s \circ P_t \quad \text{on } \mathcal{M}_b\big(\mathcal{X}^{\mathbf{a}}\big).$$

We denote $\mathbf{1} = \mathbb{1}_{\mathcal{X}^{\mathbf{a}}}$. Then $\mathbb{P}_x(T > t) = P_t\mathbf{1}(x)$ for $x \in \mathcal{X}^{\mathbf{a}}$.

Let $(\mathcal{M}_b(\mathcal{X}^{\mathbf{a}}), \|\cdot\|_\infty)$ be the Banach space, where

$$\|f\|_\infty = \sup\big\{\big|f(x)\big| : x \in \mathcal{X}^{\mathbf{a}}\big\}.$$

Note that P_t is a linear contraction on $(\mathcal{M}_b(\mathcal{X}^{\mathbf{a}}), \|\cdot\|_\infty)$,

$$\|P_t\|_\infty \leq 1 \quad \text{where } \|P_t\|_\infty := \sup\big\{\|P_t f\|_\infty : f \in \mathcal{M}_b\big(\mathcal{X}^{\mathbf{a}}\big), \|f\|_\infty \leq 1\big\}.$$

Notice that for any $g \in \mathcal{M}_b(\mathcal{X})$ and $x \in \mathcal{X}^{\mathbf{a}}$ we have

$$\mathbb{E}_x\big(g(Y_t)\big) = \mathbb{E}_x\big(g(Y_t), T > t\big) + g(Y_T)\mathbb{P}_x(T \leq t).$$

In particular, if $g(y) = 0$ $\forall y \in \partial\mathcal{X}$, we get $\mathbb{E}_x(g(Y_t)) = \mathbb{E}_x(g(Y_t), T > t)$ for all $x \in \mathcal{X}^{\mathbf{a}}$.

For any measure v on $\mathcal{X}^{\mathbf{a}}$, the set of quantities $(v(P_t\mathbb{1}_B) : B \in \mathcal{B}(\mathcal{X}^{\mathbf{a}}))$ defines a measure on $\mathcal{X}^{\mathbf{a}}$ which is denoted by $P_t^\dagger v$. So, by linearization this action can be also defined for all finite signed measures. Hence the action of the semigroup on the space of measures on $\mathcal{X}^{\mathbf{a}}$, or the space of signed finite measures on $\mathcal{X}^{\mathbf{a}}$, is defined by

$$\forall f \in \mathcal{M}_b\big(\mathcal{X}^{\mathbf{a}}\big): \quad \big(P_t^\dagger v\big)(f) = v(P_t f).$$

We note that $P_t^\dagger$ is also a semigroup acting in these spaces of measures, that is,

$$P_{t+s}^\dagger = P_t^\dagger \circ P_s^\dagger.$$

In fact,

$$P_t^\dagger\big(\big(P_s^\dagger v\big)f\big) = \big(P_s^\dagger v\big)(P_t f) = v(P_s P_t f) = v(P_{t+s} f) = \big(P_{t+s}^\dagger v\big)f.$$

For all $v \in \mathcal{P}(\mathcal{X})$ we have

$$\forall f \in \mathcal{M}_b\big(\mathcal{X}^{\mathbf{a}}\big)\ \forall t \geq 0: \quad v(P_t f) = \mathbb{E}_v\big(f(Y_t)\mathbb{1}_{T>t}\big).$$

Hence, from (2.6) and (2.7), $\nu \in \mathcal{P}(\mathcal{X})$ is a QSD if and only if it satisfies

$$\exists\theta > 0 \quad \text{such that} \quad \forall f \in \mathcal{M}_b\big(\mathcal{X}^{\mathbf{a}}\big)\ \forall t \geq 0: \quad \nu(P_t f) = e^{-\theta t}\nu(f),$$

or if and only if

$$\exists\theta > 0 \quad \text{such that} \quad \forall t \geq 0: \quad P_t^\dagger \nu = e^{-\theta t}\nu. \tag{2.18}$$

The parameter θ turns out to be equal to $\theta(\nu)$, the exponential rate of survival of ν.

Let us show that there exists some QSD when the eigenmeasure equation (2.18) is satisfied by a finite measure but for a fixed strictly positive time (result shown in Collet et al. 2011). For this purpose, the following observation is useful. Since by definition the process Y is adapted to the filtration $\mathbb{F} = (\mathcal{F}_t : t \geq 0)$, it is progressively measurable (for instance, see Revuz and Yor 1999, Proposition 4.8), that is, for all $t \geq 0$ the map

$$[0,t] \times \Omega \to \mathcal{X}, \qquad (s,\omega) \to Y_s(\omega),$$

is $\mathcal{B}([0,t]) \otimes \mathcal{F}_t$ measurable. Then also the killed process Y^T is progressively measurable with respect to the filtration $(\mathcal{F}_{t\wedge T} : t \geq 0)$ (see Revuz and Yor 1999, Proposition 4.10) and by integrating over Ω we deduce that $P_t \mathbb{1}_B$ is measurable in t.

Lemma 2.9 *Let $\widetilde{\nu} \in \mathcal{P}(\mathcal{X}^{\mathbf{a}})$ and $\beta > 0$ be such that $P_1^\dagger \widetilde{\nu} = \beta \widetilde{\nu}$. Then $\beta < 1$ and there exists a QSD ν whose exponential rate of survival is $\theta := -\log \beta > 0$.*

Proof Let $\mathbf{1} = \mathbb{1}_{\mathcal{X}^{\mathbf{a}}}$. From condition (2.3), there exists some integer $n > 0$ such that $\mathbb{P}_{\widetilde{\nu}}(T > n) < 1$. From $\beta^n = (P_n^\dagger \widetilde{\nu})\mathbf{1} = \mathbb{P}_{\widetilde{\nu}}(T > n) < 1$, we get $\beta^n < 1$, so $\beta < 1$ and $\theta := -\log \beta > 0$. We must show that there exists $\nu \in \mathcal{P}(\mathcal{X}^{\mathbf{a}})$ such that $P_t^\dagger \nu = e^{-\theta t} \nu$ for all $t \geq 0$.

Since $P_t \mathbb{1}_B$ is measurable in t, for all $B \in \mathcal{B}(\mathcal{X}^{\mathbf{a}})$, the following quantity is well-defined:

$$\nu(B) := \int_0^1 e^{\theta s} \widetilde{\nu}(P_s \mathbb{1}_B)\, ds.$$

By linearity and monotonicity, we get that ν is a finite measure on $(\mathcal{X}^{\mathbf{a}}, \mathcal{B}(\mathcal{X}^{\mathbf{a}}))$. By definition of $P_s^\dagger$, we can write

$$\nu(B) := \int_0^1 e^{\theta s} \big(P_s^\dagger \widetilde{\nu}\big)(B)\, ds.$$

Let $t \in (0,1]$. We have

$$\begin{aligned}
P_t^\dagger \nu(B) = \nu(P_t \mathbb{1}_B) &= \int_0^1 e^{\theta s} P_s^\dagger \widetilde{\nu}(P_t \mathbb{1}_B)\, ds = \int_0^1 e^{\theta s} P_{t+s}^\dagger \widetilde{\nu}(B)\, ds \\
&= \int_0^{1-t} e^{\theta s} P_{t+s}^\dagger \widetilde{\nu}(B)\, ds + \int_{1-t}^1 e^{\theta s} P_{t+s}^\dagger \widetilde{\nu}(B)\, ds \\
&= \int_t^1 e^{\theta(u-t)} P_u^\dagger \widetilde{\nu}(B)\, du + \int_1^{1+t} e^{\theta(u-t)} P_u^\dagger \widetilde{\nu}(B)\, du \\
&= e^{-\theta t} \int_t^1 e^{\theta u} P_u^\dagger \widetilde{\nu}(B)\, du + e^{-\theta t} \int_0^t e^{\theta u} e^{\theta} \big(P_u^\dagger P_1^\dagger \widetilde{\nu}\big)(B)\, du \\
&= e^{-\theta t} \nu(B),
\end{aligned}$$

where in the last equality we use $P_1^\dagger \widetilde{\nu} = e^{-\theta} \widetilde{\nu}$. (Note that the integrals $\int_0^{1-t}$ and $\int_t^1$ vanish when $t = 1$.) In particular, we proved $P_1^\dagger \nu = e^{-\theta} \nu$ and so $P_n^\dagger \nu = e^{-n\theta} \nu$.

For $t > 1$, we write $t = n + r$ with $0 \le r < 1$ and $n \in \mathbb{N} = \{1, 2, \ldots\}$. We have

$$P_t^\dagger \nu = P_r^\dagger P_n^\dagger \nu = e^{-n\theta} P_r^\dagger \nu = e^{-n\theta} e^{-r\theta} \nu = e^{-\theta t} \nu.$$

Now we normalize ν. We have shown that it verifies the QSD condition (2.18). □

2.7 Continuity Assumptions

Let us recall one of the theorems of Krein in Oikhberg and Troitsky (2005), we refer to Theorem 4 therein. Let $(B, \le, \|\cdot\|)$ be an ordered normed space and assume that there exists an element $e \in B$ that satisfies $\|e\| = 1$ and $e \ge x$ for all $x \in B$ with $\|x\| \le 1$. Then, every positive operator $R : B \to B$ is such that its adjoint R^* has a positive eigenvector v. The positivity property of $v \in B^*$ means that $v(b) \ge 0$ for all $b \ge 0$. This theorem implies the following result on the existence of QSD.

Proposition 2.10 *Assume that $\mathcal{X}^{\mathbf{a}}$ is a compact Hausdorff space and that P_1 preserves the set of continuous functions: $P_1(\mathcal{C}(\mathcal{X}^{\mathbf{a}})) \subseteq \mathcal{C}(\mathcal{X}^{\mathbf{a}})$. Then there exists a QSD.*

Proof The set $\mathcal{C}(\mathcal{X}^{\mathbf{a}})$ is an ordered normed space, endowed with the usual order between functions and the supremum $\|\cdot\|_\infty$ norm. The function $\mathbf{1}$ verifies the condition $f \le \mathbf{1}$ for all $f \in \mathcal{C}(\mathcal{X}^{\mathbf{a}})$ with $\|f\|_\infty \le 1$. Since $P_1 : \mathcal{C}(\mathcal{X}^{\mathbf{a}}) \to \mathcal{C}(\mathcal{X}^{\mathbf{a}})$ is a positive operator, the above Krein's Theorem states that $P_1^* : \mathcal{C}(\mathcal{X}^{\mathbf{a}})^* \to \mathcal{C}(\mathcal{X}^{\mathbf{a}})^*$ has a positive eigenfunction. The Riesz Representation Theorem implies that this eigenfunction is a positive finite measure v. Hence $P^* v = \beta v$ for some β, and we necessarily have $\beta > 0$. The probability measure $\nu = v(\mathbf{1})^{-1} v$ verifies $P_1^* \nu = \beta \nu$. Since $P_1^* = P_1^\dagger$ on $\mathcal{C}(\mathcal{X}^{\mathbf{a}})^*$, from Lemma 2.9 we get that ν is a QSD. □

Now we state an existence result of a QSD when $\mathcal{X}^{\mathbf{a}}$ is not necessarily compact. It can be found in Collet et al. (2011). We recall that the set of bounded continuous functions $\mathcal{C}_b(\mathcal{X}^{\mathbf{a}})$ becomes a Banach space when equipped with the supremum norm.

Theorem 2.11 *Assume that P_1 maps $\mathcal{C}_b(\mathcal{X}^{\mathbf{a}})$ into itself, $P_1 : \mathcal{C}_b(\mathcal{X}^{\mathbf{a}}) \to \mathcal{C}_b(\mathcal{X}^{\mathbf{a}})$. Also assume that the following hypothesis holds: $\exists \varphi_0 \in \mathcal{C}_b(\mathcal{X}^{\mathbf{a}})$ such that $\varphi_0 \ge 1$ and it verifies*

Hypothesis 2.1 *For any $u \ge 0$, the set $\varphi_0^{-1}([0, u])$ is compact.*

Assume also that there exist three constants $c_1 > \gamma > 0$ and $\beta > 0$ such that

$$P_1 \mathbf{1} \ge c_1 \mathbf{1}$$

and for any $h \in \mathcal{C}_b(\mathcal{X}^{\mathbf{a}})$ with $0 \le h \le \varphi_0$

$$P_1 h \le \gamma \varphi_0 + \beta.$$

Then, there exists a QSD.

Proof In the dual space $C_b(\mathcal{X}^{\mathbf{a}})^*$, we define for any real $K > 0$ the convex set $\mathscr{K}_K$ given by

$$\mathscr{K}_K = \Big\{ v \in C_b\big(\mathcal{X}^{\mathbf{a}}\big)^* : v \ge 0,\, v(1) = 1, \sup_{h \in C_b(\mathcal{X}^{\mathbf{a}}), 0 \le h \le \varphi_0} v(h) \le K \Big\}.$$

We observe that for any K large enough the set $\mathscr{K}_K$ is nonempty. It suffices to consider a Dirac measure δ_x at a point $x \in \mathcal{X}^{\mathbf{a}}$ and to take $K \ge \varphi_0(x)$.

Since for any $K \ge 0$, $\mathscr{K}_K$ is an intersection of weak* closed subsets, it is closed in the weak* topology. We now introduce the nonlinear operator R with domain $\mathscr{K}_K$ and defined by

$$Rv = \frac{P_1^* v}{v(P_1 \mathbf{1})}.$$

Note that since $P_1 \mathbf{1} > c_1 \mathbf{1}$, we have $v(P_1 \mathbf{1}) \ge c_1 v(\mathbf{1})$ and the operator R is well defined on $\mathscr{K}_K$. We obviously have $Rv(\mathbf{1}) = 1$. We now prove that R maps $\mathscr{K}_K$ into itself.

Consider $h \in C_b(\mathcal{X}^{\mathbf{a}})$ with $0 \le h \le \varphi_0$. Since

$$P_1 h \le \gamma \varphi_0 + \beta,$$

and obviously

$$0 \le P_1 h \le \gamma \frac{\| P_1 h \|}{\gamma},$$

we get

$$0 \le P_1 h \le \gamma \min\big(\varphi_0, \| P_1 h\| / \gamma\big) + \beta.$$

Therefore, since the function $h' = \min(\varphi_0, \|P_1 h\|/\gamma)$ satisfies $h' \in C_b(\mathcal{X}^{\mathbf{a}})$ and $0 \le h' \le \varphi_0$, we conclude that for $v \in \mathscr{K}_K$

$$Rv(h) \le \frac{\gamma v(h') + \beta}{c_1}.$$

From the bound $v(h') \le K$, we get

$$Rv(h) \le \frac{\gamma v(h') + \beta}{c_1} \le \frac{\gamma}{c_1} K + \frac{\beta}{c_1} \le K$$

if $K > \beta / (c_1 - \gamma)$. Therefore, for any K large enough, the set $\mathscr{K}_K$ is nonempty and mapped into itself by R.

It is easy to show that R is continuous on $\mathscr{K}_K$ in the weak* topology. This follows at once from the continuity of the operator P_1. We can now apply Tychonov's Fixed Point Theorem (see Tychonov 1935) to deduce that R has a fixed point. This implies that there is a point $v \in \mathscr{K}_K$ such that $P_1^* v = v(P_1 \mathbf{1}) v$. The proof will be finished once we prove that v can be represented by a measure ν on $\mathcal{X}^{\mathbf{a}}$, that is,

$$\forall f \in C_b\big(\mathcal{X}^{\mathbf{a}}\big): \quad v(f) = \int f \, d\nu.$$

(Since $v(1) = 1$, ν will be necessarily be a probability measure.) Let $\varpi : [0, \infty) \to [0, 1]$ be a continuous and nonincreasing with $\varpi = 1$ on the interval $[0, 1]$ and

$\varpi(2) = 0$ (so $\varpi = 0$ on $[2, \infty)$). For any integer $m \geq 1$ let υ_m be the continuous positive linear form defined on $\mathcal{C}(\mathcal{X}^{\mathbf{a}})$ by

$$\upsilon_m(f) = \upsilon\big(\varpi(\varphi_0/m)f\big).$$

This linear form has support in the set $\varphi_0^{-1}([0, 2m])$ in the sense that it vanishes on functions which vanish on this set. Since $\varphi_0^{-1}([0, 2m])$ is compact by Hypothesis 2.1, the form υ_m can be identified with a nonnegative measure ν_m on $\mathcal{X}^{\mathbf{a}}$, namely, for any $f \in \mathcal{C}_b(\mathcal{X}^{\mathbf{a}})$ we have

$$\forall f \in \mathcal{C}_b(\mathcal{X}^{\mathbf{a}}): \quad \upsilon_m(f) = \int f \, d\nu_m.$$

We now prove that this sequence of measures $(\nu_m : m \geq 1)$ is tight. Let $u > 0$ and define the set

$$\mathcal{K}_u = \varphi_0^{-1}\big([0, u]\big).$$

By Hypothesis 2.1, for any $u > 0$ this is a compact set. We now observe that

$$\mathbb{1}_{\mathcal{K}_u^c} \leq \big(\mathbf{1} - \varpi(2\varphi_0/u)\big).$$

Therefore,

$$\begin{aligned}\nu_m\big(\mathcal{K}_u^c\big) \leq \nu_m\big(\mathbf{1} - \varpi(2\varphi_0/u)\big) &= \upsilon_m\big(\mathbf{1} - \varpi(2\varphi_0/u)\big) \\ &= \upsilon\big(\varpi(\varphi_0/m)\big(\mathbf{1} - \varpi(2\varphi_0/u)\big)\big).\end{aligned}$$

We now use the following facts: the function $\varphi_0\varpi(\varphi_0/m)(\mathbf{1} - \varpi(2\varphi_0/u))$ is in $\mathcal{C}_b(\mathcal{X}^{\mathbf{a}})$ and satisfies

$$\frac{u}{2}\varpi(\varphi_0/m)\big(\mathbf{1} - \varpi(2\varphi_0/u)\big) \leq \varpi(\varphi_0/m)\big(\mathbf{1} - \varpi(2\varphi_0/u)\big)\varphi_0 \leq \varphi_0,$$

and $\upsilon \in \mathscr{H}_K$; then from definition of $\mathscr{H}_K$ we obtain that

$$\upsilon\big(\varpi(\varphi_0/m)\big(\mathbf{1} - \varpi(2\varphi_0/u)\big)\big) \leq \frac{2}{u}\upsilon\big(\varpi(\varphi_0/m)\big(\mathbf{1} - \varpi(2\varphi_0/u)\big)\varphi_0\big) \leq \frac{2K}{u}.$$

In other words, for any $u > 0$ we have for any integer $m \geq 1$

$$\nu_m\big(\mathcal{K}_u^c\big) \leq \frac{2K}{u}.$$

The sequence of measures $(\nu_m : m \geq 1)$ is therefore tight, and we denote by ν an accumulation point which is a nonnegative measure on $\mathcal{X}^{\mathbf{a}}$. We now finish the proof by showing that for any $f \in \mathcal{C}_b(\mathcal{X}^{\mathbf{a}})$ we have $\nu(f) = \upsilon(f)$. For this purpose, we write

$$\upsilon(f) = \upsilon\big(\varpi(\varphi_0/m)f\big) + \upsilon\big(\big(\mathbf{1} - \varpi(\varphi_0/m)\big)f\big).$$

We now use the inequality

$$\varphi_0 \geq \big(\mathbf{1} - \varpi(\varphi_0/m)\big)\varphi_0 \geq m\big(\mathbf{1} - \varpi(\varphi_0/m)\big),$$

and $(\mathbf{1} - \varpi(\varphi_0/m))\varphi_0 \in \mathcal{C}_b(\mathcal{X}^{\mathbf{a}})$ to conclude

$$\big|\upsilon\big(\big(\mathbf{1} - \varpi(\varphi_0/m)\big)f\big)\big| \leq \upsilon\big(\big(\mathbf{1} - \varpi(\varphi_0/m)\big)|f|\big) \leq \|f\|\upsilon\big(\mathbf{1} - \varpi(\varphi_0/m)\big) \leq \frac{K}{m}.$$

Hence, for any $f \in \mathcal{C}_b(\mathcal{X}^{\mathbf{a}})$ we have

$$\left|v(f) - \nu_m(f)\right| \leq \frac{K}{m}.$$

From the tightness, we get $\lim_{m\to\infty} \nu_m(f) = \nu(f)$ for all $f \in \mathcal{C}_b(\mathcal{X}^{\mathbf{a}})$ (see Billingsley 1968), and therefore $\nu(f) = v(f)$, which completes the proof of the theorem.

□

Chapter 3
Markov Chains on Finite Spaces

In this chapter, we introduce the main problems on QSDs for irreducible finite Markov chains. In Sect. 3.1, we show that there is a unique QSD which is the normalized left Perron–Frobenius eigenvector of the jump rates matrix restricted to the allowed states. The right eigenvector is shown to be the asymptotic ratio of survival probabilities. In Sect. 3.2, it is proved that the trajectories that survive forever form a Markov chain which is an h-process of the original one with weights given by the right eigenvector. In Sect. 3.3, we give some computations when the jump rate matrix restricted to the allowed states is symmetric. In Sect. 3.5, we compute the unique QSD for the random walk with barriers. In Sect. 3.6, we prove a Central Limit Theorem for the chain conditioned to survive when time is discrete.

3.1 Existence of a QSD

Let $Y = (Y_t : t \geq 0)$ be a Markov process taking values in a finite set I with conservative jump rate matrix $Q = (q(i, j) : i, j \in I)$. Then, the transition probability is given by,

$$\forall i, j \in I: \quad p_t(i, j) = \mathbb{P}_i(Y_t = j) = \left(e^{Qt}\right)(i, j),$$

where $e^{Qt} = \sum_{n\geq 0} \frac{Q^n t^n}{n!}$ is the exponential of the matrix Q and $e^{Qt}(i, j)$ denotes the (i, j)th entry of e^{Qt}. (In general, $M(i, j)$ denotes the (i, j)th entry of a matrix M.)

Let $\partial I \subset I$ be the set of forbidden states and $T = \inf\{t \geq 0 : Y_t \in \partial I\}$. Since the points in ∂I can be assumed to be absorbing, we have $q(i, i) = 0$ for $i \in \partial I$. Let $I^{\mathbf{a}} = I \setminus \partial I$ be the set of allowed states and $Q^{\mathbf{a}} = (q(i, j) : i, j \in I^{\mathbf{a}})$. Then, the transition probability before killing is given by

$$\forall i, j \in I^{\mathbf{a}}: \quad p_t(i, j) = \left(e^{Q^{\mathbf{a}}t}\right)(i, j). \tag{3.1}$$

We recall that $(P_t = e^{Q^{\mathbf{a}}t} : t \geq 0)$ denotes the killed semigroup. Observe that $Q^{\mathbf{a}}\mathbf{1} \leq 0$ where $\mathbf{1} = \mathbb{1}_{I^{\mathbf{a}}}$ is the unit vector in $I^{\mathbf{a}}$. We assume that $Q^{\mathbf{a}}$ is an irreducible matrix, that is,

P. Collet et al., *Quasi-Stationary Distributions*, Probability and Its Applications,
DOI 10.1007/978-3-642-33131-2_3, © Springer-Verlag Berlin Heidelberg 2013

$$\forall i, j \in I^{\mathbf{a}}, i \neq j, \exists n > 0, \exists i_1, \ldots, i_{n-1} \in I^{\mathbf{a}} \setminus \{i, j\} \quad \text{all different}$$
$$\text{such that} \quad q(i_k, i_{k+1}) > 0 \quad \forall k = 0, \ldots, n-1, \text{ where } i_0 = i, i_n = j. \tag{3.2}$$

Then $p_t(i, j) > 0$ for all $i, j \in I^{\mathbf{a}}$ and $t > 0$, that is, $(Y_t : t < T)$ is irreducible. We also assume the existence of some $i_0 \in I^{\mathbf{a}}$ that satisfies $(Q^{\mathbf{a}}\mathbf{1})(i_0) < 0$. Hence, since I is finite and $Q^{\mathbf{a}}$ is irreducible, we get that condition (2.1) is satisfied: $\mathbb{P}_i(T < \infty) = 1$ for all $i \in I^{\mathbf{a}}$. In fact, in this irreducible finite case with absorbing forbidden states, conditions (2.2) and (2.1) are equivalent.

The Perron–Frobenius theory asserts (see Theorem 1.2 in Seneta 1981) that there exists an eigenvalue $-\theta < 0$ of $Q^{\mathbf{a}}$ that has left- and right-eigenvectors ν and ψ, respectively, that can be chosen to be strictly positive. The vectors are written as column vectors, and ν' denotes the row vector, the transposed ν. We can impose the following two normalizing conditions

$$\nu'\psi = 1, \qquad \nu'\mathbf{1} = \mathbf{1},$$

where $\nu'\psi = \sum_{i \in I^{\mathbf{a}}} \psi(i)\nu(i)$. In particular, ν is a probability vector. Let us see that ν is a QSD and θ is the exponential rate of survival. From (3.1) we get

$$(\nu' P_t)(j) = (\nu' e^{Q^{\mathbf{a}} t})(j) = e^{-\theta t}\nu(j),$$

so

$$\mathbb{P}_\nu(T > t) = \nu' P_t \mathbf{1} = e^{-\theta t}. \tag{3.3}$$

Then,

$$(\nu' P_t)(j) = \nu(j)\mathbb{P}_\nu(T > t),$$

proving that ν is a QSD (see condition (2.5)).

On the other hand, as part of the Perron–Frobenius Theorem, we have

$$\exists \gamma < -\theta \quad \text{such that} \quad e^{Q^{\mathbf{a}} t} = e^{-\theta t}\psi\nu' + O(e^{\gamma t}), \tag{3.4}$$

or equivalently,

$$p_t(i, j) = e^{-\theta t}\psi(i)\nu(j) + O(e^{\gamma t}).$$

Then, θ is the exponential rate of survival for the process, that is,

$$\forall i, j \in I^{\mathbf{a}}: \quad \theta = -\lim \frac{1}{t} \log p_t(i, j).$$

Also we get that the probability of survival verifies

$$\forall i \in I^{\mathbf{a}}: \quad \mathbb{P}_i(T > t) = \sum_{j \in I^{\mathbf{a}}} p_t(i, j) = e^{-\theta t}\psi(i) + O(e^{\gamma t}). \tag{3.5}$$

Thus, the following quasi-limiting behavior is obtained:

$$\forall i \in I^{\mathbf{a}}, \ \forall j \in I^{\mathbf{a}}: \quad \lim_{t \to \infty} \mathbb{P}_i(Y_t = j \mid T > t) = \nu(j). \tag{3.6}$$

This property was shown in Darroch and Seneta (1965). When $\nu \in \mathcal{P}(I^{\mathbf{a}})$ satisfies property (3.6), it is called a quasi-limiting distribution, or a Yaglom limit.

On the other hand, (3.5) implies

$$\forall i, j \in I^{\mathbf{a}}: \quad \lim_{t\to\infty} \frac{\mathbb{P}_i(T > t)}{\mathbb{P}_j(T > t)} = \frac{\psi(i)}{\psi(j)}.$$

Then, the ratios between the components of the right-eigenvector correspond to the asymptotic ratios of the survival probabilities.

The same estimations imply

$$\forall i, j \in I^{\mathbf{a}}: \quad \lim_{t\to\infty} \frac{\mathbb{P}_i(T > t - s)}{\mathbb{P}_j(T > t)} = e^{\theta s} \frac{\psi(i)}{\psi(j)}.$$

Let us give some possible behaviors of initial distributions for which the killing time is exponentially distributed.

Proposition 3.1

(i) *Let ρ be a probability measure on $I^{\mathbf{a}}$ for which $\exists \alpha > 0$ such that $\mathbb{P}_\rho(T > t) = e^{-\alpha t}$ $\forall t \geq 0$. Then its parameter is the Perron–Frobenius eigenvalue: $\alpha = \theta$ and $\mathbb{P}_\rho(T > t) = e^{-\theta t}$ $\forall t \geq 0$.*

(ii) *If the row sums of $Q^{\mathbf{a}}$ are constant then $\psi = \mathbf{1}$ and all probability measures ρ on $I^{\mathbf{a}}$ verify the exponential distribution property $\mathbb{P}_\rho(T > t) = e^{-\theta t}$ $\forall t \geq 0$. However, by irreducibility only one of them is a QSD.*

(iii) *Assume the chain is a random walk killed at 0, that is, $I = \{0, \dots, N\}$, Q is conservative, $\partial I = \{0\}$ and for $j \in I$ we have: $q(0, j) = 0$ and $q(i, j) = 0$ if and only if $|i - j| > 1$ for $i \in I^{\mathbf{a}}$. Hence, if ρ is a probability measure for which $\exists \alpha > 0$ such that $\mathbb{P}_\rho(T > t) = e^{-\alpha t}$ $\forall t \geq 0$, then $\alpha = \theta$ and ρ is the unique QSD.*

Proof Part (i) follows straightforwardly from (3.5). Let us show part (ii). Denote by $-c$ the constant row sum of $Q^{\mathbf{a}}$, so $(Q^{\mathbf{a}})^n \mathbf{1} = (-c)^n \mathbf{1}$. Then, for all probability measures ρ, we have

$$\mathbb{P}_\rho(T > t) = \rho' e^{Q^{\mathbf{a}} t} \mathbf{1} = e^{-ct} \rho' \mathbf{1} = e^{-ct}.$$

Let us prove (iii). We have

$$\forall n \geq 1: \quad (-\alpha)^n = \frac{d^n}{dt^n} \mathbb{P}_\rho(T > t) \bigg|_{t=0} = \rho \big(Q^{\mathbf{a}}\big)^n \mathbf{1}. \tag{3.7}$$

Since the first coordinate of $Q^{\mathbf{a}}\mathbf{1}$ is the unique one which is different from 0 and in general for all $n \in \{1, \dots, N\}$ the first n coordinates of $Q^{\mathbf{a}}\mathbf{1}$ are the unique ones different from 0, there is a unique solution to (3.7). Since ρ is a probability measure, this unique solution is necessarily the QSD. □

Remark 3.2 Let $Q = (q(i, j) : i, j \in I)$ be a nonconservative irreducible jump rate matrix. Then, the associated chain looses mass at the states $bI = \{i \in I : \sum_{j \in I} q(i, j) < 0\}$. To each state $i \in bI$, we add a new state ∂_i which is absorbing, and we define $q(i, \partial_i) = -\sum_{j \in I} q(i, j)$. In this framework, $\partial = \{\partial_i : i \in bI\}$ is the set of forbidden states and $T = \inf\{T_{\partial_i} : i \in bI\}$ is the killing time. Let ν be the unique QSD of Y, the relation (2.15) in the proof of Theorem 2.6 gives

$$\mathbb{P}_\nu(Y_T = \partial_i) = \frac{\nu(i) q(i, \partial_i)}{\sum_{j \in bI} \nu(j) q(j, \partial_j)}.$$

3.2 The Chain of Trajectories That Survive Forever

Let $0 < s_1 < \cdots < s_k < t$ and $i_0, \ldots, i_k \in I^{\mathbf{a}}$. From the Markov property, we obtain

$$\mathbb{P}_i(Y_{s_1} = i_1, \ldots, Y_{s_k} = i_k, T > t) = \mathbb{P}_i(Y_{s_1} = i_1, \ldots, Y_{s_k} = i_k) \mathbb{P}_{i_k}(T > t - s_k).$$

Then,

$$\begin{aligned} &\lim_{t \to \infty} \mathbb{P}_{i_0}(Y_{s_1} = i_1, \ldots, Y_{s_k} = i_k | T > t) \\ &= \mathbb{P}_{i_0}(Y_{s_1} = i_1, \ldots, Y_{s_k} = i_k) \left(\lim_{t \to \infty} \frac{\mathbb{P}_{i_k}(T > t - s_k)}{\mathbb{P}_{i_0}(T > t)} \right) \\ &= e^{\theta s_k} \frac{\psi(i_k)}{\psi(i_0)} \mathbb{P}_{i_0}(Y_{s_1} = i_1, \ldots, Y_{s_k} = i_k) \\ &= \prod_{l=1}^{k} e^{\theta(s_l - s_{l-1})} \frac{\psi(i_l)}{\psi(i_{l-1})} \mathbb{P}_{i_{l-1}}(Y_{s_l} = i_l), \end{aligned}$$

where we put $s_0 = 0$. Therefore, the law of the process $Z = (Z_t : t \geq 0)$ starting from $i \in I^{\mathbf{a}}$ is given by

$$\mathbb{P}_i(Z_{s_1} = i_1, \ldots, Z_{s_k} = i_k) := \lim_{t \to \infty} \mathbb{P}_i(Y_{s_1} = i_1, \ldots, Y_{s_k} = i_k | T > t).$$

Then, Z is a well-defined Markov chain taking values on $I^{\mathbf{a}}$, called the survival process, and its transition probability kernel is given by

$$\mathbb{P}_i(Z_s = j) = e^{\theta s} \frac{\psi(j)}{\psi(i)} \mathbb{P}_i(Y_s = j).$$

Its jump rate matrix $\widetilde{Q} = (\tilde{q}(i, j) : i, j \in I^{\mathbf{a}})$ satisfies

$$\tilde{q}(i, j) = q(i, j) \frac{\psi(j)}{\psi(i)} \quad \text{if } i \neq j, \ \tilde{q}(i, i) = \theta + q(i, i). \tag{3.8}$$

Observe that $\widetilde{Q}$ is conservative because

$$\sum_{j \in I^{\mathbf{a}}} \tilde{q}(i, j) = \sum_{j \in I^{\mathbf{a}}} q(i, j) \frac{\psi(j)}{\psi(i)} + \theta = -\theta + \theta = 0,$$

where we used $\sum_{j \in I^{\mathbf{a}}} q(i, j) \psi(j) = -\theta \psi(i)$.

In this finite case, and under irreducibility, the chain Z is obviously positive recurrent. Moreover, the probability measure

$$\nu h = \big(\nu(i) \psi(i) : i \in I^{\mathbf{a}} \big), \tag{3.9}$$

is the stationary distribution for the process Z. In fact, it suffices to verify $(\nu h)'\widetilde{Q} = 0$, which follows from

$$\begin{aligned}\big((\nu h)'\widetilde{Q}\big)(j) &= \sum_{i\in I^{\mathbf{a}}} \nu(i)\psi(i)\tilde{q}(i,j) = \theta\nu(j)\psi(j) + \sum_{i\in I^{\mathbf{a}}} \nu(i)\psi(i)q(i,j)\frac{\psi(j)}{\psi(i)} \\ &= \theta\nu(j)\psi(j) - \theta\nu(j)\psi(j) = 0.\end{aligned}$$

Below we denote by $\widetilde{\mathbb{P}}$ the law of the process Z given by the jump rate matrix $\widetilde{Q}$ defined in (3.8).

Let us summarize what happens with some pieces of the trajectory of the process conditioned to survive. For any fixed $u > 0$ and every $\rho \in \mathcal{P}(I^{\mathbf{a}})$:

- We already proved the following convergence in distribution of the initial piece of trajectory:

$$\lim_{t\to\infty} \mathbb{P}_\rho\big((Y_s, s \in [0,u]) \in \bullet | T > t\big) = \widetilde{\mathbb{P}}_\rho\big((Z_s, s \in [0,u]) \in \bullet\big).$$

- From the quasi-limiting behavior (3.6), we have the convergence of the final piece of trajectory:

$$\lim_{t\to\infty} \mathbb{P}_\rho\big((Y_s, s \in [t-u,t]) \in \bullet | T > t\big) = \mathbb{P}_\nu\big((Y_s, s \in [0,u]) \in \bullet | T > u\big).$$

  Here ν is the unique QSD. This equality is a consequence of the following relations:

$$\lim_{t\to\infty} \mathbb{P}_\rho(Y_{t-u} = i | T > t-u) = \nu(i),$$

$$\lim_{t\to\infty} \frac{\mathbb{P}_\rho(T > t-u)}{\mathbb{P}_\rho(T > t)} = e^{\theta u} = \frac{1}{\mathbb{P}_\nu(T > u)},$$

  and

$$\begin{aligned}&\mathbb{P}_\rho\big((Y_s, s \in [t-u,t]) \in \bullet | T > t\big) \\ &\quad = \frac{\mathbb{P}_\rho(T > t-u)}{\mathbb{P}_\rho(T > t)} \sum_{i\in I^{\mathbf{a}}} \mathbb{P}_\rho(Y_{t-u} = i | T > t-u) \\ &\qquad \times \mathbb{P}_i\big((Y_s, s \in [0,u]) \in \bullet, T > u\big).\end{aligned}$$

- Let us see what happens with the renormalization of the whole trajectory. Consider the space of right continuous trajectories taking values in $I^{\mathbf{a}}$ and the time belonging to $[0, 1]$, and let $\mathcal{V}_u$ be its projections at time $u \in [0, 1]$. The following limit law is well-defined:

$$\begin{aligned}&\forall s_0 = 0 < s_1 < \cdots < s_n < 1 = s_{n+1}, i_0, i_1, \ldots, i_n, j \in I^{\mathbf{a}}: \\ &\quad \bar{\mathbb{P}}_{i_0}(\mathcal{V}_{s_1} = i_1, \ldots, \mathcal{V}_{s_1} = i_n, \mathcal{V}_1 = j) \\ &\qquad = \lim_{t\to\infty} \mathbb{P}_i(Y_{s_1 t} = i_1, \ldots, Y_{s_n t} = i_n, Y_t = j | T > t).\end{aligned}$$

  Let us compute it. Put $i_{n+1} = j$, we have

$$\bar{\mathbb{P}}_i(\mathcal{V}_{s_1}=i_1,\ldots,\mathcal{V}_{s_1}=i_n,\mathcal{V}_1=j)$$
$$=\left(\prod_{k=0}^{n}\frac{\mathbb{P}_{i_k}(Y_{(s_{k+1}-s_k)t}=i_{k+1}}{\mathbb{P}_{i_k}(T>(s_{k+1}-s_k)t)}\right)\left(\prod_{k=0}^{n}\frac{\mathbb{P}_{i_k}(T>(s_{k+1}-s_k)t)}{\mathbb{P}_{\nu}(T>(s_{k+1}-s_k)t)}\right)\frac{\mathbb{P}_{\nu}(T>t)}{\mathbb{P}_{i_0}(T>t)},$$

where we used $\mathbb{P}_{\nu}(T>t)=\prod_{k=0}^{n}\mathbb{P}_{\nu}(T>(s_{k+1}-s_k)t)$. Since $\nu(\psi)=1$, from (3.5) we get

$$\lim_{t\to\infty}\frac{\mathbb{P}_{i_k}(T>(s_{k+1}-s_k)t)}{\mathbb{P}_{\nu}(T>(s_{k+1}-s_k)t)}=\psi(i_k),$$

and so

$$\bar{\mathbb{P}}_i(\mathcal{V}_{s_1}=i_1,\ldots,\mathcal{V}_{s_1}=i_n,\mathcal{V}_1=j)=\left(\prod_{k=1}^{n}\nu(i_k)\psi(i_k)\right)\nu(j).$$

Therefore, the limit renormalized law $\bar{\mathbb{P}}_i$ is such that the projections $(\mathcal{V}_s : s\in[0,1])$ are all independent with $\mathcal{V}_0\sim\delta_i$, $\mathcal{V}_1\sim\nu$ and $\mathcal{V}_s\sim\nu\psi$ for all $s\in(0,1)$.

3.3 Symmetric Case

Assume the irreducible jump rate matrix $Q^{\mathbf{a}}$ is symmetric with respect to some measure $\pi=(\pi_i : i\in I^{\mathbf{a}})$, that is,

$$\pi_i q(i,j)=\pi_j q(j,i),\quad i,j\in I^{\mathbf{a}}.$$

Denote by $\mathcal{D}_\pi$ the diagonal matrix associated to π, so $(\mathcal{D}_\pi)_{ij}=\pi_i\delta(i,j)$ where $\delta(i,j)$ is the Kronecker's delta, $\delta(i,j)=1$ if $i=j$ and $\delta(i,j)=0$ otherwise. Then, $\mathcal{D}_\pi^{1/2}Q^{\mathbf{a}}\mathcal{D}_\pi^{-1/2}$ is a symmetric matrix.

Let $M=(m(i,j) : i,j\in I^{\mathbf{a}})$ be the orthogonal matrix formed by the normalized eigenvectors of $\mathcal{D}_\pi^{1/2}Q^{\mathbf{a}}\mathcal{D}_\pi^{-1/2}$. Then $\mathcal{D}_\pi^{1/2}Q^{\mathbf{a}}\mathcal{D}_\pi^{-1/2}=M\mathcal{D}M'$, where M' denotes the transposed matrix of M and $\mathcal{D}$ is the diagonal matrix made of the eigenvalues of $Q^{\mathbf{a}}$. These eigenvalues are denoted by $\{-d_k : k\in I^{\mathbf{a}}\}$. From (3.4) we get that all these eigenvalues are negative and dominated by the Perron–Frobenius eigenvalue $-\theta$ of $Q^{\mathbf{a}}$, so

$$\forall k\in I^{\mathbf{a}}:\quad d_k\ge\theta. \tag{3.10}$$

Then, we have

$$q(i,j)=-\sqrt{\frac{\pi_j}{\pi_i}}\sum_{k\in I^{\mathbf{a}}}d_k m(i,k)m(j,k),$$

which can be written as

$$q(i,j)=-\sqrt{\frac{\pi_j}{\pi_i}}\int_0^\infty x\,d\Gamma_{i,j}(x), \tag{3.11}$$

where $\Gamma_{i,j}$ is a signed measure, its total variation measure $\|\Gamma_{i,j}\| = \Gamma_{i,j}^+ + \Gamma_{i,j}^-$ is supported on $(0, \infty)$ and satisfies

$$\|\Gamma_{i,j}\|\big((0,\infty) \setminus \{d_k : k \in I^{\mathbf{a}}\}\big) = 0 \quad \text{and} \quad \Gamma_{i,j}\big(\{d_k\}\big) = m(i,k)m(j,k). \tag{3.12}$$

Since M is orthogonal,

$$\forall i, j \in I^{\mathbf{a}}: \quad \sum_{k \in I^{\mathbf{a}}} m(i,k)m(j,k) = \delta(i,j).$$

Hence,

$$\Gamma_{i,i}\big((0,\infty)\big) = \sum_{k \in I^{\mathbf{a}}} \Gamma_{i,i}\big(\{d_k\}\big) = \sum_{k \in I^{\mathbf{a}}} m(i,k)^2 = 1, \tag{3.13}$$

and so $\Gamma_{i,i}$ is a probability measure. From (3.10) we get that θ is the smallest value in the support of $\Gamma_{i,i}$ with weight $m(i,k_0)^2$, where k_0 is such that $-\theta = d_{k_0}$.

Notice that the Cauchy–Schwarz Inequality and the orthogonal condition of M gives

$$\forall i, j \in I^{\mathbf{a}}: \quad \Big(\sum_{k \in I^{\mathbf{a}}} \big|m(i,k)m(j,k)\big|\Big)^2 \le \Big(\sum_{k \in I^{\mathbf{a}}} m(i,k)^2\Big)\Big(\sum_{k \in I^{\mathbf{a}}} m(j,k)^2\Big) \le 1.$$

Then,

$$\forall i, j \in I^{\mathbf{a}}: \quad \|\Gamma_{i,j}\|\big([0,\infty)\big) \le 1. \tag{3.14}$$

Also the Cauchy–Schwarz Inequality and orthogonality implies that for every interval $[a,b)$,

$$\Big(\sum_{k \in I^{\mathbf{a}}: d_k \in [a,b)} \big|m(i,k)m(j,k)\big|\Big)^2 \le \Big(\sum_{k \in I^{\mathbf{a}}: d_k \in [a,b)} m(i,k)^2\Big)\Big(\sum_{k \in I^{\mathbf{a}}: d_k \in [a,b)} m(j,k)^2\Big).$$

Therefore,

$$\forall a < b: \quad \big(\|\Gamma_{i,j}\|\big([a,b)\big)\big)^2 \le \|\Gamma_{i,i}\|\big([a,b)\big)\|\Gamma_{j,j}\|\big([a,b)\big). \tag{3.15}$$

We have $e^{(\mathcal{D}_\pi^{1/2} Q^{\mathbf{a}} \mathcal{D}_\pi^{-1/2})s} = \mathcal{D}_\pi^{1/2} e^{Q^{\mathbf{a}} s} \mathcal{D}_\pi^{-1/2}$ and so

$$e^{Q^{\mathbf{a}} s} = \mathcal{D}_\pi^{-1/2} M e^{\mathcal{D} s} M' \mathcal{D}_\pi^{1/2}.$$

Then, by using (3.11), we get

$$\begin{aligned} p_s(i,j) = e^{Q^{\mathbf{a}} s}(i,j) &= \sqrt{\frac{\pi_j}{\pi_i}} \sum_{k \in I^{\mathbf{a}}} e^{-s d_k} m(i,k)m(j,k) \\ &= \sqrt{\frac{\pi_j}{\pi_i}} \int_0^\infty e^{-sx}\, d\Gamma_{i,j}(x). \end{aligned} \tag{3.16}$$

Notice that the columns of the matrix $\widetilde{M} = \mathcal{D}_\pi^{-1/2} M$ are the eigenvectors of $Q^{\mathbf{a}}$, and they are orthogonal with respect to the inner product $\langle f, g \rangle_\pi =$

$\sum_{i\in I^{\mathrm{a}}} f(i)g(i)\pi_i$ induced by π. The (i,j)th element of $\widetilde{M}$ is $\widetilde{m}(i,j)=m(i,j)/\sqrt{\pi_i}$. With this notation,

$$q(i,j)=-\pi_j\sum_{k\in I^{\mathrm{a}}} d_k\widetilde{m}(i,k)\widetilde{m}(j,k)=-\pi_j\int_0^\infty x\,d\widetilde{\Gamma}_{i,j}(x),$$

and

$$p_s(i,j)=\pi_j\sum_{k\in I^{\mathrm{a}}} e^{-sd_k}\widetilde{m}(i,k)\widetilde{m}(j,k)=\pi_j\int_0^\infty e^{-sx}\,d\widetilde{\Gamma}_{i,j}(x), \tag{3.17}$$

where $\widetilde{\Gamma}_{i,j}=(\pi_i\pi_j)^{-1/2}\Gamma_{i,j}$.

3.4 Conditions for Lumping

Let us state some conditions for lumping based upon the hypothesis that the killing time is exponentially distributed. This occurs, for instance, when the initial distribution is a QSD, see (3.3) and Proposition 3.1.

Let us introduce briefly what a lumping process is. Assume $Y=(Y_t:t\ge 0)$ is an irreducible Markov chain taking values on a finite set I with conservative jump rate matrix $Q=(q(i,j):i,j\in I)$. Let $I_0,\dots,I_s$ be a partition of I. Define $\widetilde{I}=\{0,\dots,s\}$ and take the lumping function $\xi:I\to\widetilde{I}$ given by $\xi(i)=a$ when $i\in I_a$. That is, for all $a=0,\dots,s$, the function ξ collapses all the states in I_a into a single state a. The process $\widetilde{Y}=(\widetilde{Y}_t:t\ge 0)$ defined by $\widetilde{Y}_t=\xi(Y_t)$ is said to satisfy the lumping condition if it is a Markov chain.

The usual lumping condition is

$$\forall a,b\in\{0,\dots,s\}\text{ with }a\ne b,\forall i,j\in I_a\colon\quad \sum_{k\in I_b} q(i,k)=\sum_{k\in I_b} q(j,k). \tag{3.18}$$

Since Q is conservative when (3.18) is satisfied, we also get

$$\forall i,j\in I_a\colon\quad \sum_{k\in I_a} q(i,k)=\sum_{k\in I_a} q(j,k). \tag{3.19}$$

Let us see that condition (3.18) implies that $\widetilde{Y}$ is a Markov chain and its jump rate matrix $\widetilde{Q}=(\widetilde{q}(a,b):a,b\in\widetilde{I})$ verifies

$$\forall i\in I_a\colon\quad \widetilde{q}(a,b)=\sum_{k\in I_b} q(i,k).$$

First, note that for all $a\in\{0,\dots,s\}$ the matrix $Q_{I_a\times I_a}=(q(i,j):i,j\in I_a)$ has constant row sums (see (3.19)). Then, by Proposition 3.1(ii), we get that when starting from any state $i\in I_a$ the time of leaving I_a, or equivalently, of reaching the complement $I\setminus I_a$, is exponentially distributed with parameter $\widetilde{q}(a,a)=\sum_{k\in I_a} q(i,k)$. Hence

$$\forall i\in I_a\colon\quad \mathbb{P}_i(T_{I\setminus I_a}>t)e^{-\widetilde{q}(a,a)t}=\mathbb{P}_a(\widetilde{\tau}>t),$$

where $\widetilde{\tau} = \inf\{t > 0 : \widetilde{Y}_t \neq \widetilde{Y}_0\}$ denotes the holding time for $\widetilde{Y}$. On the other hand, from (3.18) and (3.19), we also get

$$\forall i \in I_a: \quad \mathbb{P}_i(Y_{T_{I \setminus I_a}} \in I_b) = \widetilde{q}(a, b) = \mathbb{P}_a(\widetilde{Y}_{\widetilde{\tau}} = b).$$

These relations imply that for all $0 \le t_0 < \cdots < t_n$ and all $a_0, \ldots, a_{n+1} \in \{0, \ldots, s\}$ we have

$$\begin{aligned} &\mathbb{P}(\widetilde{Y}_{t_{n+1}} = a_{n+1} | \widetilde{Y}_{t_n} = a_n, \ldots, \widetilde{Y}_{t_0} = a_0) \\ &\quad = \mathbb{P}(\widetilde{Y}_{t_{n+1}} = a_{n+1} | Y_{t_n} = i_n) = \mathbb{P}(\widetilde{Y}_{t_{n+1}} = a_{n+1} | \widetilde{Y}_{t_n} = a_n), \end{aligned}$$

for all $i_n \in I_{a_n}$. Then $\widetilde{Y}$ is Markovian.

In what follows, we fix $I = \{0, \ldots, \ell\}$ with $\ell \ge 1$ and we consider a two-state lumping process, with the partition $I_0 = \{0\}$, $I_1 = \{1, \ldots, \ell\}$. So, the lumping function satisfies $\xi(0) = 0$ and $\xi(i) = 1$ for all $i \in I_1$. In our Proposition 3.3 (see below), we shall give a more general condition than (3.18) ensuring that $\widetilde{Y} = (\xi(Y_t) : t \ge 0)$ is a Markov chain. This condition is related to positive exponential rate of survival at state 0. For this purpose, let us consider $\partial I = \{0\}$ as the forbidden set, then $I^{\mathbf{a}} = I_1$ and $T = \inf\{t > 0 : Y_t = 0\}$ is the hitting time of state 0 for the process Y.

As announced, our result will contain as a particular case the condition (3.18) which in this context reduces to

$$\forall i \in I_1: \quad q(i, 0) = q(1, 0). \tag{3.20}$$

When (3.20) is satisfied, $\widetilde{Y}$ is a Markov chain with jump rate matrix $\widetilde{Q}$ given by

$$\begin{aligned} \widetilde{q}(0, 0) &= q(0, 0), & \widetilde{q}(0, 1) &= -q(0, 0), \\ \widetilde{q}(1, 0) &= q(1, 0), & \widetilde{q}(1, 1) &= -q(1, 0). \end{aligned}$$

Since $Q^{\mathbf{a}}$ has constant row sums $c = -q(1, 0)$, we get that, starting from any initial distribution on I_1, the random time T is exponentially distributed with parameter $q(1, 0)$ (see Proposition 3.1(ii)).

Let us state our generalization. The holding time of Y is denoted by $\tau = \inf\{t > 0 : Y_t \neq Y_0\}$. The transition distribution from 0 to I_1 is

$$e = (e_i : i \in I_1) \quad \text{with } e_i = \mathbb{P}_0(Y_\tau = i) = \frac{q(0, i)}{\sum_{j \in I_1} q(0, j)}.$$

We are in an irreducible finite case and we denote by θ the spectral radius. We know that there exists a unique QSD (for the process killed at $\{0\}$) and its exponential rate of survival is θ.

Proposition 3.3 *Assume that when the transition distribution from 0 to I_1 is e, the random time T is exponentially distributed. (This condition is satisfied when e is a QSD.) If the distribution of Y_0 is a convex combination of δ_0 and e then the process $\widetilde{Y} = (\widetilde{Y}_t : t \ge 0)$ is a Markov chain taking values in $\{0, 1\}$ with jump rate matrix $\widetilde{Q}$ given by*

$$\widetilde{q}(0, 0) = q(0, 0), \qquad \widetilde{q}(0, 1) = -q(0, 0), \qquad \widetilde{q}(1, 0) = \theta, \qquad \widetilde{q}(1, 1) = -\theta.$$

Proof Since the killing time T is exponentially distributed under e, from Proposition 3.1(i) we deduce that its parameter is necessarily θ. We define the distribution ζ on $\{0,1\}$ by: $\zeta(\{0\})=\mathbb{P}(Y_0=0)$, $\zeta(\{1\})=\mathbb{P}(Y_0\neq 0)=1-\zeta(\{0\})$.

There is a unique (in distribution) Markov chain $W=(W_t:t\geq 0)$ taking values in $\{0,1\}$ such that the initial distribution is ζ and the jump rate matrix $\widehat{Q}$ is given by $\widehat{q}(0,0)=-\widehat{q}(0,1)=q(0,0)$, $\widehat{q}(1,0)=-\widehat{q}(1,1)=\theta$. This process is characterized by the sequence of independent sojourn times at 0 and 1, which are exponentially distributed with parameters $-q(0,0)$ and θ, respectively.

On the other hand, the process $\widetilde{Y}$ is realized as follows. Since Y_0 is chosen at random with the distribution $\zeta(\{0\})\delta_0+\zeta(\{1\})e$, we get $\widetilde{Y}_0\sim\zeta$. If $Y_0=0$ then $\widetilde{Y}$ remains at 0 according to an exponential distribution with parameter $-q(0,0)$. If $Y_0\neq 0$ then, conditionally on this event, this initial state has the distribution e. By hypothesis, $\widetilde{Y}$ remains at 1 for a time whose distribution is exponential with parameter θ. The process $\widetilde{Y}$ then continues afresh. In this way, $\widetilde{Y}$ and W have the same distribution and the result is shown. □

3.5 Example: Finite Killed Random Walk

Let us study the QSD for the finite random walk with absorbing barriers, see Fig. 3.1. Take an integer $N>1$ and two positive values $\lambda>0$ and $\mu>0$. Let us consider the time-continuous random walk $Y=(Y_t:t\geq 0)$ on the state space $I=\{0,\dots,N\}$ with jump rate matrix $Q=(q(k,j):k,j\in I)$ given by: $q(k,k+1)=\lambda$, $q(k,k-1)=\mu$, $q(k,k)=-(\lambda+\mu)$ for $k\in\{1,\dots,N-1\}$ and such that all other values $q(k,j)$ vanish. In particular, $q(0,k)=0=q(N,k)$ for $k\in I$, so the states $\{0,N\}$ are absorbing. The forbidden set of states is $\partial I=\{0,N\}$, then $I^{\mathbf{a}}=\{1,\dots,N-1\}$. Since $(Y_{t\wedge T}:t\geq 0)$ is irreducible, there exists a unique QSD ν which is the normalized Perron–Frobenius left-eigenvector. Let us compute it.

Since $\nu=(\nu(k):k\in I^{\mathbf{a}})$ is a left-eigenvector of $Q^{\mathbf{a}}$, it must satisfy $\nu' Q^{\mathbf{a}}=-\theta\nu'$ for some $\theta>0$, so

$$
\begin{aligned}
-(\lambda+\mu+\theta)\nu(1)+\mu\nu(2)&=0,\\
\lambda\nu(k-1)-(\lambda+\mu+\theta)\nu(k)+\mu\nu(k+1)&=0,\quad k\in\{2,\dots,N-2\},\\
\lambda\nu(N-2)-(\lambda+\mu+\theta)\nu(N-1)&=0.
\end{aligned}
$$

These equations are equivalent to

$$
\begin{aligned}
&\nu(0)=0=\nu(N);\\
&\lambda\nu(k-1)-(\lambda+\mu-\theta)\nu(k)+\mu\nu(k+1)=0,\\
&\quad k\in\{1,\dots,N-1\}.
\end{aligned}
$$

This system admits a solution of the form

$$
\nu(k)=Aa_+^k+Ba_-^k,\quad k\in\{0,\dots,N\},
$$

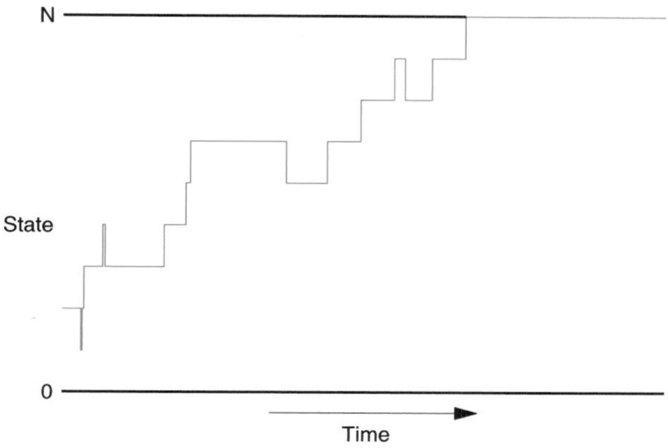

Fig. 3.1 Trajectories killed at the boundary

where

$$a_\pm = \frac{1}{2\mu}\big((\lambda+\mu-\theta)\pm\sqrt{(\lambda+\mu-\theta)^2-4\lambda\mu}\big).$$

The constants A, B must be such that $\nu(k)>0$ for $k\in\{1,\dots,N-1\}$ and they are also fixed by the boundary conditions $\nu(0)=0=\nu(N)$. From $\nu(0)=0$, we get $A+B=0$, so $B=-A$ and $A\neq 0$. From $A(a_+^N-a_-^N)=0$ and $A\neq 0$, we find the necessary condition $(\lambda+\mu-\theta)^2<4\lambda\mu$ and we assume it. Note that $(\lambda+\mu-\theta)^2=4\lambda\mu$ only for $\theta=\theta_\pm$ where $\theta_\pm=(\lambda+\mu)\pm 2\sqrt{\lambda\mu}$. Hence the condition $(\lambda+\mu-\theta)^2<4\lambda\mu$ is equivalent to $\theta\in(\theta_-,\theta_+)$.

Let us write the solutions in polar coordinates (r,φ). The modulus $r>0$ satisfies

$$r^2=\frac{1}{4\mu^2}\big((\lambda+\mu-\theta)^2+4\lambda\mu-(\lambda+\mu-\theta)^2\big)=\frac{\lambda}{\mu},$$

and the angle φ is such that

$$\varphi=\chi(\theta)\quad\text{where } \chi(\theta)=\operatorname{arctg}\left(\frac{\sqrt{4\lambda\mu-(\lambda+\mu-\theta)^2}}{(\lambda+\mu-\theta)}\right),\ \theta\in(\theta_-,\theta_+). \qquad (3.21)$$

With this notation, $a_+=re^{i\varphi}$ and $a_-=re^{-i\varphi}$. The conditions $A(a_+^N-a_-^N)=0$ and $A\neq 0$ imply $\sin(N\varphi)=0$, that is, $N\varphi=2\pi k$ for some $k\in\mathbb{Z}$. In order that the solutions be real, we take $A=Ci$ with $C\neq 0$ and real, and we can also choose $C>0$. Since $A(a_+^k-a_-^k)>0$ for $k=0,\dots,N-1$ and $A(a_+^N-a_-^N)=0$, we must necessarily have $N\varphi=2\pi$, so $\varphi=2\pi/N$. Hence $\nu(k)=2C(\sqrt{\frac{\lambda}{\mu}})^k\sin(k\varphi)$ for $k=1,\dots,N-1$.

The function $\chi(\theta)$ defined in (3.21) is such that its restrictions

$$\chi:\ (\theta_-,\lambda+\mu)\to(0,\infty)\quad\text{and}\quad \chi:\ (\lambda+\mu,\theta_+)\to(0,\infty)$$

are onto and one-to-one,

so there are exactly two values $\theta_0\in(\theta_-,\lambda+\mu)$ and $\theta_1\in(\lambda+\mu,\theta_+)$ satisfying the equality $\chi(\theta)=2\pi/N$.

On the other hand, the matrix $Q^{\mathbf{a}}$ can be written as

$$Q^{\mathbf{a}} = -(\lambda+\mu)\left(\mathbb{I} + \frac{1}{(\lambda+\mu)} M\right), \tag{3.22}$$

where $\mathbb{I}$ is the identity matrix and $M = (m(k,j) : k, j \in I^{\mathbf{a}})$ is the matrix with values $m(k, k+1) = -\lambda$, $m(k+1, k) = -\mu$ for all $k \in \{1, \dots, N-2\}$ and all other values $m(k, j)$ vanish. Since $M/(\lambda+\mu)$ is strictly substochastic, its spectral radius is strictly smaller than 1, and from (3.22) we deduce that the spectral radius of $Q^{\mathbf{a}}$ is strictly smaller than $\lambda+\mu$. Since the eigenvalue $-\theta$ having left-eigenfunction ν is minus the spectral radius of $Q^{\mathbf{a}}$, we deduce that $\theta = \theta_0$ is the unique value in the interval $(\lambda+\mu-2\sqrt{\lambda\mu}, \lambda+\mu)$ such that $\operatorname{arctg}(\sqrt{\frac{4\lambda\mu}{(\lambda+\mu-\theta_0)^2}} - 1) = 2\pi/N$. Finally, we fix the constant $C > 0$ in such a way that

$$C^{-1} = 2\sum_{k\in I^{\mathbf{a}}} (\lambda/\mu)^{\frac{k}{2}} \sin(2\pi k),$$

which implies $\sum_{k\in I^{\mathbf{a}}} \nu(k) = 1$.

3.6 The Central Limit Theorem for a QSD

Let us assume the time is discrete. So, $Y = (Y_n : n \geq 0)$ is a Markov chain taking values in I, ∂I is the set of forbidden states, $I^{\mathbf{a}} = I \setminus \partial I$ is the set of allowed states, and the time evolution is given by the stochastic kernel $P = (p(j,l) : j, l \in I)$. Let T be the hitting time of ∂I, we assume that $\mathbb{P}_i(T < \infty) = 1$ for all $i \in I^{\mathbf{a}}$, that $Y^T = (Y_n : n < T)$ is irreducible and that the strictly substochastic matrix $P^{\mathbf{a}} = (p(j,l) : j, l \in I^{\mathbf{a}})$ is aperiodic.

From the Perron–Frobenius Theorem, the spectral radius of $P^{\mathbf{a}}$ is a strictly positive simple eigenvalue, we denote it by κ and it satisfies $\kappa \in (0,1)$. Moreover, there is a unique pair of left- and right-eigenvectors ν and ψ of $P^{\mathbf{a}}$ associated to κ, that is, $\nu' P^{\mathbf{a}} = \kappa \nu'$, $P^{\mathbf{a}}\psi = \kappa\psi$, which verify the normalizing conditions

$$\nu' \mathbf{1} = 1, \qquad \nu'\psi = 1.$$

Hence ν and $\nu\psi = (\nu(j)\psi(j) : j \in I^{\mathbf{a}})$ are probability distributions on $I^{\mathbf{a}}$. The vector ν is the unique QSD, and from the aperiodicity of $P^{\mathbf{a}}$ it is also a quasi-limiting distribution, that is, it satisfies

$$\forall j, l \in I^{\mathbf{a}}: \quad \lim_{n\to\infty} \mathbb{P}_j(Y_n = l | T > n) = \nu(l).$$

Then for all $f : I^{\mathbf{a}} \to \mathbb{R}$ we have

$$\forall j \in I^{\mathbf{a}}: \quad \lim_{n\to\infty} \mathbb{E}_j\big(f(Y_n) | T > n\big) = \nu' f.$$

Also aperiodicity implies that the Markov chain $Z = (Z_n : n \geq 0)$ that gives the evolution of trajectories surviving forever has the transition matrix $\widetilde{P} = (\tilde{p}(j,k) : j, k \in I^{\mathbf{a}})$ given by

$$\tilde{p}(j,l) = \lim_{n\to\infty} \mathbb{P}_j(Y_1 = l | T > n) = \kappa^{-1} p(j,l) \frac{\psi(l)}{\psi(j)}.$$

Moreover, the distribution $\nu\psi$ is the stationary distribution associated to $\widetilde{P}$. Hence,

$$\forall j \in I^{\mathbf{a}}: \quad \lim_{l\to\infty}\lim_{n\to\infty} \mathbb{E}_j\big(f(Y_l)|T>n\big) = (\nu\psi)' f.$$

We shall prove the following Central Limit Theorem.

Theorem 3.4 *Assume $P^{\mathbf{a}}$ is aperiodic. Let $f: I^{\mathbf{a}} \to \mathbb{R}$ be a function such that $(\nu\psi)' f = 0$. Then the following limit exists*

$$\eta^2 = \lim_{N\to\infty} \frac{1}{N}\mathbb{E}_\nu\left(\left(\sum_{n=0}^{N-1} f(Y_n)\right)^2 \Bigg| T > N\right).$$

Assume that $\eta \neq 0$. Then, the following convergence in distributions holds

$$\lim_{N\to\infty} \nu\left(\frac{1}{\sqrt{N}}\sum_{n=0}^{N-1} f(Y_n) \le t \Bigg| T > N\right) = \frac{1}{\sqrt{2\pi}}\int_{-\infty}^{t} e^{-\frac{x^2}{2\eta^2}}\,dx.$$

Proof For a function $g: I^{\mathbf{a}} \to \mathbb{R}$, we denote by $\mathcal{D}_g$ the diagonal matrix indexed by $I^{\mathbf{a}} \times I^{\mathbf{a}}$ such that $\mathcal{D}_g(j,j) = g(j)$. Let $z \in \mathbb{C}$ be small in modulus and take the matrix $P(z) = P^{\mathbf{a}}\mathcal{D}_{e^{zf}}$, its coefficients are $p(z)(j,l) = p(j,l)e^{zf(j)}$. Then

$$\nu' P(z/\sqrt{N})^n \mathbf{1} = \mathbb{E}_\nu\big(e^{\frac{z}{\sqrt{N}}\sum_{n=0}^{N-1} f(Y_n)}, T > N\big).$$

On the other hand, $P(z)$ is analytic on z and $P(z) = \sum_{l\ge 0} z^l P_l$ with $P_l = \frac{1}{l!}P^{\mathbf{a}}\mathcal{D}_{f^l}$. By perturbation on z in a neighborhood around the origin, there exists a unique eigenvalue $\kappa(z)$ of maximal modulus which is simple. Let $\nu(z)$ and $\psi(z)$ be a left- and a right-eigenvectors of $P(z)$ associated to $\kappa(z)$. Also by perturbation around the origin, we get the analytic expansions

$$\kappa(z) = \sum_{l\ge 0} z^l \kappa_l, \qquad \nu(z) = \sum_{l\ge 0} z^l \nu_l, \qquad \psi(z) = \sum_{l\ge 0} z^l \psi_l,$$

for z in a neighborhood of $z = 0$, and such that $\kappa(0) = \kappa$, $\nu(0) = \nu$ and $\psi(0) = \psi$.

We can impose the following normalizing conditions on $\nu(z)$ and $\psi(z)$:

$$\nu'\psi(z) = 1 \quad \text{and} \quad \nu(z)'\mathbf{1} = 1 \quad \text{in a neighborhood of } z = 0. \tag{3.23}$$

We claim that $\kappa_1 = 0$, let us prove it. Recall $P^{\mathbf{a}}\psi = \kappa\psi$. In the equality $P(z)\psi(z) = \kappa(z)\psi(z)$, the evaluation of the coefficient of z gives the equality of functions $P_1\psi + P^{\mathbf{a}}\psi_1 = \kappa_1\psi + \kappa\psi_1$. Let us apply ν' to this equality and use $\nu' P^{\mathbf{a}} = \kappa\nu'$ to obtain $\nu'(P_1\psi) = \kappa_1\nu'\psi = \kappa_1$. But

$$\nu'(P_1\psi) = \nu'(P\mathcal{D}_f\psi) = \nu' P^{\mathbf{a}}(f\psi) = \kappa\nu'(f\psi) = 0.$$

We have shown the claim $\kappa_1 = 0$. Hence $\kappa(z) = \kappa + \kappa_2 z^2 + O(z^3)$ where $O(z^3)$ is such that $O(z^3)/z^3$ is finite for z in a neighborhood of the origin, but $z \neq 0$.

From the perturbation theory, we have $P(z) = \kappa(z)\psi(z)\nu(z) + R(z)$ with $R(z)h(z) = 0$ and $\nu(z)R(z) = 0$. Also we can assume that for some $\epsilon > 0$ small enough, the spectral radius of $R(z)$ is bounded above by some $0 < K < \kappa - 2\epsilon$ for all $|z| < \epsilon$. Hence, for N big enough and any spectral norm $\|\cdot\|$ of the matrices, we have

$$P(z)^N = \kappa(z)^N \psi(z)\nu(z) + R(z)^N, \\ \text{with } \sup\{\|R(z)^N\| : |z| < \epsilon\} \le (\kappa - \epsilon)^N. \tag{3.24}$$

Hence, for all $z \in \mathbb{C}$ and N big enough such that $|z|/\sqrt{N} < \epsilon$, we have

$$\begin{aligned} &\mathbb{E}_\nu\big(e^{\frac{z}{\sqrt{N}}\sum_{n=0}^{N-1} f(Y_n)}, T > N\big) \\ &= \nu' P(z/\sqrt{N})^N \mathbf{1} \\ &= \kappa(z/\sqrt{N})^N \nu'\psi(z/\sqrt{N})\nu(z/\sqrt{N})'\mathbf{1} + \nu' R(z/\sqrt{N})^N \mathbf{1}. \end{aligned}$$

Let s be real and N big enough such that $|s|/\sqrt{N} < \epsilon$. We have

$$\begin{aligned} &\mathbb{E}_\nu\big(e^{\frac{is}{\sqrt{N}}\sum_{n=0}^{N-1} f(Y_n)} \big| T > N\big) \\ &= \kappa^{-N}\big(\kappa(is/\sqrt{N})^N \nu'\psi(is)\nu(is/\sqrt{N})'\mathbf{1} + \nu' R(is/\sqrt{N})^N \mathbf{1}\big). \end{aligned}$$

From (3.24) the term $\kappa^{-N}\nu' R(z/\sqrt{N})^N \mathbf{1}$ goes to 0 as $n \to \infty$. Hence, from $\kappa(is/\sqrt{N}) = \kappa - \kappa_2 s^2/N + O(s^3/N^{3/2})$, we get

$$\lim_{N\to\infty} \kappa^{-N}\kappa(is)^N = \left(1 - \frac{\kappa_2 s^2}{N} + O\left(\frac{s^3}{N^{3/2}}\right)\right) = e^{-\kappa_2 s^2},$$

and then

$$\lim_{N\to\infty} \mathbb{E}_\nu\big(e^{\frac{is}{\sqrt{N}}\sum_{n=0}^{N-1} f(Y_n)} \big| T > N\big) = e^{-\kappa_2 s^2}\nu'\psi(is)\nu(is)'\mathbf{1} = e^{-\kappa_2 s^2}, \tag{3.25}$$

where in the last equality we used the normalizing conditions (3.23). Hence, if we are able to show that $\kappa_2 \ne 0$ we will prove that $\frac{1}{\sqrt{N}}\sum_{n=0}^{N-1} f(Y_n)$, conditioned to $T > N$, converges in distribution to a centered Gaussian distribution with variance $2\kappa_2$ as N goes to ∞.

By differentiating twice expression (3.25), evaluating it at $s = 0$, and by using the Dominated Convergence Theorem, we get

$$\lim_{N\to\infty} \frac{d^2}{ds^2}\big(\mathbb{E}_\nu\big(e^{\frac{is}{\sqrt{N}}\sum_{n=0}^{N-1} f(Y_n)} \big| T > N\big)\big)\big|_{s=0} = -2\kappa_2.$$

Hence,

$$\lim_{N\to\infty} \mathbb{E}_\nu\left(\frac{1}{N}\left(\sum_{n=0}^{N-1} f(Y_n)\right)^2 \middle| T > N\right) = 2\kappa_2.$$

From the hypothesis, this coefficient does not vanish, hence the result follows. □

Chapter 4
Markov Chains on Countable Spaces

The main result of this chapter is the existence theorem of a QSD when the process is exponentially killed and satisfies a condition of not coming from infinity in finite time. This is given in Theorem 4.5 in Sect. 4.4. In Theorem 4.4 of Sect. 4.3, the QSDs are characterized as solutions of an eigenvalue problem associated to the jump rate matrix of the allowed states. In Sect. 4.5, we give conditions in order that the exponential decay of the transition probabilities be the same as the exponential rate of survival. Section 4.6 is devoted to symmetric semigroups (for some measure) and we supply the spectral decomposition as a limit of finite truncations. For a monotone process, the exponential rate of convergence and the stationary distribution of the ergodic process are compared with the exponential rate of survival and the QSD, respectively, when we impose killing; this is done in Sect. 4.7.

4.1 The Minimal Process

Let I be a countable set and $Q = (q(i,j) : i, j \in I)$ be a conservative jump rate matrix, so

$$\forall i \in I^{\mathbf{a}}: \quad q(i,i) > -\infty \quad \text{and} \quad \sum_{j \in I} q(i,j) = 0.$$

Let ∂I be the set of forbidden states and $I^{\mathbf{a}} = I \setminus \partial I$. Let $Q^{\mathbf{a}} = (q(i,j) : i, j \in I^{\mathbf{a}})$ be the transition rate matrix restricted to the allowable set of states. We assume that $Q^{\mathbf{a}}$ is irreducible (it verifies property (3.2)).

We also assume

$$\exists i \in I^{\mathbf{a}}: \quad q(i,\partial I) > 0, \quad \text{where } q(i,\partial I) := \sum_{j \in \partial I} q(i,j), \tag{4.1}$$

and so $Q^{\mathbf{a}}$ is a defective jump rate matrix.

As usual we put

$$q_i = -q(i,i) \quad i \in I.$$

P. Collet et al., *Quasi-Stationary Distributions*, Probability and Its Applications,
DOI 10.1007/978-3-642-33131-2_4, © Springer-Verlag Berlin Heidelberg 2013

Since the states in ∂I are absorbing, we have $q_i = 0$ for all $i \in \partial I$.

All the notions related to a QSD only depend on the trajectory before killing, so we can assume that the set of forbidden states ∂I is a singleton. We assume this condition and we denote $\partial I = \{0\}$, where 0 denotes a state that does not belong to $I^{\mathbf{a}}$. So, $I = I^{\mathbf{a}} \cup \{0\}$ and $q(i, \partial I) = q(i, 0)$ for $i \in I$.

We will consider the semigroup $(P_t = (p_t(i, j) : i, j \in I) : t \geq 0)$ which is the minimal solution associated to Q. This is the transition probability realized with a finite (but arbitrary) number of jumps with the rates given by Q. Thus, the transition probability verifies

$$p_t(i, j) = \lim_{n \to \infty} f_t^{(n)}(i, j)$$

where $f_t^{(0)}(i, j) = \delta(i, j) e^{-q_i t}$ and the sequences $f_t^{(n)}(i, j)$, $n \geq 1$, satisfy the following backward and forward iteration equations:

$$f_t^{(n)}(i, j) = f_t^{(0)}(i, j) + \sum_{k \neq i} \int_0^t q(i, k) f_s^{(n-1)}(k, j) e^{-q_i (t-s)} \, ds, \tag{4.2}$$

$$f_t^{(n)}(i, j) = f_t^{(0)}(i, j) + \sum_{k \neq j} \int_0^t f_s^{(n-1)}(i, k) q(k, j) e^{-q_j (t-s)} \, ds. \tag{4.3}$$

We assume that the minimal solution is the unique solution to the backward equations

$$\forall i, j \in I, t \geq 0: \quad p_t'(i, j) = \sum_{k \in I} q(i, k) p_t(k, j), \tag{4.4}$$

where $p_t'(i, j) = dp_t(i, j)/dt$ is the derivative of $p_t(i, j)$ with respect to time t. This implies that $(P_t : t \geq 0)$ is honest: $\sum_{j \in I} p_t(i, j) = 1$ for all $i \in I$ and $t > 0$. See Theorem 3.3 in Anderson (1991) and discussions following it. Moreover, we also assume that the minimal solution is also the unique solution to the forward equations

$$\forall i, j \in I, t \geq 0: \quad p_t'(i, j) = \sum_{k \in I} p_t(i, k) q(k, j). \tag{4.5}$$

We denote by $Y = (Y_t : t \geq 0)$ the associated Markov chain, so $p_t(i, j) = \mathbb{P}_i(Y_t = j)$ is the transition probability. We call it, as well as its transition probability, the minimal process. By irreducibility, $p_t(i, j) > 0$ for all $i, j \in I^{\mathbf{a}}$, $t > 0$, so $(Y_t : t < T)$ is an irreducible chain.

From irreducibility and condition (4.1), we get $\mathbb{P}_i(T < \infty) > 0 \; \forall i \in I^{\mathbf{a}}$, which is condition (2.2). As announced, we will assume that the stronger condition (2.1) does hold:

$$\forall i \in I^{\mathbf{a}}: \quad \mathbb{P}_i(T < \infty) = 1.$$

4.2 The Exponential Killing Condition

As already defined, τ is the initial holding time, $\tau = \inf\{t > 0 : Y_t \neq Y_0\}$. So, when starting from $i \in I$, τ is distributed as an exponential with mean $1/q_i$. Also we define

$$\tau(t) = \inf\{s > t : Y_s \neq Y_t\}.$$

So, $\tau = \tau(0)$.

From (2.10), a necessary condition for the existence of a QSD is the existence of an exponential moment, that is,

$$\exists i \in I^{\mathbf{a}}\ \exists \theta > 0: \quad \mathbb{E}_i\big(e^{\theta T}\big) < \infty. \tag{4.6}$$

As we know from Proposition 2.4 and relation (2.9), condition (4.6) is equivalent to the exponential rate of survival condition:

$$\exists i \in I^{\mathbf{a}}\ \exists \theta' > 0: \quad \mathbb{P}_i(T > t) \leq C e^{-\theta' t}, \qquad \forall t \geq 0.$$

Let us show the following property concerning the uniform exponential moment.

Lemma 4.1 *Since* $(Y_t : t < T)$ *is irreducible, condition* (4.6) *is equivalent to the uniform exponential moment property*

$$\exists \theta > 0\ \forall i \in I^{\mathbf{a}}: \quad \mathbb{E}_i\big(e^{\theta T}\big) < \infty. \tag{4.7}$$

All $\theta > 0$ *satisfying* (4.7) *must verify*

$$\forall i \in I^{\mathbf{a}}: \quad \theta < q_i. \tag{4.8}$$

Moreover, θ_i^* *defined in* (2.9) *does not depend on* $i \in I^{\mathbf{a}}$, *that is,*

$$\theta^* := \sup\big\{\theta : \mathbb{E}_i\big(e^{\theta T}\big) < \infty\big\} = -\liminf_{t\to\infty} \frac{1}{t} \log P_i(T > t), \tag{4.9}$$

does not depend on $i \in I^{\mathbf{a}}$. *Then,*

$$\forall i \in I^{\mathbf{a}}: \quad \theta^* \leq q_i. \tag{4.10}$$

Proof The last relation (4.10) is a consequence of (4.8). The relations (4.7) and (4.8) will be consequences of the following relations. Fix $i \in I^{\mathbf{a}}$ and assume $\theta > 0$ is such that $\mathbb{E}_i(e^{\theta T}) < \infty$. From $\mathbb{P}_i(\tau \leq T) = 1$, we get $\mathbb{E}_i(e^{\theta\tau}) < \infty$. Since $\mathbb{E}_i(e^{\theta\tau}) = \infty$ when $q_i \leq \theta$, we deduce $q_i > \theta$ and $\mathbb{E}_i(e^{\theta\tau}) = q_i/(q_i - \theta)$. Now, we are able to show that

$$\big(\mathbb{E}_i\big(e^{\theta T}\big) < \infty, q(i,j) > 0\big) \quad \Rightarrow \quad \mathbb{E}_j\big(e^{\theta T}\big) < \infty. \tag{4.11}$$

This follows straightforwardly from the strong Markov property:

$$\mathbb{E}_i\big(e^{\theta T}\big) = \mathbb{E}_i\big(e^{\theta\tau} E_{Y_\tau}\big(e^{\theta T}\big)\big) = \frac{1}{q_i - \theta}\bigg(\sum_{j \in I : q(i,j) > 0} q(i,j)\mathbb{E}_j\big(e^{\theta T}\big)\bigg).$$

From irreducibility, we get that $\mathbb{E}_i(e^{\theta T}) < \infty$ implies $\mathbb{E}_j(e^{\theta T}) < \infty$ for all $j \in I^{\mathbf{a}}$, and so the equivalence between (4.6) and (4.7) is verified. We have also proved relation (4.8).

Finally, from (2.9) and (4.11), we get that θ_i^* does not depend on $i \in I^{\mathbf{a}}$, and so (4.9) holds. □

From (2.9), relation (4.7) is also equivalent to the condition

$$\exists \theta' > 0\ \forall i \in I^{\mathbf{a}}: \quad \mathbb{P}_i(T > t) \le C e^{-\theta' t}, \quad \forall t \ge 0.$$

Proposition 2.4 implies straightforwardly the following result in terms of θ^* defined in (4.9).

Proposition 4.2 *A necessary condition for the existence of a QSD is $\theta^* > 0$. Moreover, when ν is a QSD, it verifies $\theta(\nu) \le \theta^*$.*

Any QSD that verifies $\theta(\nu) = \theta^*$ will be called an extremal QSD.

Observe that since the forbidden state 0 is absorbing, we get

$$\forall i \in I^{\mathbf{a}}: \quad p_t(i, 0) = \mathbb{P}_i(T \le t).$$

Condition (2.5), in order that a probability measure ν be a QSD, is written as

$$\exists \theta > 0\ \forall j \in I^{\mathbf{a}}\ \forall t \ge 0: \quad \sum_{i \in I^{\mathbf{a}}} \nu(i) p_t(i, j) = e^{-\theta t} \nu(j).$$

Note that irreducibility implies that every QSD charges all points in $I^{\mathbf{a}}$.

4.3 The Infinitesimal Characterization of a QSD

This section is devoted to characterizing QSDs as (normalized) finite mass eigenmeasures of $Q^{\mathbf{a}}$. This result, due to Nair and Pollett (1993), will be shown in detail. The proof presented in Nair and Pollett (1993) uses strongly results and elements introduced in Tweedie (1974) and Pollett (1986). Among these elements are the subinvariant measures, a notion that will be discussed in Sect. 4.5.

First, we give the following characterization of $\theta(\nu)$, a result proved in Nair and Pollett (1993).

Proposition 4.3 *When ν is a QSD, it verifies $\theta(\nu) = \sum_{i \in I^{\mathbf{a}}} \nu(i) q(i, 0)$.*

Proof The forward equation in (4.5) computed for $j = 0$ is $p_t'(i, 0) = \sum_{k \in I} p_t(i, k) q(k, 0)$. By integrating it, we obtain

$$\forall i \in I^{\mathbf{a}}: \quad p_t(i, 0) = \sum_{k \in I} q(k, 0) \int_0^t p_u(i, k)\, du.$$

Hence

$$\sum_{i\in I^{\mathbf{a}}} \nu(i)p_t(i,0) = \sum_{k\in I} q(k,0) \int_0^t \sum_{i\in I^{\mathbf{a}}} \nu(i)p_u(i,k)\,du.$$

Since ν is a QSD, we find

$$\left(1-e^{-\theta(\nu)t}\right) = \sum_{k\in I} \nu(k)q(k,0) \int_0^t e^{-\theta(\nu)u}\,du.$$

Since $\int_0^t e^{-\theta(\nu)u}du = (1-e^{-\theta(\nu)t})/\theta(\nu)$, the result is shown. □

The following result describes the QSD in terms of the q-matrix, it is due to Nair and Pollett (1993) and it uses Tweedie (1974) and Pollett (1986). Our proof expands and simplifies some of the original arguments, but it follows closely the original proofs. We use the vector notation $(\nu' Q^{\mathbf{a}})(j) = \sum_{i\in I^{\mathbf{a}}} \nu(i)q(i,j)$. Recall our assumption that the minimal solution is honest and that it is the unique solution to the backward and the forward equations.

Theorem 4.4 *A probability measure ν is a QSD if and only if it verifies*

$$\nu' Q^{\mathbf{a}} = -\theta\nu' \quad \text{with } \theta = \sum_{i\in I^{\mathbf{a}}} \nu(i)q(i,0) \text{ and } \theta \le q_j\ \forall j\in I^{\mathbf{a}}. \tag{4.12}$$

Proof Let us first show that if ν is a QSD then it verifies (4.12). Let $\theta = \theta(\nu) = \sum_{i\in I^{\mathbf{a}}} \nu(i)q(i,0)$. Note that (4.8) implies $\theta \le q_j\ \forall j \in I^{\mathbf{a}}$. Fix $j\in I^{\mathbf{a}}$ and put $I^{\mathbf{a}}(j) = I^{\mathbf{a}}\setminus\{j\}$. The QSD property gives

$$\sum_{i\in I^{\mathbf{a}}(j)} \nu(i)p_t(i,j) = \left(e^{-\theta t} - p_t(j,j)\right)\nu(j) = \left(\left(e^{-\theta t}-1\right) + \left(1-p_t(j,j)\right)\right)\nu(j).$$

Let $I^{\mathbf{a}}(j,N)$ be an increasing sequence of finite sets converging to $I^{\mathbf{a}}(j)$ as N increases to ∞. We have

$$\sum_{i\in I^{\mathbf{a}}(j,N)} \nu(i)p_t(i,j) \le \left(\left(e^{-\theta t}-1\right) + \left(1-p_t(j,j)\right)\right)\nu(j),$$

and so by dividing by t and by taking $\lim t\to 0^+$ in the above expression, we get

$$\sum_{i\in I^{\mathbf{a}}(j,N)} \nu(i)q(i,j) \le \left(-\theta - q(j,j)\right)\nu(j).$$

Then, by making $N\to\infty$ and since the terms in the sum of the left-hand side are all positive, we deduce

$$\sum_{i\in I^{\mathbf{a}}(j)} \nu(i)q(i,j) \le \left(-\theta - q(j,j)\right)\nu(j).$$

We have shown $\sum_{i\in I^{\mathbf{a}}} \nu(i)q(i,j) \le -\theta\nu(j)$. Let us prove the other inequality. Since $\nu(i)\le 1$, we get

$$
\begin{aligned}
e^{-\theta t}\nu(j) &= \sum_{i\in I^{\mathbf{a}}(j)} \nu(i)p_t(i,j) \\
&\le \sum_{i\in I^{\mathbf{a}}(j,N)} \nu(i)p_t(i,j) + \nu(j)p_t(j,j) + \sum_{i\in I^{\mathbf{a}}(j)\setminus I^{\mathbf{a}}(j,N)} p_t(i,j) \\
&\le \sum_{i\in I^{\mathbf{a}}(j,N)} \nu(i)p_t(i,j) + \nu(j)p_t(j,j) + 1 - p_t(j,j) - p_t(j,0) \\
&\quad - \sum_{i\in I^{\mathbf{a}}(j,N)} p_t(i,j).
\end{aligned}
$$

By rearranging the above expression, by dividing by t and taking $\lim_{t\to 0^+}$, we find

$$
\big(-\theta - q(j,j)\big)\nu(j) \le \sum_{i\in I^{\mathbf{a}}(j,N)} \nu(i)q(i,j) - q(j,j) - q(j,0) - \sum_{i\in I^{\mathbf{a}}(j,N)} q(i,j).
$$

Take $N\to\infty$ in the above expression and use that the jump rate matrix Q is conservative to get $\lim_{N\to\infty}\sum_{i\in I^{\mathbf{a}}(j,N)} q(i,j) = -q(j,j) - q(j,0)$. Thus, we have proved $(-\theta - q(j,j))\nu(j) \le \sum_{i\in I^{\mathbf{a}}(j)} \nu(i)q(i,j)$, so the inequality $\sum_{i\in I^{\mathbf{a}}} \nu(i)q(i,j) \ge -\theta\nu(j)$ holds. Therefore, we have shown that if ν is a QSD then it verifies $\nu' Q^{\mathbf{a}} = -\theta\nu'$ with $\theta = \theta(\nu)$. The condition $\theta(\nu) \le q_j\ \forall j \in I^{\mathbf{a}}$ follows from $\theta(\nu) \le \theta^*$ and (4.10).

Let us now prove the converse. So assume that ν verifies $\nu' Q^{\mathbf{a}} = -\theta\nu'$ with $\theta = \sum_{i\in I^{\mathbf{a}}} \nu(i)q(i,0)$ and $\theta(\nu) \le q_j\ \forall j \in I^{\mathbf{a}}$. We must show that ν is a QSD. Recall that the transition probability of the minimal process verifies $p_t(i,j) = \lim_{n\to\infty} f_t^{(n)}(i,j)$ where $f_t^{(0)}(i,j) = \delta(i,j)e^{-q_i t}$ and $f_t^{(n)}(i,j), n\ge 1$, satisfy (4.2) and (4.3).

We must prove that ν verifies the relation

$$
\forall j \in I^{\mathbf{a}}:\quad \sum_{i\in I^{\mathbf{a}}} \nu(i)p_t(i,j) = \nu(j)e^{-\theta t}.
$$

In this purpose, define the matrix $Q^* = (q^*(i,j) : i, j \in I^{\mathbf{a}})$ by

$$
q^*(i,j) = \big(q(j,i) + \theta\delta(i,j)\big)\frac{\nu(j)}{\nu(i)}. \tag{4.13}
$$

Since $\theta(\nu) \le q_j\ \forall j \in I^{\mathbf{a}}$, the matrix Q^* is a conservative jump rate matrix. Let us consider the minimal process associated to Q^*, so its transition probability verifies $p_t^*(i,j) = \lim_{n\to\infty} f_t^{(n)*}(i,j)$ where $f_t^{(0)*}(i,j) = \delta(i,j)e^{-q_i^* t}$ and $f_t^{(n)*}(i,j)$ verifies the backward and the forward evolution equations for $n\ge 1$:

$$
f_t^{(n)*}(i,j) = f_t^{(0)*}(i,j) + \sum_{k\ne i}\int_0^t q^*(i,k) f_s^{(n-1)*}(k,j)e^{-q_i^*(t-s)}\,ds, \quad n\ge 1,
$$

$$
f_t^{(n)*}(i,j) = f_t^{(0)}(i,j) + \sum_{k\ne j}\int_0^t f_s^{(n-1)*}(i,k)q^*(k,j)e^{-q_j^*(t-s)}\,ds, \quad n\ge 1.
$$

We will show by recurrence on n that

$$
\forall n\ge 0\ \forall i,j\in I^{\mathbf{a}}:\quad \nu(i)f_t^{(n)}(i,j) = e^{-\theta t}\nu(j)f_t^{(n)*}(j,i). \tag{4.14}
$$

Since $-q_i^* = -q_i + \theta$, we get $\nu(i) f_t^{(0)}(i, j) = e^{-\theta t} \nu(j) f_t^{(0)*}(j, i)$, so (4.14) holds for $n = 0$. Let $n \geq 1$, so we assume that $\nu(i) f_t^{(n-1)}(i, j) = e^{-\theta t} \nu(j) f_t^{(n-1)*}(j, i)$ is verified for all $i, j \in I^{\mathbf{a}}$. From the backward iteration (4.2) and since 0 is an absorbing state, all $i, j \in I^{\mathbf{a}}$ verify

$$\nu(i) f_t^{(n)}(i, j) = \nu(i) f_t^{(0)}(i, j) + \sum_{\substack{k \in I^{\mathbf{a}} \\ k \neq i}} \int_0^t \nu(i) q(i, k) f_s^{(n-1)}(k, j) e^{-q_i (t-s)} \, ds. \tag{4.15}$$

From (4.13), we get $\nu(i) q(i, k) = \nu(k) q^*(k, i)$ for $i \neq k$ in $I^{\mathbf{a}}$. On the other hand, the recurrence hypothesis gives

$$\nu(k) f_s^{(n-1)}(k, j) = e^{-\theta s} \nu(j) f_t^{(n-1)*}(j, k).$$

We recall $-q_i = -q_i^* - \theta$. We put all these elements together in (4.15) to obtain

$$\begin{aligned} \nu(i) f_t^{(n)}(i, j) &= e^{-\theta t} \nu(j) f_t^{(0)*}(j, i) \\ &\quad + e^{-\theta t} \nu(j) \sum_{\substack{k \in I^{\mathbf{a}} \\ k \neq i}} \int_0^t f_s^{(n-1)*}(j, k) q^*(k, i) e^{-q_i^* (t-s)} \, ds. \end{aligned}$$

Then, we use the forward evolution equation (4.3) for $f_t^{(n)*}(j, i)$ to get $\nu(i) f_t^{(n)}(i, j) = e^{-\theta t} \nu(j) f_t^{(n)*}(j, i)$. Hence, (4.14) is verified.

We now take the limit $n \to \infty$ in (4.14) to obtain

$$\forall i, j \in I^{\mathbf{a}} \; \forall t \geq 0: \quad \nu(i) p_t(i, j) = e^{-\theta t} \nu(j) p_t^*(j, i). \tag{4.16}$$

We now claim that $(p_t^*(j, i) : j, i \in I^{\mathbf{a}}, t \geq 0)$ is honest, that is, $\forall j \in I^{\mathbf{a}}$: $\sum_{i \in I^{\mathbf{a}}} p_t^*(j, i) = 1$, Note that this property implies

$$\forall i, j \in I^{\mathbf{a}} \; \forall t \geq 0: \quad \sum_{i \in I^{\mathbf{a}}} \nu(i) p_t(i, j) = e^{-\theta t} \nu(j) \sum_{i \in I^{\mathbf{a}}} p_t^*(j, i) = e^{-\theta t} \nu(j),$$

and so ν is a QSD, which finishes the proof. Let us show the claim. Since $(p_t(i, j) : i, j \in I, t \geq 0)$ is honest and since ν is a probability vector, we have

$$\sum_{i \in I^{\mathbf{a}}} \sum_{j \in I^{\mathbf{a}}} \nu(i) p_t(i, j) = \sum_{i \in I^{\mathbf{a}}} \nu(i) \big(1 - p_t(i, 0)\big) = 1 - \sum_{i \in I^{\mathbf{a}}} \nu(i) p_t(i, 0). \tag{4.17}$$

From Fubini's Theorem and equality (4.16), we get

$$\begin{aligned} \sum_{i \in I^{\mathbf{a}}} \sum_{j \in I^{\mathbf{a}}} \nu(i) p_t(i, j) &= \sum_{i \in I^{\mathbf{a}}} \sum_{j \in I^{\mathbf{a}}} e^{-\theta t} \nu(j) p_t^*(j, i) = e^{-\theta t} \sum_{j \in I^{\mathbf{a}}} \nu(j) \Big(\sum_{i \in I^{\mathbf{a}}} p_t^*(j, i) \Big) \\ &= e^{-\theta t} \sum_{j \in I^{\mathbf{a}}} \nu(j) \big(1 - p_t^*(j, \infty)\big) \\ &= e^{-\theta t} - e^{-\theta t} \sum_{j \in I^{\mathbf{a}}} \nu(j) p_t^*(j, \infty), \end{aligned} \tag{4.18}$$

where $p_t^*(j,\infty)$ is the probability that the Markov process governed by $(p_t^*(j,i): j,i \in I^{\mathbf{a}}, t \geq 0)$ explodes for the first time before or at t. On the other hand, since $q(0,0)=0$, the forward differential equation in $j=0$ gives $p_t'(i,0) = \sum_{k\in I^{\mathbf{a}}} p_t(i,k)q(k,0)$, and so,

$$p_t(i,0) = \sum_{k\in I^{\mathbf{a}}} \left(\int_0^t p_u(i,k)\,du\right) q(k,0).$$

Then, by using again Fubini's Theorem and (4.16), we obtain

$$\begin{aligned}
\sum_{i\in I^{\mathbf{a}}} \nu(i) p_t(i,0) &= \sum_{i\in I^{\mathbf{a}}} \sum_{k\in I^{\mathbf{a}}} \left(\int_0^t \nu(i) p_u(i,k)\,du\right) q(k,0) \\
&= \sum_{i\in I^{\mathbf{a}}} \sum_{k\in I^{\mathbf{a}}} \left(\int_0^t e^{-\theta u} \nu(k) p_u^*(k,i)\,du\right) q(k,0) \\
&= \sum_{k\in I^{\mathbf{a}}} \nu(k) q(k,0) \int_0^t e^{-\theta u} \left(\sum_{i\in I^{\mathbf{a}}} p_u^*(k,i)\right) du \\
&= \sum_{k\in I^{\mathbf{a}}} \nu(k) q(k,0) \int_0^t e^{-\theta u} \big(1 - p_u^*(k,\infty)\big)\,du.
\end{aligned}$$

Now we use the equality $\theta = \theta(\nu) = \sum_{k\in I^{\mathbf{a}}} \nu(k) q(k,0)$ to get

$$\begin{aligned}
\sum_{i\in I^{\mathbf{a}}} \nu(i) p_t(i,0) &= \theta \int_0^t e^{-\theta u} - \int_0^t \left(\sum_{k\in I^{\mathbf{a}}} \nu(k) q(k,0) p_u^*(k,\infty)\right) du \\
&= \big(1 - e^{-\theta t}\big) - \int_0^t \left(\sum_{k\in I^{\mathbf{a}}} \nu(k) q(k,0) p_u^*(k,\infty)\right) du.
\end{aligned}$$

Then

$$1 - \sum_{i\in I^{\mathbf{a}}} \nu(i) p_t(i,0) = e^{-\theta t} + \int_0^t \left(\sum_{k\in I^{\mathbf{a}}} \nu(k) q(k,0) p_u^*(k,\infty)\right) du.$$

Hence, from (4.17) we deduce

$$\sum_{i\in I^{\mathbf{a}}} \sum_{j\in I^{\mathbf{a}}} \nu(i) p_t(i,j) = e^{-\theta t} + \int_0^t \left(\sum_{k\in I^{\mathbf{a}}} \nu(k) q(k,0) p_u^*(k,\infty)\right) du.$$

By comparing it with (4.18), we get that $p_u^*(k,\infty) = 0$ for all $k \in I^{\mathbf{a}}$. Therefore, $(p_t^*(j,i): j,i \in I^{\mathbf{a}}, t \geq 0)$ is honest and ν is a QSD. □

4.4 Conditions for Existence of a QSD

For countable state Markov chains, a reciprocal to the necessary condition $\theta^* > 0$ for the existence of a QSD was established in Ferrari et al. (1995a). Let us state it.

Theorem 4.5 *Assume sure killing* $\mathbb{P}_i(T < \infty) = 1$ *for some (or equivalently, for all)* $i \in I^{\mathbf{a}}$, *and*

$$\forall s > 0: \quad \lim_{i\in I^{\mathbf{a}}:\, i\to\infty} \mathbb{P}_i(T < s) = 0. \tag{4.19}$$

Under these conditions, the existence of the exponential moment property (4.6) *is necessary and sufficient for the existence of a QSD. Moreover,*

$$\theta^* = \sup\bigl\{\theta(\nu) : \nu \text{ QSD}\bigr\}, \tag{4.20}$$

and when $\mathbb{E}_i(e^{\theta^* T}) < \infty$ *for some (or equivalently, for all)* $i \in I^{\mathbf{a}}$ *then there exists an extremal QSD.*

Remark 4.6 For birth-and-death chains, discrete and continuous, the fact that the existence of an exponential moment was sufficient for the existence of a QSD was shown in Van Doorn (1991), Ferrari et al. (1995b) and Van Doorn and Schrijner (1995).

The proof follows in detail the one given in Ferrari et al. (1995a). Let us first introduce some useful probabilistic constructions done in Ferrari et al. (1995a).

4.4.1 The Resurrected Process

For $\rho \in \mathcal{P}(I^{\mathbf{a}})$, define $Q^{(\rho)} = (q^{(\rho)}(i,j) : i, j \in I^{\mathbf{a}})$ by

$$q^{(\rho)}(i,j) = q(i,j) + \rho(j)q(i,\partial I), \quad i, j \in I^{\mathbf{a}}. \tag{4.21}$$

It is easily checked that the matrix $Q^{(\rho)}$ is a conservative jump rate matrix. We will construct a Markov process $Y^{(\rho)} = (Y_t^{(\rho)} : t \ge 0)$ with jump rate matrix $Q^{(\rho)}$ starting from a distribution ρ_0 in $I^{\mathbf{a}}$. First, take a sequence of independent processes $Y^{(i)} = (Y_t^{(i)} : t \ge 0)$, each driven by the jump rate matrix Q. We assume that the first one of these copies, $Y^{(1)}$, starts from distribution ρ_0, and all the others, $Y^{(i)}$ with $i \ge 2$, start from ρ. For $i \ge 1$, denote $T^{(i)} = \inf\{t \ge 0 : Y_t^{(i)} \in \partial I\}$. Consider

$$R_0 = 0, \qquad R_i = \sum_{j=1}^{i} T^{(j)}, \quad i \ge 1.$$

Observe that $\lim_{i\to\infty} R_i = \infty$ $\mathbb{P}$-a.s. The process $Y^{(\rho)}$, defined by

$$Y_t^{(\rho)} = \sum_{i\ge 1} Y_{t-R_{i-1}}^{(i)} \mathbb{1}\bigl(t \in [R_{i-1}, R_i)\bigr), \tag{4.22}$$

fulfills the requirements: it starts from ρ_0 and its jump rate matrix is $Q^{(\rho)}$. The process $Y^{(\rho)}$ is called the resurrected process because at each time that the process hits the forbidden states it is immediately resurrected with distribution ρ.

Let us introduce an additional element in the proof of Ferrari et al. (1995a). Note that the following transformation in $\mathcal{P}(I^{\mathbf{a}})$,

$$\rho \to \varphi(\rho) \quad \text{where } \varphi(\rho)' Q^{(\rho)} = 0, \tag{4.23}$$

is well-defined for all $\rho \in \mathcal{P}(I^{\mathbf{a}})$ such that $\mathbb{E}_\rho(T) < \infty$. Indeed, in this case the process $Y^{(\rho)}$ is positive recurrent and $\varphi(\rho)$ is its (unique) stationary distribution. From Theorem 2.4.2 in Asmussen (1987), we have the explicit expression:

$$\varphi(\rho)(i) = \frac{1}{\mathbb{E}_\rho(T)} \int_0^\infty \mathbb{P}_\rho(Y_t = i)\, dt. \tag{4.24}$$

Note that by definition of $Q^{(\nu)}$ in (4.21) we have that

$$\nu' Q^{(\nu)} = 0 \quad \Leftrightarrow \quad \nu' Q^{\mathbf{a}} = -\theta(\nu)\nu'. \tag{4.25}$$

Then, following Theorem 4.4, ν is a QSD if and only if ν is a stationary measure for the process $Y^{(\nu)}$ such that $\theta(\nu) \le q_i \ \forall i \in I^{\mathbf{a}}$. We note that the last condition will be a consequence of our construction because we will show $\theta(\nu) \le \theta^*$, and so (4.10) gives the condition. Observe that (4.23) implies that the relation (4.25) can be written as: The probability measure ν on $I^{\mathbf{a}}$ is a QSD if and only if $\nu = \varphi(\nu)$, that is, if it is a fixed point for the transformation φ.

4.4.2 The Proof

The proof also requires the following elements. Let F be a distribution function concentrated on $[0, \infty)$. Consider a renewal process defined by a sequence of independent random variables distributed as F. The distribution function of the residual time of this renewal process, denoted by ΨF, is given by

$$\Psi F(s) = \frac{1}{m_1(F)} \int_0^s \big(1 - F(x)\big)\, dx,$$

where $m_1(F) = \int_0^\infty (1 - F(x))\, dx$ is the first moment of F. It is well-known and easy to check that an exponential law is a fixed point for the transformation Ψ.

Let $\rho \in \mathcal{P}(I^{\mathbf{a}})$, and let F^ρ be the distribution function of the killing time T when the process starts from law ρ. From (4.24), we get the following key equality:

$$F^{\varphi(\rho)} = \Psi\big(F^\rho\big). \tag{4.26}$$

Indeed,

$$\begin{aligned} 1 - F^{\varphi(\rho)}(s) &= \frac{1}{\mathbb{E}_\rho(T)} \int_0^\infty \sum_{i \in I^{\mathbf{a}}} \mathbb{P}_\rho(Y_t = i)\mathbb{P}_i(T > s)\, dt \\ &= \frac{1}{\mathbb{E}_\rho(T)} \int_0^\infty \mathbb{P}_\rho(T > t + s)\, dt \\ &= \frac{1}{\mathbb{E}_\rho(T)} \int_s^\infty \mathbb{P}_\rho(T > t)\, dt = 1 - \Psi\big(F^\rho\big)(s). \end{aligned}$$

Then, equality (4.26) follows from

$$\mathbb{E}_\rho(T) = \int_0^\infty \mathbb{P}_\rho(T > t)\, dt = \int_0^\infty \big(1 - F^\rho(t)\big) = m_1(F).$$

Note that the existence of the exponential moment property (4.6) implies that the sequence $\Psi^n(F^{\delta_i})$ is well-defined for all $n \geq 1$. Indeed, $F_n = \Psi^n(F^{\delta_i})$ satisfies

$$1 - F_{n+1}(s) = \frac{1}{m_1(F_n)} \int_s^\infty \big(1 - F_n(t)\big)\, dt,$$

and, by integration by parts, it is deduced that for all $k \geq 0$,

$$m_1(F_n) = \frac{m_{k+2}(F_n)}{(k+2) m_{k+1}(F_{n+1})},$$

where we denote by $m_l(G)$ the lth moment of the distribution G. So,

$$m_{k+1}(F_{n+1}) = \frac{m_{k+2}(F_n)}{(k+2) m_1(F_n)} = \frac{m_{k+3}(F_{n-1})}{(k+2)(k+3) m_1(F_n) m_1(F_{n-1})},$$

and by induction we find

$$m_{k+1}(F_{n+1}) = \frac{m_{k+n+2}(F)}{(k+2) \cdots (k+n+2) m_1(F_n) \cdots m_1(F)}.$$

Hence, by taking $k = 0$ in this formula, we obtain

$$m_1(F_{n+1}) = \frac{m_{n+2}(F)}{(n+2)! m_1(F_n) \cdots m_1(F)}.$$

The existence of an exponential moment implies $m_n(F) < \infty$ for all n, so the induction argument gives $m_1(F_n) < \infty$ for all n. Then, for all $n \geq 1$, $\Psi^n(F^{\delta_i})$ is well-defined. An important step in the proof done in Ferrari et al. (1995a) is that the existence of the exponential moment property (4.6) implies that the sequence $\Psi^n(F^{\delta_i})$ converges along subsequences to a well-defined exponential law. More precisely, the following result holds.

Lemma 4.7 *Assume that for some $C < \infty$ and $\theta_0 > 0$, the distribution F supported on $[0, \infty)$ verifies $1 - F(t) \leq Ce^{-\theta_0 t}$ $\forall t \geq 0$. Then, there exist $\theta \geq \theta_0$ and a strictly increasing subsequence $\mathcal{N} = \{n_i : i \geq 1\}$ in $\mathbb{N}$ such that $\Psi^n F$ converges to an exponential distribution with parameter θ, as $n \to \infty$, $n \in \mathcal{N}$. That is, $\forall t \geq 0 : \lim_{\substack{n \in \mathcal{N} \\ n \to \infty}} (1 - \Psi^n F(t)) = e^{-\theta t}$ $\forall t \geq 0$.*

For its proof, we refer to Ferrari et al. (1995a). The next step is the following one.

Lemma 4.8 *Assume sure killing, condition* (4.19) *and the existence of an exponential moment. Then, $\forall \theta' \in (0, \theta^*)$ there exist some $\theta \in [\theta', \theta^*]$ and $\bar{\nu} \in \mathcal{P}(I^{\mathbf{a}})$ such that when the process starts from $\bar{\nu}$ the time of killing is exponentially distributed with parameter θ, that is, $\mathbb{P}_{\bar{\nu}}(T > t) = e^{-\theta t}$ $\forall t \geq 0$.*

If $\mathbb{E}_i(e^{\theta^* T}) < \infty$ for some $i \in I$, then $\forall \theta' \in (0, \theta^*)$ the choice of $\theta = \theta^*$ fulfills the above result.

Proof Fix $i \in I^{\mathbf{a}}$. By definition of θ^*, we have that for all $\theta' \in (0, \theta^*)$ there exists $C(\theta')$ such that $\mathbb{P}_i(T > t) \le C(\theta')e^{-\theta' t}$ $\forall t \ge 0$. Therefore, Lemma 4.7 applied to $F = F^{\delta_i}$ gives the existence of a subsequence $\mathcal{N}$ and $\theta \ge \theta'$ such that $F^n = \Psi^n F^{\delta_i}$ verifies

$$\forall t \ge 0: \quad \lim_{\substack{n \in \mathcal{N} \\ n \to \infty}} \big(1 - F^n(t)\big) = e^{-\theta t}. \tag{4.27}$$

Let us see that the set of measures $\{\nu^n = \varphi^n(\delta_i) : n \in \mathcal{N}\}$ is tight. Fix $\epsilon > 0$. From (4.26) and (4.27), we deduce that there exists $t(\epsilon) > 0$ such that

$$\forall n \in \mathcal{N}: \quad 1 - F^n\big(t(\epsilon)\big) < \epsilon.$$

Now

$$1 - F^n\big(t(\epsilon)\big) = \mathbb{P}_{\nu^n}\big(T > t(\epsilon)\big) = \sum_{j \in I^{\mathbf{a}}} \nu^n(j)\mathbb{P}_j\big(T > t(\epsilon)\big),$$

and so

$$\nu^n\big(j \in I^{\mathbf{a}} : \mathbb{P}_j\big(T > t(\epsilon)\big) \ge 1/2\big) \le 2\epsilon.$$

From (4.19), we get that the set $\{j \in I^{\mathbf{a}} : \mathbb{P}_j(T \le t(\epsilon)) \ge 1/2\}$ is finite and verifies $\nu^n(j \in I^{\mathbf{a}} : \mathbb{P}_j(T \le t(\epsilon)) \ge 1/2) > 1 - 2\epsilon$. This shows that $\{\nu^n : n \in \mathcal{N}\}$ is tight. Therefore, there exists $\nu^\infty \in \mathcal{P}(I^{\mathbf{a}})$ such that for some subsequence $\mathcal{N}'$ of $\mathcal{N}$ one has $\nu^n \to \nu^\infty$, as $n \to \infty$, $n \in \mathcal{N}'$. We have

$$\begin{aligned}
\mathbb{P}_{\nu^\infty}\big(T > t(\epsilon)\big) &= \sum_{j \in I^{\mathbf{a}}} \nu^\infty(j)\mathbb{P}_j\big(T > t(\epsilon)\big) = \lim_{n \to \infty, n \in \mathcal{N}'} \sum_{j \in I^{\mathbf{a}}} \nu^n(j)\mathbb{P}_j\big(T > t(\epsilon)\big) \\
&= \lim_{\substack{n \to \infty \\ n \in \mathcal{N}'}} \big(1 - F^n\big(t(\epsilon)\big)\big) = e^{-\theta t}.
\end{aligned}$$

By definition of θ^*, we get that $\theta \in [\theta', \theta^*]$. Finally, note that the above proof also shows that if $\mathbb{E}_i(e^{\theta^* T}) < \infty$ then the above construction can be done for $\theta = \theta^*$. □

The following result will finish the proof of Theorem 4.5.

Lemma 4.9 *Assume sure killing, condition* (4.19) *and the existence of an exponential moment. Then there exists a QSD ν whose killing time is exponentially distributed with parameter θ given by Lemma* 4.8, *so* $\mathbb{P}_\nu(T > t) = e^{-\theta t}$ $\forall t \ge 0$.

Proof For any θ define the set

$$\mathcal{P}^\theta = \big\{\rho \in \mathcal{P}\big(I^{\mathbf{a}}\big) : \forall t \ge 0, \mathbb{P}_\rho(T > t) = e^{-\theta t}\big\}.$$

From the last lemma, we know that the set $\mathcal{P}^\theta$ is nonempty. It is clear that $\mathcal{P}^\theta$ is a convex set.

To avoid cumbersome notation, we assume $I^{\mathbf{a}} = \mathbb{N}$. The set $\mathcal{P}(I^{\mathbf{a}})$ is endowed with the metric $d(\rho^1, \rho^2) = \sum_{i \in \mathbb{N}} 2^{-i} |\rho^1(i) - \rho^2(i)|$. For a sequence $(\rho^n : n \ge 1)$ and ρ included in $\mathcal{P}(I^{\mathbf{a}})$, we have $d(\rho^n, \rho) \to 0$ as $n \to \infty$ if and only if the pointwise convergence holds $\rho^n(i) \to \rho(i)$ as $n \to \infty$.

Note that the argument developed in the proof of Lemma 4.8 shows that every sequence $(\rho^n : n \geq 1)$ in $\mathcal{P}^\theta$ is tight. So, it has some limit point $\rho^\infty \in \mathcal{P}(I^{\mathbf{a}})$. Obviously, $\mathbb{P}_{\rho^\infty}(T > t) = e^{-\theta t}$, and so $\rho^\infty \in \mathcal{P}^\theta$. This shows the compactness of $\mathcal{P}^\theta$.

From relation (4.26), $F^{\varphi(\rho)} = \Psi(F^\rho)$, and since an exponential distribution is invariant under Ψ, we find that $\varphi(\mathcal{P}^\theta) \subseteq \mathcal{P}^\theta$. Now, we shall prove that $\varphi : \mathcal{P}^\theta \to \mathcal{P}^\theta$ is a continuous transformation. We recall (4.24),

$$\varphi(\rho)(i) = \frac{1}{\mathbb{E}_\rho(T)} \int_0^\infty \mathbb{P}_\rho(Y_t = i)\,dt.$$

Let $(\rho^n : n \geq 1)$ be some sequence in $\mathcal{P}^\theta$ that converges to $\rho \in \mathcal{P}^\theta$. Observe that $\mathbb{E}_{\rho^n}(T) = 1/\theta = \mathbb{E}_\rho(T)$ for all $n \geq 1$. From Fatou's Lemma, we get that for all subset $L \subseteq \mathbb{N}$,

$$\begin{aligned}
\liminf_{n\to\infty} \sum_{i\in L} \varphi\big(\rho^n\big)(i) &= \liminf_{n\to\infty} \theta \int_0^\infty \sum_{j\in\mathbb{N}} \rho^n(j) \sum_{i\in L} \mathbb{P}_j(Y_t = i)\,dt \\
&\geq \theta \liminf_{n\to\infty} \int_0^\infty \sum_{j\in\mathbb{N}} \rho(j) \sum_{i\in L} \mathbb{P}_j(Y_t = i)\,dt \\
&= \sum_{i\in L} \varphi(\rho)(i).
\end{aligned} \tag{4.28}$$

Since this also holds for L^c, we get that the equality holds in (4.28). Finally, by taking $L = \{i\}$ and $L = \{i\}^c$, we can conclude

$$\liminf_{n\to\infty} \varphi\big(\rho^n\big)(i) = \varphi(\rho)(i) = \limsup_{n\to\infty} \varphi\big(\rho^n\big)(i).$$

Then $\varphi(\rho^n) \to \varphi(\rho)$ as $n \to \infty$. The continuity of φ in the metric d follows.

Since $\varphi : \mathcal{P}^\theta \to \mathcal{P}^\theta$ is a continuous transformation acting on a convex compact metric space $\mathcal{P}^\theta$, it has some fixed point ν (see Tychonov 1935). Since the equality $\varphi(\nu) = \nu$ is equivalent to ν being a QSD, the existence of a QSD is shown.

Finally, the equality $\theta^* = \sup\{\theta(\nu) : \nu \text{ is a QSD}\}$ follows from the fact that for each $\theta' \in (0, \theta^*)$ we can find some $\theta \in [\theta', \theta^*]$ for which there exists a QSD with the exponential rate of survival θ. □

Remark 4.10 In relation to quasi-limiting distributions (which satisfy (3.6)) for countable infinite Markov chains in discrete time, more general conditions for their existence were supplied in Kesten (1995).

4.5 Kingman's Parameter and Exponential Rate of Survival

We will state sufficient conditions for the exponential rate of survival to coincide with the exponential rate of the transition probabilities. Let us introduce this last quantity. For an irreducible and surely killed Markov chain, the exponential rate of the transition probabilities, denoted by θ_K and called the Kingman's parameter, is well-defined and does not depend on $i, j \in I^{\mathbf{a}}$,

$$\theta_K = -\lim_{t\to\infty} \frac{1}{t} \log p_t(i, j); \tag{4.29}$$

see Kingman (1963). This parameter verifies $\theta_K \le \inf\{q_i : i \in I^{\mathbf{a}}\}$, a subadditive argument gives $p_t(i, i) \le e^{-\theta_K t}$ $\forall i \in I^{\mathbf{a}}, t \ge 0$, and it can be shown that for all pairs $i, j \in I^{\mathbf{a}}$ there exists a finite constant C_{ij} such that $p_t(i, j) \le C_{ij} e^{-\theta_K t}$ $\forall t \ge 0$.

From $p_t(i, j) \le \mathbb{P}_i(T > t)$ for $i, j \in I^{\mathbf{a}}$, we get

$$\theta^* = \liminf_{t\to\infty} -\frac{1}{t} \log \mathbb{P}_i(T > t) \le \limsup_{t\to\infty} -\frac{1}{t} \log \mathbb{P}_i(T > t) \le \theta_K.$$

Hence, if under some conditions we are able to show $\theta^* = \theta_K$, then we will also get that the exponential rate of survival is the limit $\theta^* = \lim_{t\to\infty} -\frac{1}{t} \log \mathbb{P}_i(T > t)$ $\forall i \in I^{\mathbf{a}}$. A trivial case is $\theta_K = 0$ because it implies $\theta^* = \theta_K = 0$.

To get sufficient conditions in order that the equality $\theta^* = \theta_K$ be satisfied for some strictly positive value, let us first state some previous results on subinvariant measures.

Let μ be a measure on $I^{\mathbf{a}}$ and $\theta \ge 0$. Then, μ is said to be a θ-subinvariant measure of $Q^{\mathbf{a}}$ if it satisfies

$$\forall j \in I^{\mathbf{a}}: \quad \left(\mu' Q^{\mathbf{a}}\right)(j) \le -\theta \mu(j),$$

and it is a θ-subinvariant measure for the semigroup (P_t) if

$$\forall t \ge 0,\ \forall j \in I^{\mathbf{a}}: \quad \left(\mu' P_t\right)(j) \le e^{-\theta t} \mu(j). \tag{4.30}$$

In Tweedie (1974), it was shown that if μ is a θ-subinvariant measure of $Q^{\mathbf{a}}$ with $\theta \le \inf\{q_i : i \in I^{\mathbf{a}}\}$ then it is a θ-subinvariant measure for (P_t). (The proof of this fact can be found in the arguments and constructions supplied in the proof of Theorem 4.4.) The converse was also shown in Tweedie (1974), if μ is a θ-subinvariant measure for (P_t) then $\theta \le \inf\{q_i : i \in I^{\mathbf{a}}\}$ and μ is a θ-subinvariant measure of $Q^{\mathbf{a}}$.

If μ is a subinvariant measure for (P_t) and for some $j \in I^{\mathbf{a}}$ we have $\mu(j) = 0$, irreducibility implies $\mu = 0$, the null measure. Hence, a nontrivial subinvariant measure is strictly positive.

We can also define an analogous concept when the semigroup acts on the right. A nonnegative function $\varphi = (\varphi(j) : j \in I^{\mathbf{a}})$ is a θ-subinvariant function for the semigroup (P_t) if

$$\forall t \ge 0,\ \forall i \in I^{\mathbf{a}}: \quad (P_t \varphi)(j) \le e^{-\theta t} \varphi(j).$$

As before, if $\varphi(j) = 0$ then $\varphi = 0$, the null function. Hence, a nontrivial nonnegative θ-subinvariant function for (P_t) is strictly positive.

Let us show that for every $\theta \in (0, \theta_K]$ there exist strictly positive θ-subinvariant measures and strictly positive θ-subinvariant functions. The θ-subinvariant measures will be constructed as limits of subinvariant measures for the discrete skeletons of the original chain. First, let us fix $\delta > 0$. From (4.29), we get

$$\forall i, j \in I^{\mathbf{a}}: \quad \lim_{n\to\infty} \left(p_{n\delta}(i, j)\right)^{1/n} = e^{-\theta_K \delta}. \tag{4.31}$$

Put $\widehat{p}(i,j) := p_\delta(i,j)$, so $\widehat{p}^{(n)}(i,j) := p_{n\delta}(i,j)$. The matrix $\widehat{P} = (\widehat{p}(i,j) : i, j \in I^{\mathbf{a}})$ is the (strictly substochastic) transition matrix of the Markov chain $\widehat{Y} = (\widehat{Y}_n := Y_{n\delta} : n \geq 0)$ which is surely killed at the killing time $\widehat{T} = \inf\{n \geq 0 : \widehat{Y}_n \notin I^{\mathbf{a}}\}$. Note that $|\delta\widehat{T} - T| \leq \delta$.

Define the series $\widehat{P}_{ij}(r) := \sum_{n\geq 0} \widehat{p}^{(n)}(i,j) r^n$ for $i, j \in I^{\mathbf{a}}$ and $r \geq 0$. It is known, see Theorem 6.1 in Seneta (1981), that there exists a common convergence radius for these series,

$$R = \Big(\lim_{n\to\infty} \big(\widehat{p}^{(n)}(i,j)\big)^{1/n} \Big)^{-1},$$

so for all $i, j \in I^{\mathbf{a}}$, $\widehat{P}_{ij}(r) < \infty$ for $r \in [0, R)$ and $\widehat{P}_{ij}(r) = \infty$ for $r > R$. From (4.31), we find $R = e^{\theta_K \delta}$.

For $i \in I^{\mathbf{a}}$, denote by $\widehat{T}_i = \inf\{n > 0 : \widehat{Y}_n = i\}$ the hitting time of $i \in I^{\mathbf{a}}$ for the chain $\widehat{Y}$. Note that $\widehat{T}_i$ can be infinite. A standard proof shows that

$$\forall j \in I^{\mathbf{a}}: \quad \mathbb{E}_i \left(\sum_{n=0}^{\widehat{T}_i \wedge \widehat{T} - 1} \mathbb{1}_{\{\widehat{Y}_n = j\}} R^n \right) < \infty; \tag{4.32}$$

see Lemma 2.3 in Sect. 5.2 in Anderson (1991). Below we show the existence of subinvariant measures for (P_t) by using the existence of subinvariant measures in discrete time: a measure μ is α-subinvariant for $(\widehat{P}^n)$ if $(\mu' \widehat{P}^n)(j) \leq e^{-\alpha n} \mu(j)\ \forall j \in I^{\mathbf{a}}\ \forall n \geq 0$, and this happens if and only if it is α-subinvariant for $\widehat{P}$, that is, $(\mu' \widehat{P})(j) \leq e^{-\alpha} \mu(j)\ \forall j \in I^{\mathbf{a}}$.

The following result is stated in Kingman (1963), see Theorem 2.7 in Sect. 5.2 of Anderson (1991). Also see Theorem 6.2 in Seneta (1981) for the discrete time case.

Lemma 4.11

(i) *Let i_0 be fixed. For any $\theta \in [0, \theta_K]$, the measure $\mu^{\delta,\theta}$, defined by*

$$\forall j \in I^{\mathbf{a}}: \quad \mu^{\delta,\theta}(j) = \mathbb{E}_{i_0} \left(\sum_{n=0}^{\widehat{T}_{i_0} \wedge \widehat{T} - 1} e^{\theta \delta n} \mathbb{1}_{\{\widehat{Y}_n = j\}} \right), \tag{4.33}$$

takes strictly positive finite values, and it is $\theta\delta$-subinvariant for $(\widehat{P}^n)$:

$$\forall n \geq 0,\ \forall j \in I^{\mathbf{a}}: \quad \sum_{i \in I^{\mathbf{a}}} \mu^{\delta,\theta}(i) \widehat{p}^{(n)}(i,j) \leq e^{-\theta\delta n} \mu^{\delta,\theta}(j). \tag{4.34}$$

(ii) *For any $\theta \in [0, \theta_K]$, there exist a strictly positive measure $\mu^\theta = (\mu^\theta(j) : j \in I^{\mathbf{a}})$ and a strictly positive function $\varphi^\theta = (\varphi^\theta(j) : j \in I^{\mathbf{a}})$ which are θ-subinvariant for (P_t).*

Proof (i) Put $\mu := \mu^{\delta,\theta}$. Since $\theta \in [0, \theta_K]$, we have $e^{\theta\delta} \leq R$, so (4.32) implies that

μ is finite. Let us show that μ is $\theta\delta$-subinvariant for $\widehat{P}$. Since $\widehat{T} < \widehat{T}_{i_0}$ implies $Y_{\widehat{T}_{i_0}\wedge\widehat{T}} \notin I^{\mathbf{a}}$, we get

$$\mu(j) \geq \mathbb{E}_{i_0}\left(\sum_{n=1}^{\widehat{T}_{i_0}\wedge\widehat{T}} e^{\theta\delta n}\mathbb{1}_{\{\widehat{Y}_n=j\}}\right) = \sum_{n=1}^{\infty}\mathbb{E}_{i_0}\big(e^{\theta\delta n}\mathbb{1}_{\{\widehat{T}_{i_0}\wedge\widehat{T}\geq n\}}\mathbb{1}_{\{\widehat{Y}_n=j\}}\big).$$

By developing this last term, we find

$$\begin{aligned}
\mu(j) &\geq \sum_{n=1}^{\infty} e^{\theta\delta n}\mathbb{E}_{i_0}\big(\mathbb{E}\big(\mathbb{1}_{\{\widehat{T}_{i_0}\wedge\widehat{T}<n\}^c}\mathbb{1}_{\{\widehat{Y}_n=j\}}|\sigma(\widehat{Y}_k : k<n)\big)\big)\\
&= \sum_{n=1}^{\infty} e^{\theta\delta n}\left(\sum_{k\in I^{\mathbf{a}}}\mathbb{E}_{i_0}\big(\mathbb{1}_{\{\widehat{T}_{i_0}\wedge\widehat{T}<n\}^c}\mathbb{1}_{\{\widehat{Y}_{n-1}=k\}}\mathbb{E}_k(\mathbb{1}_{\{\widehat{Y}_1=j\}})\big)\right)\\
&= e^{\theta\delta}\sum_{k\in I^{\mathbf{a}}}\widehat{p}(k,j)\left(\mathbb{E}_{i_0}\left(\sum_{n=1}^{\infty} e^{\theta\delta(n-1)}\mathbb{1}_{\{\widehat{T}_{i_0}\wedge\widehat{T}\geq n\}}\mathbb{1}_{\{\widehat{Y}_{n-1}=k\}}\right)\right)\\
&= e^{\theta\delta}\sum_{k\in I^{\mathbf{a}}}\widehat{p}(k,j)\mu(k).
\end{aligned}$$

Then, μ is $\theta\delta$-subinvariant.

Let us prove (ii). Fix $t > 0$. Now we allow δ to vary in (4.33). We consider $\delta = 1/N$ for $N \geq 1$ and define $\mu^\theta := \liminf_{N\to\infty}\mu^{1/N,\theta}$. Let $t \geq 0$ be fixed. We take a sequence of positive integers $(n_N : N \geq 1)$ such that $t = \lim_{N\to\infty} n_N/N$.

Since for $\delta = 1/N$ we have $\widehat{p}^{(n_N)}(i,j) = p_{n_N/N}(i,j)$, by continuity of $p_s(i,j)$ on $s \in \mathbb{R}_+$ we obtain

$$\lim_{N\to\infty}\widehat{p}^{(n_N)}(i,j) = \lim_{N\to\infty} p_{n_N/N}(i,j) = p_t(i,j).$$

Take $\delta = 1/N$ and $n = n_N$ in (4.34) and compute liminf. By using Fatou's Lemma, we get for all $j \in I^{\mathbf{a}}$,

$$\begin{aligned}
\sum_{i\in I^{\mathbf{a}}}\Big(\liminf_{N\to\infty}\mu^{1/N,\theta}(i)p_{n_N/N}(i,j)\Big) &\leq \liminf_{N\to\infty}\Big(\sum_{i\in I^{\mathbf{a}}}\mu^{1/N,\theta}(i)p_{n_N/N}(i,j)\Big)\\
&\leq \liminf_{N\to\infty} e^{-\theta n_N/N}\mu^{1/N,\theta}(j).
\end{aligned}$$

We have shown

$$\forall j \in I^{\mathbf{a}}:\quad \sum_{i\in I^{\mathbf{a}}}\liminf_{N\to\infty}\mu^\theta(i)p_t(i,j) \leq e^{-\theta t}\mu^\theta(j).$$

This holds for all $t \geq 0$, so μ^θ is a θ-subinvariant measure for (P_t). Note that the measure μ^θ takes strictly positive values because from definition $\mu^{\delta,\theta}(i_0) = 1$ for all $\delta > 0$ and so $\mu^\theta(i_0) = 1$.

Now, fix $\theta \in [0,\theta_K]$. We will show that there exists a strictly positive function φ that is θ-subinvariant for (P_t). Let $\mu := \mu^\theta$ be a strictly positive θ-subinvariant measure for (P_t). Define

$$\forall t \geq 0,\ \forall i,j \in I^{\mathbf{a}}:\quad p_t^*(i,j) := \mu(j)p_t(j,i)/\mu(i).$$

It is easily checked that $(P_t^* = (p_t^*(i,j) : i, j \in I^{\mathbf{a}}) : t \geq 0)$ is a semigroup, and it is substochastic because $\sum_{j \in I^{\mathbf{a}}} p_t^*(i,j) \leq e^{-\theta t}$. Notice that

$$-\lim_{t\to\infty} \frac{1}{t} \log p_t^*(i,j) = \theta_K.$$

Then, by the part already proved, for all $\theta \in [0, \theta_K]$ there exists a strictly positive θ-subinvariant measure $\rho = (\rho(j) : j \in I^{\mathbf{a}})$ for the semigroup $(P_t^* : t \geq 0)$. Thus,

$$\forall i \in I^{\mathbf{a}}: \quad \sum_{i \in I^{\mathbf{a}}} \rho(i) p_t^*(i,j) \leq e^{-\theta t} \rho(j).$$

Since $p_t^*(i,j)\rho(i) = \mu(j) p_t(j,i)\rho(i)/\mu(i)$, the positive function φ given by $\varphi(i) = \rho(i)/\mu(i)$ is θ-subinvariant for (P_t). □

Let us prove that if the set of states connected to the killing state $\{0\}$ is finite, then $\theta^* = \theta_K$, the subinvariant measures are finite and the positive subinvariant functions are lower-bounded. This result is Theorem 3.3.2(iii) in Jacka and Roberts (1995).

Proposition 4.12 *Assume that the set* $I_0^* := \{i \in I^{\mathbf{a}} : q_{j,0} > 0\}$ *is finite.*

(i) *We have*

$$\theta^* = \theta_K \quad \textit{and} \quad \theta^* = \lim_{t\to\infty} -\frac{1}{t} \log \mathbb{P}_i(T > t). \tag{4.35}$$

(ii) *If* $\theta_K > 0$, *then for all* $\theta \in (0, \theta_K]$ *any* θ-*subinvariant measure* μ^θ *has finite mass. When* μ^{θ^*} *is normalized, we get*

$$\mathbb{P}_i(T > t) \leq \frac{e^{-\theta^* t}}{\mu^{\theta^*}(i)}. \tag{4.36}$$

Moreover, for all $\theta \in (0, \theta_K]$ *any strictly positive* θ-*subinvariant function* φ^θ *is lower-bounded by a strictly positive value.*

Proof Since $0 \leq \theta^* \leq \theta_K$, for proving (4.35) we can assume $\theta_K > 0$, and it suffices to show $\theta^* \geq \theta_K$.

We denote by $S^j = \int_0^\infty \mathbb{1}_{Y_t = j}\, dt$ the total time spent in $j \in I^{\mathbf{a}}$. We have

$$\mathbb{E}_i\big(S^j\big) = \int_0^\infty p_t(i,j)\, dt.$$

Let $T_{I_0^*} = \inf\{t \geq 0 : Y_t \in I_0^*\}$ be the hitting time of I_0^*. From the hypothesis $\mathbb{P}_i(T_{I_0^*} \leq T) = 1$ for all $i \in I^{\mathbf{a}}$, and since $\mathbb{P}_i(T < \infty) = 1$, we have $\mathbb{P}_i(T_{I_0^*} < \infty) = 1$. Obviously, $\mathbb{E}_k(S^k) \geq \mathbb{E}_k(\tau)$ for $k \in I_0^*$, where $\tau = \inf\{t \geq 0 : Y_t \neq Y_0\}$ is the holding time. These relations imply

$$\begin{aligned}\forall k \in I^{\mathbf{a}}: \quad & \sum_{j \in I_0^*} \int_0^\infty p_t(k,j)\, dt = \mathbb{E}_k\Big(\sum_{j \in I_0^*} S^j\Big) = \mathbb{E}_k\Big(\mathbb{E}_{Y_{T_{I_0^*}}}\Big(\sum_{j \in I_0^*} S^j\Big)\Big) \\ & \geq \mathbb{E}_k\big(\mathbb{E}_{Y_{T_{I_0^*}}}(\tau)\big) \geq \min\big\{\mathbb{E}_j(\tau) : j \in I_0^*\big\} = \min\big\{q_j^{-1} : j \in I_0^*\big\}.\end{aligned} \tag{4.37}$$

Let $\theta \in (0, \theta_K]$ and let μ be a θ-subinvariant measure. Let us prove that μ has finite mass. By integrating equations (4.30) with respect to time, we get

$$\forall j \in I^{\mathbf{a}}: \quad \sum_{k \in I^{\mathbf{a}}} \mu(k) \int_0^\infty p_t(k, j)\, dt \le \frac{\mu(j)}{\theta}.$$

Then,

$$\sum_{k \in I^{\mathbf{a}}} \mu(k) \biggl(\sum_{j \in I_0^*} \int_0^\infty p_t(k, j)\, dt \biggr) \le \frac{1}{\theta} \biggl(\sum_{j \in I_0^*} \mu^\theta(j) \biggr).$$

From (4.37), we have $\sum_{j \in I_0^*} \int_0^\infty p_t(k, j)\, dt \ge \min\{q_j^{-1} : j \in I_0^*\}$ for all $k \in I^{\mathbf{a}}$, and then we obtain

$$\sum_{k \in I^{\mathbf{a}}} \mu(k) \le \frac{\sum_{j \in I_0^*} \mu^\theta(j)}{\theta \min\{q_j^{-1} : j \in I_0^*\}} < \infty,$$

proving that μ has finite mass. We normalize it, so by using (4.30), we get

$$\begin{aligned} \mu(i)\mathbb{P}_i(T > t) &\le \sum_{j \in I^{\mathbf{a}}} \mu(j)\mathbb{P}_j(T > t) = \sum_{j \in I^{\mathbf{a}}} \biggl(\sum_{k \in I^{\mathbf{a}}} \mu(j) p_t(j, k) \biggr) \\ &\le e^{-\theta t} \sum_{j \in I^{\mathbf{a}}} \mu(j) = e^{-\theta t}. \end{aligned}$$

By taking $\theta = \theta_K$, we obtain $\mathbb{P}_i(T > t) \le e^{-\theta_K t}/\mu(i)$. Then $\theta^* \ge \theta_K$ and both relations (4.35) and (4.36) are shown.

The only thing left to prove is that for $\theta \in (0, \theta_K]$, all θ-subinvariant strictly positive functions φ are lower-bounded by a strictly positive constant. Note that

$$\sum_{j \in I_0^*} \int_0^\infty p_t(i, j)\varphi(j)\, dt \le \sum_{j \in I^{\mathbf{a}}} \int_0^\infty p_t(i, j)\varphi(j)\, dt \le \int_0^\infty e^{-\theta t} \varphi(i)\, dt = \frac{\varphi(i)}{\theta}.$$

From (4.37), $\sum_{j \in I_0^*} \int_0^\infty p_t(i, j)\, dt \ge \min\{q_j^{-1} : j \in I_0^*\}$, so we obtain the result $\varphi(i) \ge \theta \min\{q_j^{-1} : j \in I_0^*\}\ \forall i \in I^{\mathbf{a}}$. □

4.6 Symmetric Case

Let us assume there exists some measure $\pi = (\pi_i : i \in I^{\mathbf{a}})$ such that $Q^{\mathbf{a}}$ is symmetric with respect to π, that is,

$$\forall i, j \in I^{\mathbf{a}}: \quad \pi_i q(i, j) = \pi_j q(j, i).$$

Let $(I_N : N \ge 1)$ be an increasing sequence of finite subsets such that $I_N \nearrow I^{\mathbf{a}}$ as $N \nearrow \infty$. Let $Q^{(N)} = (q(i, j) : i, j \in I_N)$ be the truncated jump rate matrix and let

$$p_t^{(N)}(i, j) = \bigl(e^{Q^{(N)} t}\bigr)(i, j) = \mathbb{P}_i(Y_t = j, T_{I^{\mathbf{a}} \setminus I_N} > t) \tag{4.38}$$

be the truncated kernel. The matrix $Q^{(N)}$ is symmetric with respect to the restriction of π to I_N. Let $\{d_k^{(N)} : k \in I_N\}$ be the set of eigenvalues of $Q^{(N)}$, which are all negative and denote by $\mathcal{D}^{(N)}$ the diagonal matrix whose elements are $\{d_k^{(N)} : k \in I_N\}$. Recall the decomposition given in (3.16),

$$p_t^{(N)}(i,j) = \sqrt{\frac{\pi_j}{\pi_i}} \int_0^\infty e^{-tx}\, d\Gamma_{i,j}^{(N)}(x),$$

where $\Gamma_{i,j}^{(N)}$ is a signed measure supported on $(0,\infty)$. When $i=j$, $\Gamma_{i,i}^{(N)}$ is a probability measure. Let

$$\Gamma_{i,j}^{(N)} = \Gamma_{i,j}^{(N)+} - \Gamma_{i,j}^{(N)-}, \qquad \|\Gamma_{i,j}^{(N)}\| = \Gamma_{i,j}^{(N)+} + \Gamma_{i,j}^{(N)-}.$$

From (3.12) and (3.13), we get that the total variation measure $\|\Gamma_{i,j}^{(N)}\|$ satisfies

$$\|\Gamma_{i,j}^{(N)}\|\big((0,\infty)\setminus\{-d_k^{(N)} : k \in I^{\mathbf{a}}\}\big) = 0,$$
$$\Gamma_{i,j}^{(N)}\big(\{-d_k^{(N)}\}\big) = m^{(N)}(i,k)m^{(N)}(j,k),$$

where $M^{(N)} = (m^{(N)}(i,j) : i,j \in I^{\mathbf{a}})$ is the orthogonal matrix formed by the normalized eigenvectors of $(\mathcal{D}_\pi^{(N)})^{1/2} Q^{\mathbf{a}(N)} (\mathcal{D}_\pi^{(N)})^{-1/2}$.

From (3.14), we get that the masses of the positive measures $\Gamma_{i,j}^{(N)+}$ and $\Gamma_{i,j}^{(N)-}$ are bounded by 1. The Helly's Theorem implies the existence of an infinite subsequence $(N_l : l \ge 1)$ and of measures $\Gamma_{i,j}^+$, $\Gamma_{i,j}^-$ such that

$$\Gamma_{i,j}^{(N_l)+} \to \Gamma_{i,j}^+ \quad \text{and} \quad \Gamma_{i,j}^{(N_l)-} \to \Gamma_{i,j}^- \quad \text{vaguely as } l \to \infty. \tag{4.39}$$

Therefore, the signed measure

$$\Gamma_{i,j} = \Gamma_{i,j}^+ - \Gamma_{i,j}^- \tag{4.40}$$

is well-defined, and (3.14) implies that the mass of the total variation measure $\|\Gamma_{i,j}\| = \Gamma_{i,j}^+ + \Gamma_{i,j}^-$ is bounded by 1:

$$\forall i,j \in I^{\mathbf{a}}: \quad \|\Gamma_{i,j}\|\big([0,\infty)\big) \le 1.$$

When $i=j$, $\Gamma_{i,i}$ is a measure whose mass is bounded by 1. Moreover,

$$\int_0^\infty e^{-tx}\, d\Gamma_{i,j}^+(x) = \lim_{l\to\infty} \int_0^\infty e^{-tx}\, d\Gamma_{i,j}^{(N_l)+}(x),$$
$$\int_0^\infty e^{-tx}\, d\Gamma_{i,j}^-(x) = \lim_{l\to\infty} \int_0^\infty e^{-tx}\, d\Gamma_{i,j}^{(N_l)-}(x).$$

Put

$$\theta_N^* = \liminf_{t\to\infty} -\frac{1}{t} \log \mathbb{P}_i(T > t, T_{I^{\mathbf{a}}\setminus I_N} \le t),$$

we have that the spectrum of $Q^{(N)}$ satisfies $\theta_N^* = \inf\{d_k^{(N)} : k \in 1,\dots,N\}$. From

$$\liminf_{t\to\infty} -\frac{1}{t}\log \mathbb{P}_i(T>t) \le \liminf_{t\to\infty} -\frac{1}{t} \log \mathbb{P}_i(T > t, T_{I^{\mathbf{a}}\setminus I_N} \le t),$$

we get $\{d_k^{(N)} : k \in I^{\mathbf{a}}\} \subset [\theta^*, \infty)$.

Remark 4.13 From $\Gamma_{i,j}^{(N)+}(\{0\}) = 0 = \Gamma_{i,j}^{(N)+}(\{0\})$ $\forall N$, we get $\|\Gamma_{i,j}^{+}\|(\{0\}) = \|\Gamma_{i,j}^{-}\|(\{0\}) = 0$. Therefore, from

$$\|\Gamma_{i,j}\|\big(\{0\}\big) = 0 \quad \forall i, j \in I^{\mathbf{a}} \quad \text{and} \quad \big\{d_k^{(N)} : k \in I^{\mathbf{a}}\big\} \subset \big[\theta^*, \infty\big),$$

we get $\int_0^\infty d\Gamma_{i,j} = \int_{\theta^*}^\infty d\Gamma_{i,j}$, where the domain of integration is the set $[\theta^*, \infty) \cap (0, \infty)$.

Recall that the closed support of a measure v, denoted by Support(v), is the smallest closed set in $\mathbb{R}$ such that its complement has no mass, $v(\text{Support}(v)^c) = 0$. We have $\theta_0 = \inf \text{Support}(v)$, a finite number, if and only if $v((-\infty, \theta_0)) = 0$ and for all $\epsilon > 0$ we have $v([\theta_0, \theta_0 + \epsilon)) > 0$. The support of a signed measure v is by definition the support of the total variation measure $\|v\| = v^+ + v^-$.

Proposition 4.14 *Assume that there exists an increasing sequence of finite subsets* $(I_N : N \geq 1)$ *such that* $I_N \nearrow I^{\mathbf{a}}$ *as* $N \nearrow \infty$ *and*

$$\forall t \geq 0,\ \forall i \in I^{\mathbf{a}}: \quad \lim_{N\to\infty} \mathbb{P}_i(T_{I^{\mathbf{a}} \setminus I_N} \leq t) = 0. \tag{4.41}$$

Then $\Gamma_{i,i}$ *is a probability measure* $\forall i \in I^{\mathbf{a}}$. *Moreover,* $\theta_K = \inf \text{Support}(\Gamma_{i,j})$ $\forall i, j \in I^{\mathbf{a}}$ *and*

$$p_t(i, j) = \sqrt{\frac{\pi_j}{\pi_i}} \int_{\theta_K}^\infty e^{-tx}\, d\Gamma_{i,j}(x).$$

If, in addition to condition (4.41), *the set* $I_0^* = \{k \in I^{\mathbf{a}} : q(k, 0) > 0\}$ *is finite, then* $\theta^* = \inf \text{Support}(\Gamma_{i,j})$, *so*

$$p_t(i, j) = \sqrt{\frac{\pi_j}{\pi_i}} \int_{\theta^*}^\infty e^{-tx}\, d\Gamma_{i,j}(x). \tag{4.42}$$

Furthermore, the following relation is satisfied:

$$\mathbb{P}_i(T > t) = \sum_{k \in I^{\mathbf{a}}} q(k, 0) \sqrt{\frac{\pi_k}{\pi_i}} \int_{\theta^*}^\infty \frac{e^{-tx}}{x}\, d\Gamma_{i,k}(x). \tag{4.43}$$

Proof From (4.38) and condition (4.41), we deduce

$$\forall t \geq 0,\ \forall i, j \in I^{\mathbf{a}}: \quad \lim_{N\to\infty} p_t^{(N)}(i, j) = p_t(i, j).$$

Hence,

$$p_t(i, j) = \sqrt{\frac{\pi_j}{\pi_i}} \int_0^\infty e^{-tx}\, d\Gamma_{i,j}(x),$$

If $i = j$, we evaluate this formula at $t = 0$ to get that $\Gamma_{i,i}$ is a probability measure.

Let us show that for all $i, j \in I^{\mathbf{a}}$ we have $\theta_K = \inf \text{Support}(\Gamma_{i,j})$. We first prove it for $i = j$. Fix $i \in I^{\mathbf{a}}$ and denote by $\theta_0 = \inf \text{Support}(\Gamma_{i,i})$ which is nonnegative. For all $\epsilon > 0$,

$$p_t(i, i) \geq e^{-t(\theta_0 + \epsilon)}\, \Gamma_{i,i}\big([\theta_0, \theta_0 + \epsilon]\big).$$

From (4.29), we deduce $\theta_K \le \theta_0 + \epsilon$ for all $\epsilon > 0$, so $\theta_K \le \theta_0$. Then, if $\theta_0 = 0$, we obtain the equality $\theta_K = \theta_0 = 0$. Assume $\theta_0 > 0$. In this case, we find

$$p_t(i,i) \le e^{-t\theta_0} \Gamma_{i,i}\big([\theta_0, \infty)\big) \le e^{-t\theta_0},$$

and so $\theta_K \ge \theta_0$. We have proved $\theta_K = \inf \mathrm{Support}(\Gamma_{i,i})$. Then, in the case $\theta_K > 0$, we have shown that we necessarily have $\|\Gamma_{i,i}\|([0,\theta_K)) = 0\ \forall i \in I^{\mathbf{a}}$.

Let us show the result for a pair i, j. Denote $\theta_0 = \inf \mathrm{Support}(\Gamma_{i,j})$, we first prove $\theta_0 \ge \theta_K$. This obviously holds if $\theta_K = 0$, and in the case $\theta_K > 0$ we use (3.15) with $[0, \theta_K)$ to obtain

$$\forall i, j: \quad \big(\|\Gamma_{i,j}\|\big([0,\theta_K)\big)\big)^2 \le \|\Gamma_{i,i}\|\big([0,\theta_K)\big)\|\Gamma_{j,j}\|\big([0,\theta_K)\big) = 0.$$

This implies $\theta_0 \ge \theta_K$. In the case $\theta_0 > \theta_K$, we deduce that for all $\theta \in (\theta_K, \theta_0)$ the following relation is satisfied:

$$e^{\theta t} p_t(i,j) = \sqrt{\frac{\pi_j}{\pi_i}} \int_{\theta_0}^{\infty} e^{-t(x-\theta)}\, d\Gamma_{i,j}(x) \to 0 \quad \text{as } t \to \infty,$$

which contradicts (4.29). We conclude $\theta_K = \inf \mathrm{Support}(\Gamma_{i,j})\ \forall i, j \in I^{\mathbf{a}}$.

Under the additional hypothesis that I_0^* is finite, we get $\theta^* = \theta_K$, and then relation (4.42) is satisfied. Let us show (4.43). We first evaluate the forward equations (4.5) at the killing state $j = 0$. We have

$$\forall s \ge 0,\ \forall i \in I^{\mathbf{a}}: \quad p_s'(i,0) = \sum_{k \in I} p_s(i,k) q(k,0).$$

Since $q(k,0) = 0$ except for $k \in I_0^*$, the last sum is finite, and it runs over $k \in I_0^*$. By integrating this equation in the interval (t, ∞) and using the fact that the process is surely killed, we get

$$\mathbb{P}_i(T > t) = 1 - p_t(i,0) = \sum_{k \in I_0^*} q(k,0) \int_t^{\infty} p_s(i,k).$$

Then, by using equality (4.42) and Fubini's Theorem, we obtain

$$\begin{aligned} \mathbb{P}_i(T > t) &= \sum_{k \in I_0^*} q(k,0) \sqrt{\frac{\pi_k}{\pi_i}} \int_t^{\infty} \left(\int_{\theta^*}^{\infty} e^{-sx}\, d\Gamma_{i,k}(x) \right) ds \\ &= \sum_{k \in I_0^*} q(k,0) \sqrt{\frac{\pi_k}{\pi_i}} \int_{\theta^*}^{\infty} \frac{e^{-tx}}{x}\, d\Gamma_{i,k}(x). \end{aligned}$$

Hence, relation (4.43) is proved. □

Remark 4.15 Since by hypothesis the process Y is nonexplosive, when from each state the chain can only jump to a bounded number of states, then there does always exist a sequence of finite sets $(I_N : N \ge 1)$ that verifies (4.41).

Proposition 4.16 *Assume conditions* (4.41) *and that* $I_0^* = \{k \in I^{\mathbf{a}} : q(k,0) > 0\}$ *is finite. Then,*

$$\lim_{t \to \infty} e^{-\theta^* t} p_t(i,i) = \Gamma_{i,i}\big(\{\theta^*\}\big). \tag{4.44}$$

Moreover, we have the following solidarity property: If $\Gamma_{i_0,i_0}(\{\theta^*\}) > 0$ *for some* $i_0 \in I^{\mathbf{a}}$ *then* $\Gamma_{i,i}(\{\theta^*\}) > 0\ \forall i \in I^{\mathbf{a}}$.

The exponential killing condition $\theta^* > 0$ *is necessary in order that* $\exists i_0 \in I^{\mathbf{a}}$ *such that* $\lim_{t\to\infty} e^{-\theta^* t} p_t(i,i) > 0$.

Proof From (4.42), we have $p_t(i,i) = \int_{\theta^*}^{\infty} e^{-tx}\, d\Gamma_{i,i}(x)$. By the Dominated Convergence Theorem, we have that for all $\delta > 0$,

$$\lim_{t\to\infty} e^{-\theta^* t} \int_{\theta^*+\delta}^{\infty} e^{-tx}\, d\Gamma_{i,i}(x) = 0.$$

Hence

$$\begin{aligned}\Gamma_{i,i}\big(\{\theta^*\}\big) &\le \liminf_{t\to\infty} \int_{\theta^*}^{\theta^*+\delta} e^{-t(x-\theta^*)}\, d\Gamma_{i,i}(x) \le \limsup_{t\to\infty} \int_{\theta^*}^{\theta^*+\delta} e^{-t(x-\theta^*)}\, d\Gamma_{i,i}(x)\\ &\le \Gamma_{i,i}\big[\theta^*, \theta^* + \delta\big].\end{aligned}$$

Then, the first part is shown. To prove the solidarity property of the second part, an argument based upon irreducibility suffices. This is given by the inequality $p_{h+t+s}(i,i) \ge p_h(i,i_0)p_t(i_0,i_0)p_s(i_0,i)$ for fixed $s, h > 0$. The inequalities $p_h(i,i_0) > 0$ and $p_s(i_0,i) > 0$ give the second part.

The final part is a simple consequence of $\Gamma_{i,i}\{0\} = 0$. □

A state $i \in I^{\mathbf{a}}$ is called R-positive if $\lim_{t\to\infty} e^{-\theta_K t} p_t(i,i) > 0$ (see Anderson 1991, Sect. 5.2, therein it is also shown that the limit exists). Now, under the assumptions of Proposition 4.16, we have $\theta_K = \theta^*$, so we can rephrase the result as: A state i is R-positive if and only if the spectral measure $\Gamma_{i,i}$ gives positive measure to the bottom of the spectrum.

From the solidarity property, we can refer to the R-positive chain when some state is R-positive. When this happens, it can be shown that there exists a unique θ^*-invariant measure. If this measure is finite, it can be proved that its normalization is a QSD (θ^* being its exponential rate of survival). For these results, see Seneta and Vere-Jones (1996) and Proposition 2.10 in Anderson (1991), Sect. 5.2.

4.7 Comparison of Exponential Rates for a Monotone Process

We will give some comparisons between the stationary distribution and the QSD, and between the exponential rates of survival and the exponential ergodic coefficients. This will be done in the framework of monotone processes. Here $Y = (Y_t : t \ge 0)$ is an irreducible positive recurrent Markov chain taking values in the countable set I.

As before $p_t(i,j) = \mathbb{P}_i(Y_t = j)$, and denote by $\overline{P}_t = (p_t(i,j) : i, j \in I)$, $t \ge 0$, the associated semigroup. (Recall that notation (P_t) is reserved to the killed semigroup). We assume that the Markov chain Y is positive recurrent with stationary distribution $\pi = (\pi_i : i \in I)$, so

$$\pi' \overline{P}_t = \pi' \quad \text{for all } t \ge 0.$$

Consider the exponential rate of convergence to equilibrium starting from the distribution δ_i for $i \in I$,

$$\alpha^*(i) = \sup\{\alpha : \exists C < \infty, \left\| \delta_i' \overline{P}_t - \pi' \right\|_{\mathrm{TV}} \le C e^{-\alpha t}\}, \tag{4.45}$$

where $\|\bullet\|_{\mathrm{TV}}$ is the total variation metric.

Let $\preceq$ be a partial order on I. A function $f : I \to \mathbb{R}$ is called monotone if $i \preceq j$ implies $f(i) \le f(j)$. It induces the following stochastic order relation between probability measures: For $\rho_1, \rho_2 \in \mathcal{P}(I)$, we put $\rho_1 \preceq \rho_2$ if $\rho_1(f) \le \rho_2(f)$ for all monotone functions f (where $\rho(f) = \sum_{i \in I} f(i)\rho(i)$). The Markov semigroup $(\overline{P}_t : t \ge 0)$ is said to be monotone with respect to $\preceq$ if

$$i \preceq j \quad \Rightarrow \quad \delta_i' \overline{P}_t \preceq \delta_j' \overline{P}_t \quad \forall t \ge 0.$$

Proposition 4.17 *Assume that the Markov semigroup $(\overline{P}_t : t \ge 0)$ defined by Y is monotone. In addition, suppose that the countable state space I has a strictly smallest element 0 in I, that is, $0 \prec j$ for all $j \in I \setminus \{0\}$. Consider the process $Y^T = (Y_{t \wedge T} : t \ge 0)$ with hitting time $T = \inf\{t > 0 : Y_t = 0\}$. Let ν be a QSD for Y^T, then (a) $\pi \preceq \nu$ and (b) $\theta(\nu) \le \alpha^*(0)$.*

Proof We will repeat the proof done in Martínez and Ycart (2001) which uses coupling arguments.

(a) Since the process is monotone and since $\delta_0 \preceq \nu$, from the Strassen's Theorem (see Lindvall 1992), we can construct a Markov process (W^1, W^2) with right continuous trajectories taking values in $I \times I$ starting from the distribution $\delta_0 \otimes \nu$ such that each coordinate evolves with semigroup $(\overline{P}_t)$ and $W^1 \preceq W^2$ $\mathbb{P}$-a.s. Let $T^2 = \inf\{t \ge 0 : W_t^2 = 0\}$. Since $W^1 \preceq W^2$, we get $W_{T^2}^1 = W_{T^2}^2 = 0$. We redefine the process at time T^2 by resurrecting W^2 with distribution ν. Since $W_{T^2}^1 = 0$ and the coupling is Markovian, at T^2 we are as at the initial time $t = 0$ with distribution $\delta_0 \otimes \nu$, so the same construction can be done once again. This iterative procedure gives a monotone coupling $(\overline{W}^1, \overline{W}^2)$, $\overline{W}^1 \preceq \overline{W}^2$, for which $\overline{W}^1$ evolves with semigroup $(\overline{P}_t)$ and $\overline{W}^2$ evolves with semigroup $(P_t^{(\nu)})$ (see the construction of the resurrected process (4.22)). Since ν is a stationary distribution for $(P_t^{(\nu)})$ (recall (4.25)) and $\lim_{t \to \infty} \|\delta_0' \overline{P}_t - \pi'\|_{\mathrm{TV}} = 0$, we conclude $\pi \preceq \nu$.

(b) Let (U^1, U^2) be a Markovian coupling, both components evolving with semigroup $(\overline{P}_t)$ and such that $U_0^1 \sim \delta_0$ and $U_0^2 \sim \pi$. Also from the Strassen's Theorem, we can assume $U^1 \preceq U^2$ a.s. Let $T' = \inf\{t \ge 0 : U_t^1 = U_t^2\}$ and $T'' = \inf\{t \ge 0 : U_t^2 = 0\}$. Since $T' \le T''$, we get $\mathbb{P}(T' > t) \le \mathbb{P}(T'' > t) = \mathbb{P}_\pi(T > t)$. From Doeblin's Inequality, we find $\mathbb{P}(T' > t) \ge \|\delta_0' \overline{P}_t - \pi'\|_{\mathrm{TV}}$. Then

$$\mathbb{P}_\pi(T > t) \ge \left\| \delta_0' \overline{P}_t - \pi' \right\|_{\mathrm{TV}}.$$

Now, let (V^1, V^2) be a Markovian coupling, both components evolving with semigroup $(\overline{P}_t)$, and such that $V_0^1 \sim \pi$ and $V_0^2 \sim \nu$. Let $T^{(i)} = \inf\{t \ge 0 : V^i(t) = 0\}$. From part (a) we can assume $T^{(1)} \le T^{(2)}$. Then

$$\left\| \delta_0' \overline{P}_t - \pi' \right\|_{\mathrm{TV}} \le \mathbb{P}_\pi(T > t) = \mathbb{P}\big(T^{(1)} > t\big) \le \mathbb{P}\big(T^{(2)} > t\big) = \mathbb{P}_\nu(T > t) = e^{-\theta(\nu)t}.$$

The result follows. □

We point out that part (b) can be also deduced from (4.20) and Theorem 2.1 in Lund et al. (1996). Indeed, in this work, it was shown that $\theta^* \le \alpha^*(0)$ under the same hypotheses we had on the order relations, including the existence of a smallest element. Since $\theta(\nu) \le \theta^*$ for all QSDs ν, the part (b) follows directly. Other order relations involving QSDs can be found in Collet et al. (2003).

Chapter 5
Birth-and-Death Chains

In this chapter, we study QSDs on birth-and-death chains. In Sect. 5.1, we describe the quantities relevant for this study. The main tools for describing the QSDs are the Karlin–McGregor decomposition and the dual process, which are supplied in Sect. 5.2.

Among the main results of the chapter is the description of the set of QSDs done in Theorem 5.4 of Sect. 5.3: exponential killing implies that there exist QSDs; if ∞ is a natural boundary, there is a continuum of QSDs; when ∞ is an entrance boundary, there is a unique one. In Theorem 5.8 of Sect. 5.4, it is shown that the extremal QSD is a quasi-limiting distribution. The process of survival trajectories is studied in Sect. 5.5 and its classification is given in Theorem 5.12. In Sect. 5.6, it is proved that the QSDs are stochastically ordered, the extremal QSD is the minimal one. In Sect. 5.7, explicit conditions on the birth-and-death parameters ensuring discrete spectrum are supplied. Some examples of killed birth-and-death chains can be found in Sect. 5.9. In the last Sect. 5.10, we study QSDs for a population model where each individual is characterized by some trait, and it can die or gives birth to a clonal or a mutated individual.

5.1 Parameters for the Description of the Chain

Let $Y = (Y_t : t \geq 0)$ be a birth-and-death chain taking values in $\mathbb{Z}_+ = \{0, 1, \ldots\}$. Its jump rate matrix Q is such that $q(i, i-1) = \mu_i$ (death rates), $q(i, i+1) = \lambda_i$ (birth rates), $q(i, i) = -(\lambda_i + \mu_i)$ and all the other coefficients vanish. The chain is killed at $\partial I = \{0\}$, and the set of allowed states is $\mathbb{N} = \{1, 2, \ldots\}$. We assume that the birth and death rates $(\lambda_i, \mu_i : i \in \mathbb{N})$ are strictly positive. Obviously, $\mu_0 = 0$. Since we impose killing at 0, $\lambda_0 = 0$. We recall the notation $p_t(i, j) = \mathbb{P}_i(Y_t = j)$.

We assume that the chain is surely killed at 0,

$$\forall i \in \mathbb{N}: \quad \mathbb{P}_i(T < \infty) = 1 \quad \text{with } T = \inf\{t \geq 0 : Y_t = 0\}.$$

P. Collet et al., *Quasi-Stationary Distributions*, Probability and Its Applications,
DOI 10.1007/978-3-642-33131-2_5,

This property implies

$$\forall i \in \mathbb{N}: \quad \mathbb{P}_i(T_\infty < \infty) = 0,$$

where

$$T_\infty = \lim_{N\to\infty} T_N, \quad \text{with } T_N = \inf\{t \geq 0 : Y_t = N\}.$$

Indeed, as already stated, sure killing implies there is no explosion.

Note that the backward and forward equations (4.4) and (4.5) have respectively the following form. For all $i, j \geq 0$ and all $t \geq 0$,

$$p_t'(i,j) = \mu_i p_t(i-1,j) - (\lambda_i + \mu_i) p_t(i,j) + \lambda_i p_t(i+1,j), \tag{5.1}$$

$$p_t'(i,j) = \lambda_{j-1} p_t(i,j-1) - (\lambda_j + \mu_j) p_t(i,j) + \mu_{j+1} p_t(i,j+1), \tag{5.2}$$

where we put $\lambda_{-1} = 0$.

Consider $\pi = (\pi_i : i \in \mathbb{N})$ with the coefficients

$$\pi_1 = 1 \quad \text{and} \quad \pi_i = \prod_{j=1}^{i-1} \frac{\lambda_j}{\mu_{j+1}}, \quad i > 1. \tag{5.3}$$

The relations $\pi_{i+1}/\pi_i = \lambda_i/\mu_{i+1}$ $\forall i \in \mathbb{N}$, imply that the birth-and-death chain is reversible with respect to π, that is,

$$\forall i, j \in \mathbb{N}: \quad \pi_i q(i,j) = \pi_j q(j,i). \tag{5.4}$$

(Relation that only needs to be verified for $j = i + 1$.) Put

$$\begin{aligned} A &= \sum_{i\geq 1} (\lambda_i \pi_i)^{-1}, \qquad B = \sum_{i\geq 1} \pi_i, \qquad C = \sum_{i\geq 1} (\lambda_i \pi_i)^{-1} \sum_{j=1}^{i} \pi_j, \\ D &= \sum_{i\geq 2} (\mu_i \pi_i)^{-1} \sum_{j\geq i} \pi_j = \sum_{i\geq 1} (\lambda_i \pi_i)^{-1} \sum_{j\geq i+1} \pi_j = \sum_{i\geq 2} \pi_i \sum_{j=1}^{i-1} (\lambda_j \pi_j)^{-1}. \end{aligned} \tag{5.5}$$

We have

$$C + D = AB, \qquad (A = \infty \Rightarrow C = \infty), \qquad (B = \infty \Rightarrow D = \infty). \tag{5.6}$$

It is well known that the assumption on sure killing, $\mathbb{P}_i(T < \infty) = 1$ $\forall i \in \mathbb{N}$, is equivalent to $A = \infty$. Indeed, the quantities $u_i = \mathbb{P}_i(T < \infty)$, $i \geq 0$, verify the set of equations

$$u_i = \frac{\lambda_i}{\lambda_i + \mu_i} u_{i+1} + \frac{\mu_i}{\lambda_i + \mu_i} u_{i-1}, \quad i \geq 1, \ u_0 = 1.$$

By iterating these quantities, it is shown that $u_1 = 1$ is equivalent to $A = \infty$. So, the assumption of sure killing is $A = \infty$. Hence, we also have $C = \infty$. These assumptions are taken all along the chapter.

Let us now consider the quantities $w_i = \mathbb{E}_i(T)$, $i \geq 0$. They satisfy the equations

$$w_i = \frac{1}{\lambda_i + \mu_i} + \frac{\lambda_i}{\lambda_i + \mu_i} w_{i+1} + \frac{\mu_i}{\lambda_i + \mu_i} w_{i-1}, \quad i \geq 1,\ w_0 = 1.$$

The analysis of these quantities in the case $A = \infty$ allows proving that $B < \infty$ holds if and only if $\mathbb{E}_i(T) < \infty$ for all $i \geq 1$. Analysis for both A and B can be found in Karlin and Taylor (1975, Theorem 7.1, Chap. 4).

Let us give meaning to C and D. We follow the approach developed in Anderson (1991, Sect. 8.1). Let us do it first for D. For each $n \geq 1$, consider the chain $Y^{(n)} = (Y_t^{(n)} : t \geq 0)$ reflected at n, so taking values on $\{0, 1, \ldots, n\}$. Its parameters $(\lambda_i^{(n)}, \mu_i^{(n)} : i = 0, 1, \ldots, n)$ coincide with the original ones $\lambda_i^{(n)} = \lambda_i$, $\mu_i^{(n)} = \mu_i$, with the unique exception $\lambda_n^{(n)} = 0$. So, when the chain is at n, it waits for an exponential time of parameter μ_n before jumping to $n - 1$. Observe that this chain is also surely killed at 0. Denote by $T_i^{(n)}$ the first time that the chain $Y^{(n)}$ hits i, and denote $y_i = \mathbb{E}_i(T_{i-1}^{(n)})$ for $i \in \{1, \ldots, n\}$. From the Markov property, these quantities verify the equations

$$y_i = \frac{1}{\lambda_i + \mu_i} + \frac{\lambda_i}{\lambda_i + \mu_i}(y_{i+1} + y_i) \quad \text{for } 1 \leq i < n \text{ and } y_n = \frac{1}{\mu_n}.$$

It is not difficult to show that $y_i = (\lambda_i \pi_i)^{-1} \sum_{j=i}^{n} \pi_j$ for $i = 1, \ldots, n$. Then $D = \lim_{n \to \infty} \sum_{i=1}^{n} y_i$. The Markov property gives

$$D = \lim_{n \to \infty} \mathbb{E}_n\big(T_0^{(n)}\big).$$

Regarding the quantity C, the analysis is done on the reflected birth-and-death chain $Y^{[r]} = (Y_t^{[r]} : t \geq 0)$ taking values in $\mathbb{N}$ with the same parameters as for Y with only one exception: if the chain $Y^{[r]}$ is at state 1, it waits for an exponential time of parameter λ_1 before jumping to state 2. In this chain, let $T_j^{[r]}$ be the first time that the process $Y^{[r]}$ hits j and define $z_i = \mathbb{E}_i(T_{i+1}^{[r]})$, $i \geq 1$. The Markov property implies that

$$z_i = \frac{1}{\lambda_i + \mu_i} + \frac{\mu_i}{\lambda_i + \mu_i}(z_{i-1} + z_i) i \geq 2, \quad z_1 = \frac{1}{\lambda_1}.$$

It is easy to show that $z_i = (\lambda_i \pi_i)^{-1} \sum_{j=1}^{i} \pi_j$ for $i \geq 1$. Hence $C = \sum_{i \geq 1} z_i$. From the Markov property, we find

$$C = \lim_{n \to \infty} \mathbb{E}_1\big(T_n^{[r]}\big).$$

So, $C = \infty$ is the condition for having infinite mean time of explosion for the reflected chain.

Let us note that $D < \infty$ implies $B < \infty$. From Karlin and Taylor (1975, Theorem 7.1, Chap. 4), we also get

$$B + D = \lim_{n\to\infty} \mathbb{E}_n(T). \tag{5.7}$$

Since we are assuming $A = \infty$ (and so $C = \infty$) there are only two possibilities left for ∞ in the Feller classification:

- If $D = \infty$ then ∞ is a natural boundary;
- If $D < \infty$ then ∞ is an entrance boundary.

In both cases, the minimal solution is the unique solution to the backward and forward equation, see Anderson (1991, Sect. 8.1, p. 262).

5.2 Polynomials, Representation and Duality

In the analysis of the birth-and-death chains, a main role is played by the family of polynomials $(\psi_i(\theta) : i \geq 0)$ defined by $\psi_0(\theta) \equiv 0$ and the following recurrence relations:

$$\psi_1(\theta) \equiv 1, \qquad \lambda_i \psi_{i+1}(\theta) - (\lambda_i + \mu_i - \theta)\psi_i(\theta) + \mu_i \psi_{i-1}(\theta) = 0, \quad i \in \mathbb{N}. \tag{5.8}$$

The polynomial $\psi_i(\theta)$ is of degree $i - 1$ when $i \geq 1$. By definition, the sequence $\psi(\theta) := (\psi_i(\theta) : i \in \mathbb{N})$ verifies

$$Q^{\mathbf{a}} \psi(\theta) = -\theta \psi(\theta). \tag{5.9}$$

Also notice that $f = \psi(\theta)$ is the unique solution to the equation $Q^{\mathbf{a}} f = -\theta f$ up to a multiplicative constant.

It can be shown that for $i \geq 2$ the roots of the polynomial $\psi_i(\theta)$ are all real; moreover, they are simple and strictly positive. Let us denote them by $a_{1,i} < \cdots < a_{i-1,i}$. When i varies, they verify the interlacing property (see Karlin and McGregor 1957b and Theorem 5.3 in Chihara 1972),

$$a_{k,i+1} < a_{k,i} < a_{k+1,i+1} \quad \forall k = 1, \ldots, i. \tag{5.10}$$

Therefore, there exists

$$\xi_k = \lim_{i\to\infty} a_{k,i}, \quad k \geq 1. \tag{5.11}$$

The interlacing property of the roots also gives

$$\xi_1 = \sup\{\theta \geq 0 : \psi_i(\theta) > 0 \ \forall i \geq 1\}. \tag{5.12}$$

In the rest of this chapter, we put $I_N = \{1, \ldots, N\}$. For all $N \geq 1$, we denote by

$$\psi^{(N)}(\theta) := \psi(\theta)|_{I_N} = (\psi_1(\theta), \ldots, \psi_N(\theta))$$

the restriction of $\psi(\theta)$ to I_N.

From definition of the function $\psi_i(\theta)$ and the truncated matrix $Q^{(N)}$, it is easy to see that for each $k \in I_N$ the vector $\psi^{(N)}(a_{k,N+1})$ is an eigenvector of $Q^{(N)}$,

$$Q^{(N)}\psi^{(N)}(a_{k,N+1}) = -a_{k,N+1}\psi^{(N)}(a_{k,N+1}).$$

Hence, $(\psi^{(N)}(a_{k,N+1}) : k \in I_N)$ is a set of N independent eigenvectors of $Q^{(N)}$. The eigenvalues of $Q^{(N)}$ are minus the zeros of ψ_{N+1}, and so

$$\left\{d_k^{(N)} : k \in I_N\right\} = \{a_{k,N+1} : k \in I_N\}.$$

Therefore, we find for $\Gamma_{1,1}^{(N)}$ defined in (3.12) and (3.13)

$$\forall N \geq 1: \quad a_{1,N} = \inf \text{Support}\bigl(\Gamma_{1,1}^{(N)}\bigr). \tag{5.13}$$

Let us give more insight to the spectral decomposition of the truncated matrix $Q^{(N)}$. Let $\langle f, g\rangle_\pi = \sum_{j\in I^{\mathbf{a}}} f(i)g(i)\pi_i$ be the inner product defined by π in $\mathbb{R}^N$ and $\|\cdot\|_\pi = \langle\cdot,\cdot\rangle_\pi^{1/2}$ be the π-norm. The matrix $Q^{(N)}$ is symmetric and negative definite with respect to $\langle\cdot,\cdot\rangle_\pi$. Since $(\psi^{(N)}(a_{k,N+1}) : k \in I_N)$ is a complete set of eigenfunctions of $Q^{(N)}$, it is a basis of orthogonal functions of $(\mathbb{R}^N, \langle\cdot,\cdot\rangle_\pi)$. Then for all $f \in \mathbb{R}^N$,

$$\begin{aligned}
Q^{\mathbf{a}} f &= -\sum_{k\in I^{\mathbf{a}}} \frac{a_{k,N+1}}{\|\psi^{(N)}(a_{k,N+1})\|_\pi^2}\bigl\langle f, \psi^{(N)}(a_{k,N+1})\bigr\rangle_\pi \psi^{(N)}(a_{k,N+1}) \\
&= -\int_0^\infty \theta\bigl\langle f, \psi^{(N)}(\theta)\bigr\rangle_\pi \psi^{(N)}(\theta)\, d\Gamma^{(N)}(\theta),
\end{aligned}$$

where $\Gamma^{(N)} = \sum_{n\in I_N} \delta_{x_{k,N}}$ with $x_{k,N} = a_{k,N+1}/\|\psi^{(N)}(\theta)\|_\pi^2$. Then

$$e^{Q^{(N)}t} f = \int_0^\infty e^{-\theta t}\bigl\langle f, \psi^{(N)}(\theta)\bigr\rangle_\pi \psi^{(N)}(\theta)\, d\Gamma^{(N)}(\theta).$$

The transition probability $p_t^{(N)}(i, j)$ of the semigroup generated by $Q^{(N)}$ is given by

$$p_t^{(N)}(i, j) = e^{Q^{(N)}t}(i, j) = \bigl(e^{Q^{(N)}t}\mathbf{1}_{\{j\}}\bigr)(i),$$

where $\mathbf{1}_{\{j\}}$ is the function that verifies $\mathbf{1}_{\{j\}}(i) = \delta(i, j)$. Then, we obtain the following formula:

$$\forall i, j \in I_N: \quad p_t^{(N)}(i, j) = \pi_j \int_0^\infty e^{-\theta t}\psi_i^{(N)}(\theta)\psi_j^{(N)}(\theta)\, d\Gamma^{(N)}(\theta). \tag{5.14}$$

From $\psi_1^{(N)}(\theta) \equiv 1$, $\pi_1 = 1$, we get

$$p_t^{(N)}(1, 1) = \pi_1 \int_0^\infty e^{-\theta t}\, d\Gamma^{(N)}(\theta) = \int_0^\infty e^{-\theta t}\, d\Gamma^{(N)}(\theta),$$

and so $\Gamma^{(N)} = \Gamma^{(N)}_{1,1} = \widetilde{\Gamma}^{(N)}_{1,1}$, where $\Gamma^{(N)}_{1,1}$ and $\widetilde{\Gamma}^{(N)}_{1,1}$ are the measures defined in (3.16) and (3.17).

Formula (5.14) admits a limit when $N \to \infty$, giving the so-called Karlin–McGregor representation. In this formula, it is shown (see Karlin and McGregor 1957b) that there exists a measure Γ supported by $[\theta^*, \infty) \cap (0, \infty)$ such that

$$\forall i, j \in \mathbb{N}: \quad p_t(i, j) = \pi_j \int_0^\infty e^{-t\theta} \psi_i(\theta) \psi_j(\theta)\, d\Gamma(\theta). \tag{5.15}$$

In Anderson (1991, Theorem 2.1 in Sect. 8.2), the Karlin–McGregor representation is shown, and its proof is based upon a recurrence argument on i, j. Below we give the constructions that are required and the main steps of the proof.

The class of sets $(I_N : N \geq 1)$ verifies condition (4.41) because the chain Y is assumed to be nonexplosive and from each state in $\mathbb{N}$ a birth-and-death chain can jump only to two states (see Remark 4.15). Besides, the set $I_0^* = \{k \in \mathbb{N} : q(k, 0) > 0\} = \{1\}$ is a singleton. Then we can apply Proposition 4.14 to have

$$\forall i, j \in \mathbb{N}: \quad p_t(i, j) = \sqrt{\frac{\pi_j}{\pi_i}} \int_0^\infty e^{-t\theta}\, d\Gamma_{i,j}(\theta) = \pi_j \int_0^\infty e^{-t\theta}\, d\widetilde{\Gamma}_{i,j}(\theta).$$

Here $\Gamma_{i,j}$ are signed measures whose total variation measures have mass smaller or equal to 1 and $\widetilde{\Gamma}_{i,j} = (\pi_i \pi_j)^{-1/2} \Gamma_{i,j}$. We recall $\Gamma_{i,i}$ are probability measures. Let us put

$$\Gamma := \Gamma_{1,1} = \widetilde{\Gamma}_{1,1}, \tag{5.16}$$

so

$$p_t(1, 1) = \int_0^\infty e^{-t\theta}\, d\Gamma(\theta).$$

From Proposition 4.14, the construction of $\Gamma_{i,i}$ done in (4.39) and (4.40), relation (5.13) and definition (5.11), we also get

$$\xi_1 \leq \theta^* = \inf \text{Support}(\Gamma). \tag{5.17}$$

Since the measure Γ is supported in the domain $[\theta^*, \infty) \cap (0, \infty)$, instead of the integral $\int_0^\infty \cdot d\Gamma$ we can also put $\int_{\theta^*}^\infty \cdot d\Gamma$.

Let us prove that the representation (5.15) is verified for $j = 1$, that is,

$$\forall i \geq 1: \quad p_t(i, 1) = \int_0^\infty e^{-t\theta} \psi_i(\theta)\, d\Gamma(\theta). \tag{5.18}$$

By definition $\Gamma = \Gamma_{1,1}$, so the formula holds for $i = 1$. For the induction step, we assume it has been proved up to i, we will show that it is verified for $i + 1$. By using the Dominated Convergence Theorem, we differentiate this equality to get

$$p_t'(i, 1) = -\int_0^\infty \theta e^{-t\theta} \psi_i(\theta)\, d\Gamma(\theta). \tag{5.19}$$

From the backward equation (5.1),

$$\lambda_i p_t(i+1,1) = p_t'(i,1) - \mu_i p_t(i-1,1) + (\lambda_i + \mu_i) p_t(i,1).$$

By writing $p_t(i-1,1)$ and $p_t(i,1)$ as in (5.18) and $p_t'(i,1)$ as in (5.19), and using the recurrence relation (5.8), we obtain

$$\lambda_i \psi_{i+1}(\theta) = -\theta \psi_i(\theta) - \mu_i \psi_{i-1}(\theta) + (\lambda_i + \mu_i)\psi_i(\theta),$$

and we find

$$\lambda_i p_t(i+1,1) = \int_0^\infty e^{-t\theta} \lambda_i \psi_{i+1}(\theta)\, d\Gamma(\theta).$$

Then, the induction follows. We have shown that (5.18) is verified. Let us now prove the general case in (5.15),

$$\forall i, j \geq 1: \quad p_t(i,j) = \pi_j \int_0^\infty e^{-t\theta} \psi_i(\theta)\psi_j(\theta)\, d\Gamma(\theta),$$

by induction on j. It holds for $j = 1$, assume that it holds up to j and let us prove it for $j+1$. By using the Dominated Convergence Theorem, we differentiate the above equality and get

$$p_t'(i,j) = -\pi_j \int_0^\infty \theta e^{-t\theta} \psi_i(\theta)\psi_j(\theta)\, d\Gamma(\theta), \quad i, j \geq 1.$$

From the forward equation (5.2),

$$\mu_{j+1} p_t(i, j+1) = p_t'(i,j) - \lambda_{j-1} p_t(i, j-1) + (\lambda_j + \mu_j) p_t(i,j),$$

and the recurrence relation (5.8),

$$\pi_j \lambda_j \psi_{j+1}(\theta) = -\pi_j \theta \psi_j(\theta) - \pi_j \mu_j \psi_{j-1}(\theta) + \pi_j(\lambda_j + \mu_j)\psi_j(\theta),$$

we find

$$\mu_{j+1} p_t(i, j+1) = \pi_j \lambda_j \int_0^\infty e^{-t\theta} \psi_i(\theta)\psi_{j+1}(\theta)\, d\Gamma(\theta).$$

Since $\pi_j \lambda_j / \mu_{j+1} = \pi_{j+1}$, (5.15) is verified for $(i, j+1)$. This shows formula (5.15) and finishes the proof of the Karlin–McGregor representation.

Let $\langle f, g\rangle_\Gamma = \int f g\, d\Gamma$ be the inner product defined in $L^2(\Gamma)$, which is the set of real functions f with domain $\mathbb{R}_+$ and such that $\|f\|_\Gamma^2 := \langle f, f\rangle_\Gamma$ is finite. Observe that when we evaluate (5.15) at $t = 0$, we get that the polynomials are orthogonal,

$$\forall i, j \in \mathbb{N}: \quad \pi_j \langle \psi_i, \psi_j\rangle_\Gamma = \pi_j \int_0^\infty \psi_i(\theta)\psi_j(\theta)\, d\Gamma(\theta) = \delta(i,j).$$

Also $\psi_i \in L^2(\Gamma)$ with $\|\psi_i\|_\Gamma = 1/\sqrt{\pi_i}$.

We recall that from Proposition 4.14 we have $\theta_K = \theta^*$, and so

$$\forall i \in \mathbb{N}: \quad \theta^* = -\lim_{t\to\infty} \frac{1}{t} \log P_i(T > t), \tag{5.20}$$

and $\forall i, j \in \mathbb{N}$, $\theta^* = -\lim_{t\to\infty} \frac{1}{t} \log p_t(i, j)$.

Also, from Proposition 4.16, we get $\lim_{t\to\infty} e^{-\theta^* t} p_t(1, 1) = \Gamma_{1,1}(\{\theta^*\})$. Then $\Gamma(\{\theta^*\})$ is the necessary and sufficient condition in order that the chain be R-positive, $\theta^* > 0$ being a necessary condition in order for this to happen.

Let us obtain the following expression for the survival probabilities:

$$\forall i \geq 1: \quad \mathbb{P}_i(T > t) = \mu_1 \int_{\theta^*}^{\infty} \frac{e^{-t\theta}}{\theta} \psi_i(\theta)\, d\Gamma(\theta). \tag{5.21}$$

It is shown in a completely similar way as when proving relation (4.43) in Proposition 4.14. In fact, from the forward equation (5.2), we get

$$\forall i \in \mathbb{N}, s \geq 0: \quad p'_s(i, 0) = \mu_1 p_s(i, 1).$$

By integrating this equation in the interval (t, ∞), by using the fact that the process is surely killed and from expression (5.18), we obtain

$$\mathbb{P}_i(T > t) = 1 - p_t(i, 0) = \mu_1 \int_t^{\infty} p_s(i, 1)\, ds = \mu_1 \int_t^{\infty} \int_{\theta^*}^{\infty} e^{-t\theta} \psi_i(\theta)\, d\Gamma(\theta)\, ds.$$

Hence, Fubini's Theorem implies equality (5.21). By evaluating this expression at $i = 1$ and $t = 0$, we obtain

$$\mu_1 \int_{\theta^*}^{\infty} \frac{1}{\theta}\, d\Gamma(\theta) = 1. \tag{5.22}$$

Proposition 5.1 *We have* $\theta^* = \xi_1$.

Proof From (5.17), we are reduced to showing the converse, $\xi_1 \geq \theta^*$. Fix N and put $T_{\geq N} = \inf\{t \geq 0 : Y_t \geq N\}$. By definition of the truncated kernel $Q^{(N)}$, there exists some $c_N > 0$ that verifies

$$\forall t \geq 0: \quad \mathbb{P}_1(Y_t = 1, t < T \wedge T_{\geq N}\} = \big(e^{Q^{(N)} t}\big)(1, 1) \geq c_N e^{-a_{1,N+1} t}.$$

By definition of θ^*, we deduce $\theta^* \leq a_{1,N+1}$. The result is shown. □

Let us introduce the dual birth-and-death chain defined in Karlin and McGregor (1957a). All the parameters and quantities related to it will be written with a superscript $\widehat{}$. The dual chain takes values in $\mathbb{N} = \{1, 2, \ldots\}$ and its birth and death parameters are given by

$$\widehat{\lambda}_i = \mu_i, \qquad \widehat{\mu}_i = \lambda_{i-1}, \quad i \in \mathbb{N},$$

in particular, $\widehat{\mu}_1 = 0$. Then, the coefficients $\widehat{\pi}_i$ verify

$$\widehat{\pi}_1 = 1, \qquad \widehat{\pi}_i = \prod_{r=1}^{i-1} \frac{\widehat{\lambda}_r}{\widehat{\mu}_{r+1}} = \prod_{r=1}^{i-1} \frac{\mu_r}{\lambda_r} = \frac{\mu_1}{\mu_i \pi_i}, \qquad i \in \mathbb{N}. \tag{5.23}$$

Also, the associated family of polynomials $(\widehat{\psi}_i(\theta) : i \geq 0)$ verify $\widehat{\psi}_0(\theta) \equiv 0$ and

$$\widehat{\psi}_1(\theta) \equiv 1, \qquad \mu_i \widehat{\psi}_{i+1}(\theta) = (\mu_i + \lambda_{i-1} - \theta)\widehat{\psi}_i(\theta) - \lambda_{i-1}\widehat{\psi}_{i-1}(\theta), \quad i \in \mathbb{N}. \tag{5.24}$$

Let us prove that the following relations between both classes of polynomials is satisfied (see Karlin and McGregor 1957a, Van Doorn 1985):

$$\forall i \in \mathbb{Z}_+: \quad \psi_{i+1}(\theta) - \psi_i(\theta) = \widehat{\pi}_{i+1}\widehat{\psi}_{i+1}(\theta) \tag{5.25}$$

and

$$\forall i \in \mathbb{N}: \quad \widehat{\psi}_i(\theta) - \widehat{\psi}_{i+1}(\theta) = \mu_1^{-1}\pi_i \theta \psi_i(\theta). \tag{5.26}$$

First, let us show (5.25). By using

$$\mu_i \widehat{\pi}_{i+1}^{-1} = \lambda_i \widehat{\pi}_i^{-1} \quad \text{and the shifted relation} \quad \lambda_{i-1}\widehat{\pi}_{i-1}^{-1} = \mu_{i-1}\widehat{\pi}_i^{-1},$$

we find that for $i \in \mathbb{N}$,

$$\begin{aligned} 0 = {} & \mu_i \widehat{\pi}_{i+1}^{-1}\big(\psi_{i+1}(\theta) - \psi_i(\theta)\big) - (\mu_i + \lambda_{i-1} - \theta)\widehat{\pi}_i^{-1}\big(\psi_i(\theta) - \psi_{i-1}(\theta)\big) \\ & + \lambda_{i-1}\widehat{\pi}_{i-1}^{-1}\big(\psi_{i-1}(\theta) - \psi_{i-2}(\theta)\big). \end{aligned}$$

From the uniqueness of the family of polynomials satisfying (5.24), we deduce equalities (5.25). We note that (5.25) holds for $i = 0$ because $\widehat{\pi}_1 = 1$.

Let us now prove (5.26). From (5.25) and (5.8), we get

$$\begin{aligned} \widehat{\psi}_{i+1}(\theta) - \widehat{\psi}_i(\theta) &= \frac{1}{\mu_i \widehat{\pi}_i}\big(\lambda_i\big(\psi_{i+1}(\theta) - \psi_i(\theta)\big) - \mu_i\big(\psi_i(\theta) - \psi_{i-1}(\theta)\big)\big) \\ &= \frac{1}{\mu_i \widehat{\pi}_i}\theta \psi_i(\theta). \end{aligned}$$

Then, (5.23) shows the result. Note that the equality (5.26) implies

$$\forall i \in \mathbb{N}: \quad \widehat{\psi}_i(0) = 1. \tag{5.27}$$

It can be shown (see Theorem 7.1 in Chihara 1972 and Kijima et al. 1997) that the zeros of both families of polynomials $(\widehat{\psi}_i(\theta) : i \geq 0)$ and $(\widehat{\psi}_i(\theta) : i \geq 0)$ verify the set of inequalities,

$$0 < \widehat{a}_{1,i} < a_{1,i} < \cdots < \widehat{a}_{k,i} < a_{k,i}, \quad k = 1, \ldots, i. \tag{5.28}$$

The above relations allow one to get the following increasing property, which is a stronger version of (5.12).

Proposition 5.2 *Assume $\theta^* > 0$. Then,*

$$\theta \le \theta^* \quad \Rightarrow \quad \big(\psi_i(\theta) > 0 \text{ and } \psi_i(\theta) < \psi_{i+1}(\theta)\ \forall i \in \mathbb{N}\big). \tag{5.29}$$

In particular, $\theta \le \theta^*$ implies $\psi_i(\theta) \ge 1$ for all $i \in \mathbb{N}$.

Proof The last relation follows from (5.29) and $\psi_1(\theta) \equiv 1$. Let us show (5.29).

Recall $\xi_1 = \theta^*$. From (5.28), we get $\widehat{\xi}_k \le \xi_k$ for all $k \ge 1$, in particular, $\widehat{\xi}_1 \le \xi_1$. We are assuming $A = \infty$ and $\theta^* > 0$. Since this condition implies $B < \infty$, we are under the hypotheses that in Kijima et al. (1997) allow showing $\widehat{\xi}_1 = \xi_1$ (see Case 4.1 in this reference). Since $\widehat{\psi}_i(\theta) > 0$ when $\theta \le \widehat{\xi}_1$, from (5.25) we deduce (5.29). □

We introduce the following measure

$$d\Gamma^d(\theta) = \frac{\mu_1}{\theta}\, d\Gamma(\theta), \tag{5.30}$$

which is supported on $[\theta^*, \infty) \cap (0, \infty)$. From (5.22), Γ^d is a probability measure. Equality (5.21) can be written as

$$\forall i \in \mathbb{N}: \quad \mathbb{P}_i(T > t) = \int_{\theta^*}^{\infty} e^{-t\theta} \psi_i(\theta)\, d\Gamma^d(\theta).$$

In particular, when we evaluate at $t = 0$, we find that

$$\forall i \in \mathbb{N}: \quad \int_{\theta^*}^{\infty} \psi_i(\theta)\, d\Gamma^d(\theta) = 1. \tag{5.31}$$

We observe that $\text{Support}(\Gamma^d) = \text{Support}(\Gamma)$.

Let us show that the family of polynomials $(\widehat{\psi}_i(\theta) : i \in \mathbb{N})$ is orthogonal with respect to the measure Γ^d. From (5.25) and (5.26), we find for $j > i + 1$:

$$\begin{aligned}\int_{\theta^*}^{\infty} \big(\widehat{\psi}_i(\theta) - \widehat{\psi}_{i+1}(\theta)\big)\widehat{\psi}_j(\theta)\, d\Gamma^d(\theta) &= \frac{\pi_i}{\widehat{\pi}_j} \int_{\theta^*}^{\infty} \psi_i(\theta)\big(\psi_j(\theta) - \psi_{j-1}(\theta)\big)\, d\Gamma(\theta)\\ &= 0.\end{aligned}$$

Let us take $j \ge 2$. For all $l \le j - 1$, we have

$$\int_{\theta^*}^{\infty} \sum_{i=1}^{l-1} \big(\widehat{\psi}_{i+1}(\theta) - \widehat{\psi}_i(\theta)\big)\widehat{\psi}_j(\theta)\, d\Gamma^d(\theta) = 0.$$

Then, since $\widehat{\psi}_1 \equiv 1$,

$$\int_{\theta^*}^{\infty} \widehat{\psi}_l(\theta)\widehat{\psi}_j(\theta)\, d\Gamma^d(\theta) = \int_{\theta^*}^{\infty} \widehat{\psi}_1(\theta)\widehat{\psi}_j(\theta)\, d\Gamma^d(\theta) = \int_{\theta^*}^{\infty} \widehat{\psi}_j(\theta)\, d\Gamma^d(\theta).$$

Now, we use equalities (5.25) and (5.31) to obtain

$$\int_{\theta^*}^{\infty} \widehat{\psi}_j(\theta)\, d\Gamma^d(\theta) = \frac{1}{\widehat{\pi}_j} \int_{\theta^*}^{\infty} \big(\psi_j(\theta) - \psi_{j-1}(\theta)\big)\, d\Gamma^d(\theta) = 0.$$

Hence, $\int_{\theta^*}^{\infty} \widehat{\psi}_l(\theta)\widehat{\psi}_j(\theta)\, d\Gamma^d(\theta) = 0$ when $l \le j-1$, and so by the symmetry of the argument, we show it for all $l \ne j$. The orthogonality is shown.

Observe that from (5.15) we have $\pi_i \int_{\theta^*}^{\infty} \psi_i^2(\theta)\, d\Gamma(\theta) = 1$. We claim that the following set of equalities hold:

$$\forall i \ge 1: \quad \widehat{\pi}_i \int_{\theta^*}^{\infty} \widehat{\psi}_i^2(\theta) = 1.$$

For $i = 1$, it follows from $\widehat{\pi}_1 \int_{\theta^*}^{\infty} \widehat{h}_1^2(\theta)\, d\Gamma^d(\theta) = \int_{\theta^*}^{\infty} d\Gamma^d(\theta) = 1$. Now, with the same computations as above, we obtain

$$\begin{aligned}
&\int_{\theta^*}^{\infty} \big(\widehat{\psi}_i(\theta) - \widehat{\psi}_{i+1}(\theta)\big)\widehat{\psi}_{i+1}(\theta)\, d\Gamma^d(\theta) \\
&\quad = \frac{\pi_i}{\widehat{\pi}_{i+1}} \int_{\theta^*}^{\infty} \psi_i(\theta)\big(\psi_{i+1}(\theta) - \psi_i(\theta)\big)\, d\Gamma(\theta) \\
&\quad = \frac{1}{\widehat{\pi}_{i+1}}.
\end{aligned}$$

Since $\int_{\theta^*}^{\infty} \widehat{\psi}_i(\theta)\widehat{\psi}_{i+1}(\theta)\, d\Gamma^d(\theta) = 0$, the claim is shown.

5.3 Structure of the Set of QSDs

As we saw, each vector $\psi(\theta) = (\psi_i(\theta) : i \in \mathbb{N})$ is a right-eigenvector of $Q^{\mathbf{a}}$ associated to the eigenvalue $-\theta$ and it is unique up to a multiplicative constant. From the equality $\theta^* = \xi_1$, the eigenvector $\psi(\theta)$ is strictly positive when $\theta \le \theta^*$. The birth-and-death chain is reversible with respect to π. Then if we choose $C(\theta) > 0$, any strictly positive constant for $\theta \le \theta^*$, the measure $\nu_\theta = (\nu_\theta(i) : i \in \mathbb{N})$ defined by

$$\forall i \in \mathbb{N}: \quad \nu_\theta(i) = C(\theta)\psi_i(\theta)\pi_i$$

is a strictly positive left-eigenvector of $Q^{\mathbf{a}}$ associated to the eigenvalue $-\theta$. So, it verifies $\nu_\theta Q^{\mathbf{a}} = -\theta\nu_\theta$. Observe that by evaluating at $i = 1$, and since $\psi_1(\theta) = 1 = \pi_1$, we find $C(\theta) = \nu_\theta(1)$, so we can write

$$\forall i \in \mathbb{N}: \quad \nu_\theta(i) = \nu_\theta(1)\psi_i(\theta)\pi_i. \tag{5.32}$$

On the other hand, in Proposition 4.2, it was proved that $\theta^* > 0$ is a necessary condition for the existence of QSDs for Markov chains on countable states; and from Proposition 4.3 and Theorem 4.4, we get that a QSD ν must necessarily verify $\nu' Q^{\mathbf{a}} = -\theta(\nu)\nu'$ with

$$\theta(\nu) = \sum_{i\in\mathbb{N}} \nu(i)q(i,0) = \nu(1)\mu_1. \tag{5.33}$$

Hence, ν is a QSD if it verifies the following system of equations

$$\forall i \in \mathbb{N}: \quad \lambda_{i-1}\nu(i-1) + \mu_{i+1}\nu(i+1) = \big(\lambda_i + \mu_i - \nu(1)\mu_1\big)\nu(i), \tag{5.34}$$

with $\nu(0) = 0$. Observe that (5.33) can be written as $\theta = \nu_\theta(1)\mu_1$ when ν_θ is a QSD. Then, for $\theta \leq 0$, ν_θ is a positive measure but it cannot be a QSD.

Remark 5.3 For instance, in the case $\theta = 0$, the left-eigenfunction ν_θ verifies

$$\nu_0(i) = \nu_0(1)\pi_i \left(1 + \sum_{j=1}^{i-1}\prod_{l=1}^{j}\frac{\mu_l}{\lambda_l}\right),$$

but as shown it cannot be a QSD.

Let us assume

$$\theta^* > 0.$$

The candidates to be QSDs are the measures ν_θ for $\theta \in (0, \theta^*]$. Note that we proved that $\nu_\theta = (\nu_\theta(i) : i \geq 1)$ is a QSD if and only if it is a probability distribution on $\mathbb{N}$, $\sum_{i\geq 1}\nu_\theta(i) = 1$, and if it verifies (5.34). This result was shown in Theorem 3.1 in Van Doorn (1991).

The following characterization of the set of QSDs is given in Theorem 3.2 in Van Doorn (1991). (For the last part of the statement also see Cavender 1978.)

Theorem 5.4 *Let $A = \infty$. Assume $\theta^* > 0$. If $D < \infty$, then ν_{θ^*} is the unique QSD. When $D = \infty$, the set of QSDs is given by the continuum $(\nu_\theta : \theta \in (0, \theta^*])$.*

Proof Let us consider $\theta \in (0, \theta^*]$. From relations (5.32) and (5.26), we have $\nu_\theta(i) = \nu_\theta(1)\psi_i(\theta)\pi_i$ and $\widehat{\psi}_i(\theta) - \widehat{\psi}_{i+1}(\theta) = \mu_1^{-1}\pi_i\theta\psi_i(\theta)$ for $i \geq 1$. Hence, by using $\theta = \mu_1\nu_\theta(1)$, we get the expression

$$\nu_\theta(i) = \widehat{\psi}_i(\theta) - \widehat{\psi}_{i+1}(\theta).$$

Therefore, by using $\widehat{\psi}_1 \equiv 1$, we get

$$\sum_{i=1}^{N}\nu_\theta(i) = 1 - \widehat{\psi}_{N+1}(\theta).$$

Hence $\sum_{i\in\mathbb{N}} \nu_\theta(i) = 1 \Leftrightarrow \lim_{N\to\infty} \widehat{\psi}_N(\theta) = 0$, and we get the equivalence

$$\forall\theta \in \big(0, \theta^*\big]: \quad \nu_\theta \text{ is a QSD} \quad \Leftrightarrow \quad \lim_{N\to\infty} \widehat{\psi}_N(\theta) = 0. \tag{5.35}$$

Recall that for $i \geq 2$, the polynomial $\widehat{\psi}_i(\theta)$ is of degree $i-1$ and all its roots $(\widehat{a}_{j,i} : j = 1, \ldots, i-1)$ are strictly positive and simple. Let $\theta \in (0, \theta^*)$. Since $\widehat{\psi}_i(0) = 1$ for all $i \geq 1$ (see (5.27)), we get

$$\widehat{\psi}_i(\theta) = \frac{\widehat{\psi}_i(\theta)}{\widehat{\psi}_i(0)} = \prod_{j=1}^{i-1}\left(1 - \frac{\theta}{\widehat{a}_{j,i}}\right). \tag{5.36}$$

From the interlacing property (5.10), we find

$$\widehat{\psi}_i(\theta) = \prod_{j=1}^{i-1}\left(1 - \frac{\theta}{\widehat{a}_{j,i}}\right) > \prod_{j=1}^{i-1}\left(1 - \frac{\theta}{\widehat{a}_{j,i+1}}\right)\left(1 - \frac{\theta}{\widehat{a}_{j,i+1}}\right) = \widehat{\psi}_{i+1}(\theta).$$

Then, the sequence $\widehat{\psi}_i(\theta)$ is decreasing and it converges to the limit:

$$\forall\theta \in \big(0, \theta^*\big): \quad \lim_{i\to\infty} \widehat{\psi}_i(\theta) = \prod_{j\geq 1}\left(1 - \frac{\theta}{\widehat{\xi}_j}\right). \tag{5.37}$$

Recall that the sequence $\widehat{\xi}_j$ is increasing, $\widehat{\xi}_j = \xi_j$ and $\xi_1 = \theta^* > 0$.

From (5.37), we obtain $\lim_{i\to\infty} \widehat{\psi}_i(\theta^*) = \lim_{i\to\infty} \widehat{\psi}_i(\xi_1) = 0$, and then ν_{θ^*} is always a QSD.

On the other hand, (5.37) also implies

$$\forall\theta \in \big(0, \theta^*\big): \quad \lim_{i\to\infty} \widehat{\psi}_i(\theta) > 0 \quad \Leftrightarrow \quad \sum_{j\geq 1} \xi_j^{-1} < \infty. \tag{5.38}$$

Since the right-hand term does not depend on $\theta \in (0, \theta^*)$, we get from (5.35) and (5.38) that:

- If $\sum_{j\geq 1} \xi_j^{-1} < \infty$ then the unique QSD is ν_{θ^*};
- If $\sum_{j\geq 1} \xi_j^{-1} = \infty$ then $(\nu_\theta : \theta \in (0, \theta^*])$ is the set of all QSDs.

The above argument was given in Chihara (1972) and it was further developed in Van Doorn (1986) in the framework of QSDs.

The only thing left to prove is the equivalence

$$D < \infty \quad \Leftrightarrow \quad \sum_{j\geq 1} \xi_j^{-1} < \infty. \tag{5.39}$$

Since $\widehat{a}_{j,i} \searrow \xi_j$ for all $j \geq 1$, we get

$$\sum_{j=1}^{i-1} \widehat{a}_{j,i}^{-1} \nearrow \sum_{j\geq 1} \xi_j^{-1}. \tag{5.40}$$

From (5.36), we have that

$$m_i = -\sum_{j=1}^{i-1} \widehat{a}^{-1}_{j,i} \tag{5.41}$$

is the coefficient of order 1 (in θ) of the polynomial $\widehat{\psi}_i(\theta)$. From equation (5.24) and since the constant term is $\widehat{\psi}_i(0) = 1$, we find

$$\mu_i m_{i+1} = (\mu_i + \lambda_{i-1}) m_i - 1 - \lambda_{i-1} m_{i-1},$$

where $m_0 = m_1 = 0$ and $m_2 = -1/\mu_1$. By iteration, we get

$$m_{i+1} - m_i = -\sum_{j=1}^{i} \frac{1}{\mu_i} \left(\prod_{l=j}^{i-1} \frac{\lambda_l}{\mu_l} \right) = -\pi_i \sum_{l=1}^{i} (\lambda_l \pi_l)^{-1}.$$

Hence

$$m_{i+1} = -\frac{1}{\mu_1} \pi_i + \pi_i \sum_{j=1}^{i} \pi_j \sum_{l=1}^{j} (\lambda_j \pi_j)^{-1}.$$

This expression, together with (5.40) and (5.41), gives

$$\sum_{j\ge 1} \xi_j^{-1} = \frac{1}{\mu_1} \sum_{j\ge 1}^{\infty} + \sum_{j\ge 1} \pi_j \sum_{l=1}^{j} (\lambda_j \pi_j)^{-1} = \frac{1}{\mu_1} B + D.$$

Since $D < \infty$ implies $B < \infty$, the last equality shows the equivalence (5.39). □

Recall that the relation (5.7) is $B + D = \lim_{n\to\infty} \mathbb{E}_n(T)$. Again by using that $D < \infty$ implies $B < \infty$, from the last theorem we get that there exists a unique QSD only when there is an exponential moment and the mean killing time is uniformly bounded. The conditions $A = \infty$, $D < \infty$ and $B < \infty$ imply that ∞ is an entrance boundary, then Theorem 5.4 can be rephrased in the following way.

Theorem 5.5 *Assume the birth-and-death process is exponentially killed. Then the extremal QSD ν_{θ^*} does exist.*

If ∞ is an entrance boundary then ν_{θ^*} is the unique QSD, and when ∞ is a natural boundary, the set of QSDs is the continuum $(\nu_\theta : \theta \in (0, \theta^*])$.

Let us show that for birth-and-death chains the QSDs are determined by the exponential distribution property of the killing time.

Proposition 5.6 *Let ν be a probability measure on $\mathbb{N}$ for which $\exists \theta > 0$ such that $\mathbb{P}_\nu(T > t) = e^{-\theta t}$ $\forall t \ge 0$. Then ν is a QSD.*

Proof By integrating the forward equations (5.2) for $j = 0$, we find $p_t(i, 0) = \mu_1 \int_0^\infty p_u(i, 1)\, du$. Multiplying by $\nu(i)$, summing over $\mathbb{N}$ and using $\mathbb{P}_\nu(T > t) = e^{-\theta t}$, we get

$$1 - e^{-\theta t} = \mu_1 \int_0^t \mathbb{P}_\nu(Y_u = 1)\, du. \tag{5.42}$$

By differentiating at $t = 0$, we obtain $\theta = \mu_1 \nu(1)$. Now, we integrate (5.2) for $j > 0$ to get

$$p_t(i, j) = \lambda_{j-1} \int_0^t p_u(i, j-1)\, du - (\lambda_j + \mu_j) \int_0^t p_u(i, j)\, du$$
$$+ \mu_{j+1} \int_0^t p_u(i, j+1)\, du.$$

In a similar way as before, we obtain

$$\mathbb{P}_\nu(Y_t = j) = \lambda_{j-1} \int_0^t \mathbb{P}_\nu(Y_u = j-1)\, du - (\lambda_j + \mu_j) \int_0^t \mathbb{P}_\nu(Y_u = j)\, du$$
$$+ \mu_{j+1} \int_0^t \mathbb{P}_\nu(Y_u = j+1)\, du. \tag{5.43}$$

From relations (5.42) and (5.43), we obtain a system of equations that must be satisfied by $(\nu(i) : i \in \mathbb{N})$ in terms of $\theta = \mu_1 \nu(1)$. Moreover, this system is a linear recurrence on $n \in \mathbb{N}$ so it has a unique solution. But this system of equations is the same as the one satisfied by a QSD, thus ν must satisfy (5.34). Since ν is a probability measure, we deduce that it must be a QSD. □

5.4 The Quasi-limiting Distribution

The following result, useful in the study of ratio limit behavior, can be found in Van Doorn (1991).

Lemma 5.7 *Define*

$$M(t) = \int_{\theta^*}^\infty e^{-t(\theta - \theta^*)}\, d\Gamma^d(\theta).$$

Let $w : [\theta^*, \infty) \to \mathbb{R}$ *be a measurable and* Γ^d*-integrable real function, continuous at* θ^* *and vanishing at* θ^*, $w(\theta^*) = 0$. *Then, it verifies*

$$\lim_{t \to \infty} M(t)^{-1} \int_{\theta^*}^\infty e^{-t(\theta - \theta^*)} w(\theta)\, d\Gamma^d(\theta) = 0. \tag{5.44}$$

Proof (i) Suppose $\Gamma^d((\theta^*, \theta^* + \delta)) = 0$ for some $\delta > 0$, so necessarily $\Gamma^d(\{\theta^*\}) > 0$ (because $\theta^* = \inf \text{Support}(\Gamma^d)$). Then, $\lim_{t\to\infty} M(t) = \Gamma^d(\{\theta^*\})$. On the other hand, from $w(\theta^*)=0$, w is Γ^d-integrable, and by using the Dominated Convergence Theorem, we find

$$\lim_{t\to\infty}\int_{\theta^*}^{\infty} e^{-t(\theta-\theta^*)}w(\theta)\,d\Gamma^d(\theta)=\lim_{t\to\infty}\int_{\theta^*+\delta}^{\infty} e^{-t(\theta-\theta^*)}w(\theta)\,d\Gamma^d(\theta)=0.$$

Thus the result.

(ii) Assume $\Gamma^d((\theta^*,\theta^*+\delta))>0$ for all $\delta>0$, so θ^* is an accumulation point of the support. Fix $\epsilon>0$ and let $\delta>0$ be such that $|w(\theta)|<\epsilon$ for all $\theta\in[\theta^*,\theta^*+\delta)$. Then

$$\begin{aligned}\left|\int_{\theta^*}^{\infty} e^{-t(\theta-\theta^*)}w(\theta)\,d\Gamma^d(\theta)\right| &\le \epsilon\int_{\theta^*}^{\theta^*+\delta} e^{-t(\theta-\theta^*)}\,d\Gamma^d(\theta)\\ &\quad+\int_{\theta^*+\delta}^{\infty} e^{-t(\theta-\theta^*)}\big|w(\theta)\big|\,d\Gamma^d(\theta).\end{aligned}$$

We have

$$M(t)^{-1}\epsilon\left(\int_{\theta^*}^{\theta^*+\delta} e^{-t(\theta-\theta^*)}\,d\Gamma^d(\theta)\right)\le\epsilon.$$

On the other hand, $\Gamma^d([\theta^*,\theta^*+\delta/2])>0$. We have

$$M(t)\ge\int_{\theta^*}^{\theta^*+\delta/2} e^{-t(\theta-\theta^*)}\,d\Gamma^d(\theta),$$

and then the result will be shown once we prove

$$\left(\int_{\theta^*}^{\theta^*+\delta/2} e^{-t(\theta-\theta^*)}\,d\Gamma^d(\theta)\right)^{-1}\left(\int_{\theta^*+\delta}^{\infty} e^{-t(\theta-\theta^*)}\big|w(\theta)\big|\,d\Gamma^d(\theta)\right)\to 0$$

as $t\to\infty$.

Define the function

$$\forall\theta\in\big[\theta^*+\delta,\infty\big):\quad g_t(\theta):=\left(\int_{\theta^*}^{\theta^*+\delta/2} e^{-t(\theta-\theta^*)}\,d\Gamma^d(\theta)\right)^{-1} e^{-t(\theta-\theta^*)}\big|w(\theta)\big|.$$

It verifies

$$\forall\theta\in\big[\theta^*+\delta,\infty\big):\quad g_t(\theta)\le\Gamma\big(\big[\theta^*,\theta^*+\delta/2\big]\big)^{-1}e^{-t\delta/2}\big|w(\theta)\big|.$$

So the Dominated Convergence Theorem gives the result. □

In Good (1968), it was shown that the quasi-limiting distribution (3.6) exists and that it is the extremal QSD ν_{θ^*}. Let us prove it.

Theorem 5.8 *Assume $A=\infty$ and $\theta^*>0$. Then the following limit exists and verifies*

$$\forall i\in\mathbb{N},\ \forall j\in\mathbb{N}:\quad \lim_{t\to\infty}\mathbb{P}_i(Y_t=j|T>t)=\nu_{\theta^*}(j).$$

Proof We follow the proof in Van Doorn (1991). From relations (5.15), (5.21) and definition (5.30), we find

$$\mathbb{P}_i(Y_t=j|T>t)=\frac{p_t(i,j)}{\mathbb{P}_i(T>t)}=\frac{\pi_j}{\mu_1}\frac{\int_{\theta^*}^{\infty}\theta e^{-t\theta}\psi_i(\theta)\psi_j(\theta)\,d\Gamma^d(\theta)}{\int_{\theta^*}^{\infty}e^{-t\theta}\psi_i(\theta)\,d\Gamma^d(\theta)}.$$

Recall the definition of $M(t)$ and the relation (5.44) stated in Lemma 5.7. We observe that $\psi_i(\theta^*)>0\ \forall i\ge 1$, then we can consider $w(\theta)=\psi_i(\theta)-\psi_i(\theta^*)$ in (5.44), to obtain

$$\lim_{t\to\infty}M(t)^{-1}\int_{\theta^*}^{\infty}e^{-t(\theta-\theta^*)}\psi_i(\theta)\,d\Gamma^d(\theta)=\psi_i(\theta^*). \tag{5.45}$$

Since $\theta^*>0$, we can also take $w(\theta)=\theta\psi_i(\theta)\psi_j(\theta)-\theta^*\psi_i(\theta^*)\psi_j(\theta^*)$ to get

$$M(t)^{-1}\int_{\theta^*}^{\infty}e^{-t(\theta-\theta^*)}\theta\psi_i(\theta)\psi_j(\theta)\,d\Gamma^d(\theta)=\theta^*\psi_i(\theta^*)\psi_j(\theta^*).$$

Now, by putting all the elements together and by using (5.33) and (5.32), we conclude

$$\begin{aligned}\mathbb{P}_i(Y_t=j|T>t)&=\frac{\pi_j}{\mu_1}\frac{\theta^*\psi_i(\theta^*)\psi_j(\theta^*)}{\psi_i(\theta^*)}=\frac{\theta^*}{\mu_1}\pi_j\psi_j(\theta^*)=\nu_{\theta^*}(1)\pi_j\psi_j(\theta^*)\\&=\nu_{\theta^*}(j).\end{aligned}$$

The result is shown. □

5.5 The Survival Process and Its Classification

Let us give the following result for the process of survival trajectories as an h-process. See Jacka and Roberts (1995), and Martínez and Vares (1995) for the discrete time case. We assume $\theta^*>0$, which implies the existence of a QSD.

Proposition 5.9 *Let $\theta^*>0$. The process $Z=(Z_t:t\ge 0)$ whose law starting from $i\in\mathbb{N}$ is given by*

$$\mathbb{P}_i(Z_{s_1}=i_1,\dots,Z_{s_k}=i_k):=\lim_{t\to\infty}\mathbb{P}_i(Y_{s_1}=i_1,\dots,Y_{s_k}=i_k|T>t)$$

is a Markov chain and its transition kernel is

$$\forall i,j\in\mathbb{N}:\quad \mathbb{P}_i(Z_s=j)=e^{\theta^*s}\frac{\psi_j(\theta^*)}{\psi_i(\theta^*)}\mathbb{P}_i(Y_s=j). \tag{5.46}$$

Proof Observe that when we take $w(\theta)=e^{s(\theta-\theta^*)}-1$ and apply (5.44), we get

$$\lim_{t\to\infty}\frac{M(t-s)}{M(t)}=M(t)^{-1}\left(\int_{\theta^*}^{\infty}e^{-t(\theta-\theta^*)}\left(e^{s(\theta-\theta^*)}-1\right)d\Gamma^d(\theta)\right)+1=1. \quad (5.47)$$

Now, we have

$$\begin{aligned}&\lim_{t\to\infty}\mathbb{P}_{i_0}(Y_{s_1}=i_1,\ldots,Y_{s_k}=i_k|T>t)\\ &\quad=\mathbb{P}_{i_0}(Y_{s_1}=i_1,\ldots,Y_{s_k}=i_k)\lim_{t\to\infty}\frac{\mathbb{P}_{i_k}(T>t-s_k)}{\mathbb{P}_{i_0}(T>t)}.\end{aligned}$$

On the other hand, from formulas (5.21), (5.30), (5.45) and (5.47), we get

$$\begin{aligned}&\lim_{t\to\infty}\frac{\mathbb{P}_j(T>t-s)}{\mathbb{P}_i(T>t)}\\ &\quad=\lim_{t\to\infty}\frac{\int_{\theta^*}^{\infty}e^{-(t-s)\theta}\psi_j(\theta)\,d\Gamma^d(\theta)}{\int_{\theta^*}^{\infty}e^{-t\theta}\psi_i(\theta)\,d\Gamma^d(\theta)}\\ &\quad=\lim_{t\to\infty}e^{s\theta^*}\frac{M(t-s)^{-1}\int_{\theta^*}^{\infty}e^{-(t-s)(\theta-\theta^*)}\psi_j(\theta)\,d\Gamma^d(\theta)}{M(t)^{-1}\int_{\theta^*}^{\infty}e^{-t(\theta-\theta^*)}\psi_i(\theta)\,d\Gamma^d(\theta)}\left(\lim_{t\to\infty}\frac{M(t-s)}{M(t)}\right)\\ &\quad=e^{s\theta^*}\frac{\psi_j(\theta^*)}{\psi_i(\theta^*)}.\end{aligned}$$

Then, the result follows. □

So, the extremal QSD ν_{θ^*} is the significant one for describing the survival process Z because it is the quasi-limiting distribution.

From (5.46), we get that the process Z is a birth-and-death chain taking values in $\mathbb{N}$ (so reflected at 1), and its birth and death parameters $(\overline{\lambda}_i : i\in\mathbb{N})$, $(\overline{\mu}_i : i\in\mathbb{N})$ verify

$$\forall i\in\mathbb{N}:\quad \overline{\lambda}_i=\frac{\psi_{i+1}(\theta^*)}{\psi_i(\theta^*)}\lambda_i=\frac{\nu_{\theta^*}(i+1)}{\nu_{\theta^*}(i)}\mu_{i+1}, \quad (5.48)$$

$$\forall i\in\mathbb{N}:\quad \overline{\mu}_i=\frac{\psi_{i-1}(\theta^*)}{\psi_i(\theta^*)}\mu_i=\frac{\nu_{\theta^*}(i-1)}{\nu_{\theta^*}(i)}\lambda_{i-1}. \quad (5.49)$$

Observe that, as expected, these formulas give $\overline{\mu}_1=0$. Also it holds that

$$\overline{\lambda}_i+\overline{\mu}_i=\lambda_i+\mu_i-\theta^*.$$

The classification of the stopped process Y^T was introduced in Vere-Jones (1962). Based upon the results already found for the survival process Z, the classification can be presented in the following form. The chain Y^T is said to be θ^*-transient (respectively, θ^*-null recurrent, or θ^*-positive recurrent) if Z is transient (respectively, null recurrent, or positive recurrent). To show which of these properties is verified at each case, it is useful to compute the vector $\overline{\pi} = (\overline{\pi}_i : i \ge 1)$ analogous to (5.3): $\overline{\pi}_1 = 1$ and

$$\overline{\pi}_i = \prod_{r=1}^{i-1} \frac{\overline{\lambda}_r}{\overline{\mu}_{r+1}} = \prod_{r=1}^{i-1} \left(\frac{\psi_{r+1}(\theta^*)}{\psi_r(\theta^*)} \right)^2 \frac{\lambda_r}{\mu_{r+1}} = \psi_i(\theta^*)^2 \pi_i, \quad i > 1. \tag{5.50}$$

The constants $\overline{A}, \overline{B}, \overline{C}, \overline{D}$ are the analogs of A, B, C, D in (5.5), but computed with $\overline{\pi}$, $\overline{\lambda}$ and $\overline{\mu}$ instead of π, λ and μ. This gives:

$$\overline{A} = \sum_{i \ge 1} \left(\psi_{i+1}(\theta^*)\psi_i(\theta^*)\lambda_i \pi_i\right)^{-1}, \qquad \overline{B} = \sum_{i \ge 1} \psi_i(\theta^*)^2 \pi_i,$$

$$\overline{C} = \sum_{i \ge 1} \left(\psi_{i+1}(\theta^*)\psi_i(\theta^*)\lambda_i \pi_i\right)^{-1} \sum_{j=1}^{i} \psi_j(\theta^*)^2 \pi_j,$$

$$\overline{D} = \sum_{i \ge 1} \left(\psi_{i+1}(\theta^*)\psi_i(\theta^*)\lambda_i \pi_i\right)^{-1} \sum_{j=i+1}^{\infty} \psi_j(\theta^*)^2 \pi_j.$$

We have that the process Y^T is θ^*-transient if $\overline{A} < \infty$ and θ^*-positive recurrent when it is θ^*-recurrent and $\overline{B} < \infty$.

From now on, we assume $D = \infty$, so ∞ is a natural boundary and there is a continuum of QSDs.

Lemma 5.10 *Under the assumption $D = \infty$, we have that Y^T is θ^*-positive recurrent if and only if $\overline{B} < \infty$.*

Proof We claim that $\overline{D} \ge D$. In fact, from Proposition 5.2, we have that $\psi_i(\theta^*)$ is increasing with $i \ge 1$, so

$$\begin{aligned}\overline{D} &\ge \sum_{i \ge 1} \psi_{i+1}(\theta^*)\psi_i(\theta^*)\left(\psi_{i+1}(\theta^*)\psi_i(\theta^*)\lambda_i \pi_i\right)^{-1} \sum_{j=i+1}^{\infty} \pi_j \\ &= \sum_{i \ge 1} (\lambda_i \pi_i)^{-1} \sum_{j=i+1}^{\infty} \pi_j = D.\end{aligned}$$

Then $\overline{D} = \infty$. From (5.6), we get $\infty = \overline{C} + \overline{D} = \overline{A}\overline{B}$. Hence $\overline{B} < \infty$ implies $\overline{A} = \infty$, and the result follows. □

We will develop an approach based on the behavior of the sequence of the exponential rates of survival to classify the limit Markov chain Z. For this consider the following sequence of killing times that expresses more and more restrictions. Let $a \ge 1$ be fixed and consider the hitting time of the set of states $\{0, \dots, a-1\}$,

$$T_{a-1} = \inf\{t > 0 : Y_t \le a - 1\}.$$

So $T_{a-1} = T$ when $a = 1$. We denote by $Y^{T_{a-1}} = (Y_{t \wedge T_{a-1}} : t \geq 0)$ the process stopped at $\{0, \dots, a-1\}$. It is always assumed that the initial distribution of $Y^{T_{a-1}}$ is concentrated on the set of states $\{a, a+1, \dots\}$. Note that when dealing with $Y^{T_{a-1}}$ it can be assumed that $\lambda_i = 0 = \mu_i$ for $i < a$. We observe that $A = \infty$ also implies $\mathbb{P}_i(T_{a-1} < \infty) = 1$ for some $i \geq a$ (or for all $i \geq a$).

From (5.20), there exists the limit exponential rate of survival of the chain $Y^{T_{a-1}}$, which is

$$\theta^{*(a)} = -\lim_{t\to\infty} \frac{1}{t} \log P_i(T_{a-1} > t), \quad i \geq a.$$

By solidarity property, this limit does not depend on the initial condition $i \geq a$. The sequence $(\theta^{*(a)} : a \in \mathbb{N})$ is nondecreasing in a. Note that $\theta^{*(1)} = \theta^*$ is the exponential rate of killing of the original chain.

The following property is verified.

Lemma 5.11 *We have the equivalence*

$$\theta^* > 0 \quad \Leftrightarrow \quad \left[\exists a \geq 1 : \theta^{*(a)} > 0\right].$$

Proof If $\theta^* > 0$, it is obvious that $\theta^{*(a)} > 0 \ \forall a \geq 1$. Let us prove the converse.

Assume that $\theta^{*(a_0)} > 0$ for some $a_0 \geq 1$. We can assume $a_0 > 1$; for if not, it is trivial. Note that it suffices to show that starting from any $a \geq 1$ the time of first visit to $a-1$ has a finite exponential moment, that is, there exists $\theta_a > 0$ such that $\mathbb{E}_a(e^{\theta_a T_{a-1}}) < \infty$, where $T_{a-1} = \inf\{t > 0 : Y_t = a-1\}$. In fact, this proves that T has a finite exponential moment because, from the Markov property, $\mathbb{E}_a(e^{\theta T}) = \prod_{j=1}^{a} \mathbb{E}_j(e^{\theta T_{j-1}})$ is finite for all $0 < \theta \leq \min\{\theta_j : j = 1, \dots, a\}$.

We need to show that for all $a \in \{1, \dots, a_0\}$ the exponential moment $\mathbb{E}_a(e^{\theta T_{a-1}}) < \infty$ is finite for some $\theta > 0$. This property will be shown by induction on $a = a_0, a_0 - 1, \dots, 1$. It holds for $a = a_0$ because, from our assumption and by taking $\theta_{a_0} \in (0, \theta^{*(a_0)})$, we get

$$\forall a > a_0,\ \forall \theta \leq \theta_{a_0}: \ \ \mathbb{E}_a\left(e^{\theta T_{a-1}}\right) \leq \mathbb{E}_a\left(e^{\theta T_{a_0-1}}\right) < \infty.$$

Now we show the induction step. Assume that for some $a \in \{2, \dots, a_0\}$ $\exists \theta_a > 0$ such that $\mathbb{E}_{a+1}(e^{\theta_a T_a}) < \infty$, then we will get the existence of $\theta_{a-1} > 0$ such that $\mathbb{E}_a(e^{\theta_{a-1} T_{a-1}}) < \infty$. This will be done by studying the sequence of times of successive visits to a, starting from a, before visiting $a-1$. Let $(E_m : m \geq 1)$ be a sequence of i.i.d. random variables distributed exponentially with parameter $\lambda_a + \mu_a$. These random variables represent the time spent at a in each successive visit.

On the other hand, consider $(L_m : m \geq 1)$, a sequence of i.i.d. random variables, where the common distribution is the one of the hitting time at a, under $\mathbb{P}_{a+1}$. We also take them independent from $(E_m : m \geq 1)$. The last ingredient we need is M, a random variable geometrically distributed with parameter $0 < \epsilon = \frac{\lambda_a}{\lambda_a + \mu_a} < 1$, that is, $\mathbb{P}(M = m) = (1-\epsilon)\epsilon^m$ for all $m \in \mathbb{Z}_+$. M represents the number of excursions from a to a, prior to visiting $a-1$ for the first time. The variable M is taken independent from $(E_m : m \geq 1)$, $(L_m : m \geq 1)$.

It is straightforward to show that under $\mathbb{P}_a$:

$$T_{a-1} \sim \sum_{m=1}^{M} E_m + \sum_{m=1}^{M-1} L_m.$$

From $\sum_{m=1}^{M} E_m + \sum_{m=1}^{M-1} L_m \le \sum_{m=1}^{M}(E_m + L_m)$, we get

$$\forall \theta > 0\colon \quad \mathbb{E}_a\big(e^{\theta T_{a-1}}\big) \le \sum_{m=1}^{\infty} \big(\mathbb{E}\big(e^{\theta(E_1+L_1)}\big)\big)^m (1-\epsilon)\epsilon^m.$$

We notice that for $0 \le \theta < \min(\theta_a, \lambda_a + \mu_a)$ one has $\mathbb{E}(e^{\theta(E_1+L_1)}) < \infty$. From the Monotone Convergence Theorem, we have

$$\lim_{\theta \searrow 0} \mathbb{E}\big(e^{\theta(E_1+L_1)}\big) = 1.$$

Therefore, there exists $\theta_{a-1} > 0$ such that for all $\theta \in (0, \theta_{a-1})$

$$\mathbb{E}\big(e^{\theta(E_1+L_1)}\big) < \frac{1}{\epsilon},$$

and then $\mathbb{E}_a(e^{\theta_{a-1} T_{a-1}}) < \infty$. Hence, the result follows by induction. □

Let us write the parameters, measures and processes already obtained for $Y^{T_{a-1}}$. We put

$$\pi_a^{(a)} = 1 \quad \text{and} \quad \pi_i^{(a)} = \prod_{j=a}^{i-1} \frac{\lambda_j}{\mu_{j+1}}, \qquad i > a.$$

On the other hand, since we assumed $D = \infty$ and $\theta^{*(a)} > 0$ $\forall a \ge 1$, there is a continuum of QSDs $(\nu_\theta^{(a)} : \theta \in [0, \theta^*))$, the extremal one being $\nu_{\theta^{*(a)}}^{(a)}$. Each one of these QSDs, $\nu_\theta^{(a)} = (\nu_\theta^{(a)}(i) : i \ge a)$, verifies

$$\lambda_{i-1}\nu_\theta^{(a)}(i-1) + \mu_{i+1}\nu_\theta^{(a)}(i+1) = (\lambda_i + \mu_i - \theta)\nu_\theta^{(a)}(i), \quad i \ge a,$$

where $\nu_\theta^{(a)}(i) = 0$ for $i < a$. Let $\theta \in [0, \theta^*)$. Then the function

$$\psi^{(a)}(\theta)\colon \quad \{i \in \mathbb{N} : i \ge a\} \to (0, \infty) \quad \text{given by} \quad \psi_i^{(a)}(\theta) = \frac{\nu_\theta^{(a)}(i)}{\pi_i \nu_\theta^{(a)}(1)}$$

verifies $Q^{\mathbf{a}(a)}\psi^{(a)}(\theta) = -\theta\psi^{(a)}(\theta)$, where $Q^{\mathbf{a}(a)}$ is the jump rates' matrix $Q^{\mathbf{a}}$ restricted to $\{a, a+1, \dots\}$, more precisely,

$$\psi_a^{(a)}(\theta) \equiv 1, \qquad \lambda_i \psi_{i+1}^{(a)}(\theta) - (\lambda_i + \mu_i - \theta)\psi_i^{(a)}(\theta) + \mu_i \psi_{i-1}^{(a)}(\theta) = 0, \quad i \ge a, \tag{5.51}$$

where $\psi_{a-1}^{(a)}(\theta) \equiv 0$.

The process $Z^{(a)} = (Z_t^{(a)} : t \geq 0)$ conditioned to survival starting from $i \geq a$,

$$\mathbb{P}\big(Z_{s_1}^{(a)} = i_1, \ldots, Z_{s_k}^{(a)} = i_k\big) := \lim_{t\to\infty} \mathbb{P}_i(Y_{s_1} = i_1, \ldots, Y_{s_k} = i_k | T_{a-1} > t),$$

is a birth-and-death chain taking values in $\{a, a+1, \ldots\}$ reflected at a. Its birth and death parameters $(\overline{\lambda}_i : i \geq a)$, $(\overline{\mu}_i : i \geq a)$ verify $\overline{\mu}_a^{(a)} = 0$, and for $i > a$,

$$\overline{\lambda}_i^{(a)} = \frac{\psi_{i+1}^{(a)}(\theta^{*(a)})}{\psi_i^{(a)}(\theta^{*(a)})}\lambda_i = \frac{\nu_{\theta^{*(a)}}^{(a)}(i+1)}{\nu_{\theta^{*(a)}}^{(a)}(i)}\mu_{i+1},$$

$$\overline{\mu}_i^{(a)} = \frac{\psi_{i-1}^{(a)}(\theta^{*(a)})}{\psi_i^{(a)}(\theta^{*(a)})}\mu_i = \frac{\nu_{\theta^{*(a)}}^{(a)}(i-1)}{\nu_{\theta^{*(a)}}^{(a)}(i)}\lambda_{i-1}.$$

The following equality holds $\overline{\lambda}_i^{(a)} + \overline{\mu}_i^{(a)} = \lambda_i + \mu_i - \theta^{*(a)}$.

The constants $\overline{A}^{(a)}$, $\overline{B}^{(a)}$, $\overline{C}^{(a)}$, $\overline{D}^{(a)}$ are the analogs of $\overline{A}$, $\overline{B}$, $\overline{C}$, $\overline{D}$, but for a instead of $a = 1$.

The chain $Y^{T_{a-1}}$ is $\theta^{*(a)}$-transient (respectively, $\theta^{*(a)}$-null recurrent, or $\theta^{*(a)}$-positive recurrent) if $Z^{(a)}$ is transient (respectively, null recurrent, or positive recurrent). Based on the behavior of the sequence of survival rates, in Theorem 1 in Hart et al. (2003), the following classification result was shown.

Theorem 5.12 *Let* $1 \leq a \leq b$, *then*:

(a) *If* $\theta^{*(a)} < \theta^{*(b)}$ *for some* $b > a$ *then* $Y^{T_{a-1}}$ *is* $\theta^{*(a)}$*-positive recurrent. When* $a \geq 2$ *we also get* $\theta^{*(a-1)} < \theta^{*(a)}$.

In particular, $\theta^{*(a)} < \theta^{*(b)}$ *for* $b > a$ *implies that* Y *is* θ^**-positive recurrent.*

(b) *If* $\theta^{*(a)} = \theta^{*(b)}$ *for some* $b > a$ *then* $Y^{T_{b-1}}$ *is* $\theta^{*(b)}$*-transient and* $\theta^{*(b)} = \theta^{*(b+1)}$.

For the proof of Theorem 5.12, we need to establish some notation and some intermediate results.

Lemma 5.13 *Let* $\beta > 0$ *and* $p > a$. *Assume that* $\psi_i^{(a)}(\beta)$ *is increasing in* $i \in \{a, \ldots, p\}$. *Then, so is* $\psi_i^{(a)}(\beta')$ *for* $\beta' \in [0, \beta]$. *Moreover, we have* $\psi_i^{(a)}(\beta') > \psi_i^{(a)}(\beta)$ *for* $i \in \{a+1, \ldots, p\}$.

Proof Since $\psi_i^{(a)}(\beta)$ is increasing, (5.25) implies that $\widehat{\psi}_i^{(a)}(\beta)$ is positive when $i \in \{a, \ldots, p\}$. Hence $\beta \leq \widehat{a}_{1,p}^{(a)}$ the first zero of $\widehat{\psi}_p^{(a)}$. Then, $\widehat{\psi}_i^{(a)}(x)$ and $\psi_i^{(a)}(x)$ are both strictly decreasing on $[0, \beta]$ for all $i \in \{a, \ldots, p\}$. Hence, $\psi_i^{(a)}(\beta') > \psi_i^{(a)}(\beta)$ for $i \in \{a+1, \ldots, p\}$. □

Now, we introduce the discrete Wronskian. For two vectors $u, v \in \mathbb{R}^{\mathbb{N}}$, we define $w := w(u, v) \in \mathbb{R}^{\mathbb{N}}$ by

$$w_i = (u_{i+1} - u_i)v_i - u_i(v_{i+1} - v_i) = u_{i+1}v_i - u_i v_{i+1}.$$

Lemma 5.14 *Let $1 \le a < b$. Let $\theta \le \theta'$. Fix the vectors u and v by $u_i = \psi_i^{(a)}(\theta)$ and $v_i = \psi_i^{(b)}(\theta')$, $i \ge 1$. Then the Wronskian $w(u,v)$ verifies*

$$w_i = \frac{1}{\lambda_i \pi_i^{(b)}} \left((\theta' - \theta) \sum_{r=b}^{i} u_r v_r \pi_r^{(b)} - \mu_b u_{b-1} \right), \quad i \ge b. \tag{5.52}$$

Proof From definition, relation (5.51) and by development, we find for $i \ge b$,

$$\begin{aligned} w_i - w_{i-1} &= (u_{i+1} + u_{i-1}) v_i - u_i (v_{i+1} + v_{i-1}) \\ &= \frac{\mu_i}{\lambda_i} (-u_{i-1} v_i + u_i v_{i-1}) - (u_i v_{i-1} - u_{i-1} v_i) + \frac{(\theta' - \theta)}{\lambda_i} u_i v_i, \end{aligned}$$

then we obtain

$$w_i = \frac{\mu_i}{\lambda_i} w_{i-1} + \frac{(\theta' - \theta)}{\lambda_i} u_i v_i. \tag{5.53}$$

By noting that $w_{b-1} = u_b v_{b-1} - u_{b-1} v_b = -u_{b-1}$, we iterate (5.53) to get the result. □

Lemma 5.15 *Let $a < b$ and assume $\theta^{*(a)} < \theta^{*(b)}$. Then, for all $\beta \in (\theta^{*(a)}, \theta^{*(b)}]$, we have*

$$\frac{\psi_{i+1}^{(a)}(\theta^{*(a)})}{\psi_i^{(a)}(\theta^{*(a)})} < \frac{\psi_{i+1}^{(b)}(\beta)}{\psi_i^{(b)}(\beta)}, \quad i \ge b. \tag{5.54}$$

Moreover,

$$\left(\theta^{*(b)} - \theta^{*(a)}\right) \sum_{r=b}^{\infty} \psi_r^{(a)}\left(\theta^{*(a)}\right) \psi_r^{(b)}\left(\theta^{*(b)}\right) \pi_r^{(b)} = \mu_b \psi_{b-1}^{(a)}\left(\theta^{*(a)}\right). \tag{5.55}$$

Proof Assume that $w_p(\psi_\bullet^{(a)}(\theta^{*(a)}), \psi_\bullet^{(b)}(\beta)) > 0$ for some $p \ge b$. We know that there exists $\bar\theta > \theta^{*(a)}$ such that $\psi_i^{(a)}(\bar\theta)$ is increasing in $i \in \{a, \dots, p\}$. So, from formula (5.52), we get

$$(\beta - \bar\theta) \sum_{r=b}^{p} \psi_r^{(a)}(\bar\theta) \psi_r^{(b)}(\beta) \pi_r^{(b)} > \mu_b \psi_{b-1}^{(a)}(\bar\theta).$$

Thus $w_p(\psi_\bullet^{(a)}(\bar\theta), \psi_\bullet^{(b)}(\beta)) > 0$. Then, since $\psi_i^{(b)}(\beta)$ is positive and strictly increasing for all $i \ge b$ and $\psi_p^{(a)}(\bar\theta) > 0$, we deduce that $\psi_{p+1}^{(a)}(\bar\theta) > \psi_p^{(a)}(\bar\theta) > 0$, so that $\psi_i^{(a)}(\bar\theta)$ is increasing in $i \in \{a, \dots, p, p+1\}$. Hence, also from (5.52), we deduce that $w_{p+1}(\psi_\bullet^{(a)}(\bar\theta), \psi_\bullet^{(b)}(\beta)) > 0$.

By applying this argument in an inductive way, we find that $\psi_i^{(a)}(\bar\theta)$ is increasing in the variable $i \ge a$, but this is a contradiction to $\bar\theta > \theta^{*(a)}$. Then, we deduce that

$w_p(\psi_\bullet^{(a)}(\bar{\theta}), \psi_\bullet^{(b)}(\beta)) \le 0$ for all $p \ge b$. From (5.52), we get that this inequality must be strict for all $p \ge b$. Then formula (5.54) holds.

Let us now show (5.55). We have proved $w_p(\psi_\bullet^{(a)}(\theta^{*(a)}), \psi_\bullet^{(b)}(\theta^{*(a)})) \le 0$, then (5.52) gives us the inequality $\le$ in (5.55). So, we are left with proving the opposite inequality.

Let $p > b$ be as large as necessary. We can always find $\beta' > \theta^{*(b)}$ such that $\psi_i^{(b)}(\beta')$ is strictly increasing for $i \in \{b, \dots, p\}$. Since $\beta' > \theta^{*(b)}$, there exists $s > p$ such that $\psi_{s+1}^{(b)}(\beta') \le \psi_s^{(b)}(\beta')$ and we fix s as the smallest value bigger than p satisfying this condition. We get $w_s(\psi_\bullet^{(a)}(\theta^{*(a)}), \psi_\bullet^{(b)}(\beta')) > 0$. Now, by definition of s, $\psi_i^{(b)}(\beta')$ is strictly increasing for $i \in \{b, \dots, s\}$, with $s \ge p$. Since Lemma 5.13 implies $\psi_r^{(b)}(\theta^{*(b)}) > \psi_r^{(b)}(\beta')$, we find

$$\begin{aligned}
&\left(\beta' - \theta^{*(a)}\right) \sum_{r=b}^{\infty} \psi_r^{(a)}\left(\theta^{*(a)}\right) \psi_r^{(b)}\left(\theta^{*(b)}\right) \\
&\quad > \left(\beta' - \theta^{*(a)}\right) \sum_{r=b}^{s} \psi_r^{(a)}\left(\theta^{*(a)}\right) \psi_r^{(b)}\left(\beta'\right) \pi_r^{(b)} \pi_r^{(b)} \ge \mu_b \psi_{b-1}^{(a)}\left(\theta^{*(a)}\right).
\end{aligned}$$

Letting $\beta' \searrow \theta^{*(b)}$ allows one to get the inequality $\ge$ in (5.55). This proves the lemma. □

Let us now finish the proof of Theorem 5.12. Let us show part (a). So assume that for some $a < b$ we have $\theta^{*(a)} < \theta^{*(b)}$. We must prove that $Y^{T_{a-1}}$ is $\theta^{*(a)}$-positive recurrent. From Lemma 5.10, we only need to verify $\overline{B}^{(a)} < \infty$. From relation (5.54), we have $\psi_i^{(a)}(\theta^{*(a)}) \le \psi_b^{(a)}(\theta^{*(a)}) \psi_i^{(b)}(\theta^{*(b)})$ (recall $\psi_b^{(b)}(\theta^{*(b)}) = 1$). Then, from (5.55), we find

$$\begin{aligned}
\overline{B}^{(a)} - \sum_{i=a}^{b-1} \left(\psi_i^{(a)}\left(\theta^{*(a)}\right)\right)^2 \pi_i^{(a)} &= \sum_{i=a}^{\infty} \left(\psi_i^{(a)}\left(\theta^{*(a)}\right)\right)^2 \pi_i^{(a)} \\
&\le \psi_b^{(a)}\left(\theta^{*(a)}\right) \pi_b^{(a)} \sum_{i=a}^{\infty} \psi_i^{(a)}\left(\theta^{*(a)}\right) \psi_i^{(b)}\left(\theta^{*(b)}\right) \pi_i^{(b)} \\
&= \left(\theta^{*(b)} - \theta^{*(a)}\right)^{-1} \mu_b \pi_b^{(a)} \psi_b^{(a)}\left(\theta^{*(a)}\right) \psi_{b-1}^{(a)}\left(\theta^{*(a)}\right) \\
&< \infty.
\end{aligned}$$

We have proved $\overline{B}^{(a)} < \infty$, so the first part of (a) follows.

Let us show the second part of (a), that is, $1 < a < b$ and $\theta^{*(a)} < \theta^{*(b)}$ imply $\theta^{*(a-1)} < \theta^{*(a)}$. Suppose the contrary, that is, $\theta^{*(a-1)} = \theta^{*(a)}$. Thus, from (5.52), we get

$$w_i := w_i\left(\psi_i^{(a-1)}\left(\theta^{*(a-1)}\right), \psi_a^{(a)}\left(\theta^{*(a)}\right)\right) = -\frac{\mu_a \psi_{a-1}^{(a-1)}(\theta^{*(a-1)})}{\lambda_i \pi_i^{(a)}} = -\frac{\mu_a}{\lambda_i \pi_i^{(a)}}.$$

For $p > a$, the last equality implies

$$\frac{\psi_p^{(a-1)}(\theta^{*(a-1)})}{\psi_p^{(a)}(\theta^{*(a)})} - \psi_a^{(a-1)}\big(\theta^{*(a-1)}\big) = \sum_{r=a}^{p-1} \frac{w_k}{\psi_{r+1}^{(a)}(\theta^{*(a)}\psi_r^{(a)}(\theta^{*(a)})}$$
$$= -\mu_a \sum_{r=a}^{p-1} \frac{1}{\psi_{r+1}^{(a)}(\theta^{*(a)})\psi_r^{(a)}(\theta^{*(a)})\lambda_r \pi_r^{(a)}}.$$

Then,

$$\psi_p^{(a-1)}\big(\theta^{*(a-1)}\big)$$
$$= \psi_p^{(a)}\big(\theta^{*(a)}\big)\left(\psi_a^{(a-1)}\big(\theta^{*(a-1)}\big) - \mu_a \sum_{r=a}^{p-1} \frac{1}{\psi_{r+1}^{(a)}(\theta^{*(a)})\psi_r^{(a)}(\theta^{*(a)})\lambda_r \pi_r^{(a)}}\right). \tag{5.56}$$

Since $Y^{T_{a-1}}$ is $\theta^{*(a)}$-positive recurrent, we have $\overline{A}^{(a)} = \infty$, and so there exists some large p' for which

$$\psi_a^{(a-1)}\big(\theta^{*(a-1)}\big) < \mu_a \sum_{r=a}^{p'-1} \frac{1}{\psi_{r+1}^{(a)}(\theta^{*(a)})\psi_r^{(a)}(\theta^{*(a)})\lambda_r \pi_r^{(a)}}.$$

Hence $\psi_a^{(a-1)}(\theta^{*(a-1)}) < 0$, which is a contradiction, and we conclude $\theta^{*(a-1)} < \theta^{*(a)}$. Thus, part (a) of Theorem 5.12 has been proved.

Let us show part (b) of Theorem 5.12. Assume $\theta^{*(a)} = \theta^{*(b)}$ for some $b > a$. From equality (5.56), we deduce that $\overline{A}^{(b)} < \infty$, so Y^{T_b} is $\theta^{*(b)}$-transient. On the other hand, if $\theta^{*(b)} = \theta^{*(b+1)}$, we should deduce from part (a) that Y^{T_b} is $\theta^{*(b)}$-positive recurrent, which is a contradiction. This finishes the proof.

5.6 Comparison of QSDs

We recall that for two probability vectors $\rho = (\rho(i) : i \in \mathbb{N})$ and $\rho' = (\rho'(i) : i \in \mathbb{N})$ we put $\rho' \preceq \rho$ and said that ρ' is stochastically smaller than ρ if $\rho'(f) \le \rho(f)$ for any increasing real function f with domain $\mathbb{N}$. It is easy to check that $\rho' \preceq \rho$ if and only if all $j \in \mathbb{N}$ satisfy $\sum_{i=1}^{j} \rho'(i) \ge \sum_{i=1}^{j} \rho(i)$.

A stronger order relation between probability vectors is the following one. We put $\rho' \precsim \rho$ if and only if the sequence $(\rho(i)/\rho'(i) : i \in \mathbb{N})$ is increasing. To show that this order is stronger than $\preceq$, observe that, since ρ' and ρ are probability vectors, $\rho' \precsim \rho$ implies the existence of some $k \in \mathbb{N}$ such that

$$\cdots \le \rho(k-1)/\rho'(k-1) < 1 \le \rho(k)/\rho'(k) \le \cdots.$$

From this set of inequalities, we find that all $j \in \mathbb{N}$ necessarily satisfy $\sum_{i=1}^{j} \rho'(i) \geq \sum_{i=1}^{j} \rho(i)$.

The next result can be found in Cavender (1978).

Proposition 5.16 *Let $\theta, \theta' \in (0, \theta^*]$, $\theta < \theta'$. Then $\nu_{\theta'} \lesssim \nu_\theta$, and in particular $\nu_{\theta'} \preceq \nu_\theta$.*

Proof Recall that the family of QSDs verifies (5.34), with $\theta \in (0, \theta^*]$:

$$\forall i \in \mathbb{N}: \quad \lambda_{i-1}\nu_\theta(i-1) - \big(\lambda_i + \mu_i - \nu_\theta(1)\mu_1\big)\nu_\theta(i) + \mu_{i+1}\nu_\theta(i+1) = 0.$$

Let $0 < \theta < \theta' \leq \theta^*$. Define the sequences ρ and ρ' by $\rho(i) = \nu_\theta(i)$ and $\rho(i) = \nu_{\theta'}(i)$ for $i \in \mathbb{N}$. If $(\rho(i)/\rho'(i) : i \in \mathbb{N})$ is not increasing in $i \in \mathbb{N}$, there will exist some $k \geq 1$ (we take the smallest one) such that

$$\rho(k-1)/\rho'(k-1) \leq \rho(k)/\rho'(k) > \rho(k+1)/\rho'(k+1).$$

Note that $k = 1$ if and only if $\rho(1)/\rho'(1) > \rho(2)/\rho'(2)$. Since ρ' is a QSD and the coefficients of ρ and ρ' are strictly positive, we have

$$\forall i \in \mathbb{N}: \quad \lambda_{i-1}\rho'(i-1)\frac{\rho(i)}{\rho'(i)} - \big(\lambda_i + \mu_i - \rho'(1)\mu_1\big)\rho'(i)\frac{\rho(i)}{\rho'(i)} + \mu_{i+1}\rho'(i+1)\frac{\rho(i)}{\rho'(i)} = 0.$$

Then

$$\lambda_{k-1}\rho'(k-1)\frac{\rho(k-1)}{\rho'(k-1)} - \big(\lambda_k + \mu_k - \rho'(1)\mu_1\big)\rho'(k)\frac{\rho(k)}{\rho'(k)} + \mu_{k+1}\rho'(k+1)\frac{\rho(k+1)}{\rho'(k+1)} < 0.$$

Hence $\lambda_{k-1}\rho(k-1) - (\lambda_k + \mu_k)\rho(k) + \mu_{k+1}\rho'(k) + \rho'(1)\mu_1\rho(k) < 0$, which is a contradiction to the fact that ρ verifies the QSD property. □

The next result was shown in Ferrari et al. (1995b).

Proposition 5.17 *Assume $\forall i \geq 1 : \lambda_i + \mu_i = c$ is a constant. Then for all $\theta, \theta' \in (0, \theta^*]$ with $\theta < \theta'$, $\nu_{\theta'}$ has a tail exponentially lighter than ν_θ, more precisely,*

$$\forall i \in \mathbb{N}: \quad \frac{\nu_{\theta'}(i)}{\nu_\theta(i)} \leq \frac{\theta'}{\theta}\left(\frac{c - \theta'}{c - \theta}\right)^{i-1}.$$

Proof Let $\theta \in (0, \theta^*]$. We know from relation (4.8) that $c = \lambda_i + \mu_i \geq \theta$. On the other hand, if $c = \theta$ the right-hand side in (5.34) will vanish and there cannot exist the QSD ν_θ. Therefore, we can assume $c > \theta$. Let us define

$$z_i(\theta) = \frac{\mu_{i+1}\nu_\theta(i+1)}{(c-\theta)\nu_\theta(i)}. \tag{5.57}$$

So, the QSD equation is equivalent to

$$z_1(\theta) = 1 \text{ and } \forall i \in \mathbb{N}: \quad z_{i+1}(\theta) = 1 - \frac{\lambda_{i-1}\mu_i}{(c-\theta)^2 z_i(\theta)}.$$

Since $z_i(\theta) > 0$, the above evolution equation implies that $z_{i+1}(\theta) \in (0,1)$ $\forall i \in \mathbb{N}$. An inductive argument gives that for all $i \in \mathbb{N}$, $z_i(\theta)$ decreases with $\theta \in (0, \theta^*]$. In fact, if $0 < \theta \le \theta' \le \theta^*$ $z_i(\theta') \le z_i(\theta)$, then $(c-\theta')^2 z_i(\theta') \le (c-\theta)^2 z_i(\theta)$, and so $z_{i+1}(\theta') \le z_{i+1}(\theta)$.

By iterating expression (5.57), we get

$$\nu_\theta(i+1) = \nu_\theta(1)\left(\prod_{j=1}^{i} \frac{z_j(\theta)}{\mu_{j+1}}\right)(c-\theta)^i,$$

and then

$$\frac{\nu_{\theta'}(i+1)}{\nu_\theta(i+1)} = \frac{\nu_{\theta'}(1)}{\nu_\theta(1)}\left(\prod_{j=1}^{i} \frac{z_j(\theta')}{z_j(\theta)}\right)\left(\frac{c-\theta'}{c-\theta}\right)^i.$$

For $0 < \theta \le \theta' \le \theta^*$, we have $\prod_{j=1}^{i} \frac{z_j(\theta')}{z_j(\theta)} \le 1$. Since $\nu_\theta(1) = \theta/\mu_1$, the result follows. □

Since in Proposition 5.17 we assume $\lambda_i + \mu_i = c$ is a constant, it applies to discrete Markov chains. In this case, in the QSD equation, λ_i and μ_i are respectively the jump probabilities to go to the right and to the left from site i, and so their sums are equal to $c = 1$.

5.7 Action on ℓ^2_π and Discrete Spectrum

We will prove that $Q^{\mathbf{a}}$ admits a spectral decomposition on $\ell^2_\pi = \{(f_n : n \in \mathbb{N}) \in \mathbb{C}^{\mathbb{N}} : \sum_{n\in\mathbb{N}} |f_n|^2 \pi_n < \infty\}$. Afterwards we give conditions in order for the $Q^{\mathbf{a}}$ to have a discrete spectrum, that is, pure point spectrum without a finite accumulation point.

In the case $\theta^* > 0$ and $\sum_{i\in\mathbb{N}} \pi_i < \infty$, we study the conditional limiting behavior when the spectrum has a gap after θ^*.

We recall that $Q^{\mathbf{a}} = (q(i,j) : i, j \in \mathbb{N})$ verifies $\pi_i q(i,j) = \pi_j q(j,i)$, see (5.4). Let $\langle f, g\rangle_\pi = \sum_{n\in\mathbb{N}} f(n)g(n)\pi_n$ be the inner product on ℓ^2_π and let $\|f\|_\pi = (\langle f, f\rangle_\pi)^{1/2}$ be the associated norm.

Notice that for every function $f : \mathbb{N} \to \mathbb{R}$, $Q^{\mathbf{a}} f$ can be defined pointwise by

$$\forall n \in \mathbb{N}: \quad Q^{\mathbf{a}} f(n) = \mu_n f(n-1) - (\lambda_n + \mu_n) f(n) + \lambda_n f(n+1). \tag{5.58}$$

Consider the following dense linear domain in ℓ^2_π:

$$\mathcal{D}(Q^{\mathbf{a}}) = \left\{ f \in \ell^2_\pi : \sum_{n\in\mathbb{N}} \bigl(\mu_n f(n-1) - (\lambda_n + \mu_n) f(n) + \lambda_n f(n+1)\bigr)^2 \pi_n < \infty \right\}.$$

We have

$$\mathcal{D}(Q^{\mathbf{a}}) = \{ f \in \ell^2_\pi : Q^{\mathbf{a}} f \in \ell^2_\pi \}, \tag{5.59}$$

and $Q^{\mathbf{a}} : \mathcal{D}(Q^{\mathbf{a}}) \to \ell^2_\pi$ is well-defined.

We recall that we defined $\Gamma = \Gamma_{1,1}$ in (5.16) in Sect. 5.2.

Theorem 5.18 *The operator $Q^{\mathbf{a}} : \mathcal{D}(Q^{\mathbf{a}}) \to \ell^2_\pi$ has spectral decomposition*

$$Q^{\mathbf{a}} = -\int_{\theta^*}^{\infty} \theta \, dE_\theta, \tag{5.60}$$

where $(E_\theta : \theta \in [\theta^*, \infty))$ is a right-continuous increasing family of orthogonal projections and the integral $-\int_{\theta^*}^{\infty}$ runs over the domain $[\theta^*, \infty) \cap (0, \infty)$ (if needed, we put $E_\theta \equiv 0$ for $\theta < \theta^*$). Moreover,

$$\forall t > 0: \quad e^{Q^{\mathbf{a}} t} := \int_{\theta^*}^{\infty} e^{-\theta t} \, dE_\theta$$

is an analytic semigroup, and we have the equality

$$P_t = \int_{\theta^*}^{\infty} e^{-\theta t} \, dE_\theta \tag{5.61}$$

on the set of bounded functions in $\mathcal{D}(Q^{\mathbf{a}})$. In particular,

$$p_t(i,j) = \pi_i^{-1} \int_{\theta^*}^{\infty} e^{-\theta t} \, d\langle E_\theta \mathbf{1}_{\{j\}}, \mathbf{1}_{\{i\}} \rangle = \pi_i^{-1} \int_{\theta^*}^{\infty} e^{-\theta t} \, d\mathcal{E}_{i,j}(\theta), \tag{5.62}$$

where $\mathcal{E}_{i,j}(B) = \int_B d\langle E_\theta \mathbf{1}_{\{j\}}, \mathbf{1}_{\{i\}} \rangle_\pi$ for $B \in \mathcal{B}(\mathbb{R}_+)$ is a finite signed measure on $(\mathbb{R}_+, \mathcal{B}(\mathbb{R}_+))$, and for $j = i$, $\mathcal{E}_{i,i}$ is a probability measure. Furthermore,

$$\Gamma = \mathcal{E}_{1,1}, \tag{5.63}$$

and

$$\mathcal{E}_{i,j} \ll \mathcal{E}_{1,1} \quad \textit{with} \quad \frac{d\mathcal{E}_{i,j}}{d\mathcal{E}_{1,1}}(\theta) = \pi_j \psi_i(\theta) \psi_j(\theta), \ \mathcal{E}_{1,1}\text{-a.s.}$$

Proof We first state some properties of the operator $Q^{\mathbf{a}} : \mathcal{D}(Q^{\mathbf{a}}) \to \ell^2_\pi$, allowing one to get a spectral decomposition.

If $(f^k : k \in \mathbb{N}) \subset \mathcal{D}(Q^{\mathbf{a}})$ is such that $f^k \to g \in \ell^2_\pi$ and $Q^{\mathbf{a}} f^k \to h \in \ell^2_\pi$ as $k \to \infty$, then necessarily for all $n \in \mathbb{N}$,

$$h(n) = \lim_{k\to\infty} \bigl(\mu_n f^k(n-1) - (\lambda_n + \mu_n) f^k(n) + \lambda_n f^k(n+1)\bigr) = Q^{\mathbf{a}} g(n).$$

So, $g \in \mathcal{D}(Q^{\mathbf{a}})$ and $Q^{\mathbf{a}} g = h$. Then, $Q^{\mathbf{a}} : \mathcal{D}(Q^{\mathbf{a}}) \to \ell^2_\pi$ is a closed operator.

Let $I_N = \{1, \dots, N\}$. We denote by $\ell^{2(N)}_\pi$ the set of functions $f \in \ell^2_\pi$ such that $f|_{I^{\mathbf{a}} \setminus I_N} \equiv 0$. We identify $\ell^{2(N)}_\pi$ with $\mathbb{R}^N$ which is the space where the truncated matrix $Q^{(N)} = (q(i,j) : i, j \in I_N)$ acts. Notice that the set $\bigcup_{N \in \mathbb{N}} \ell^{2(N)}_\pi$ is dense in ℓ^2_π.

Since $Q^{(N)}$ is symmetric with respect to π and $Q^{\mathbf{a}}$ is closed, we find $\langle Q^{\mathbf{a}} f, g\rangle_\pi = \langle f, Q^{\mathbf{a}} g\rangle_\pi$ $\forall f, g \in \mathcal{D}(Q^{\mathbf{a}})$. We claim that $Q^{\mathbf{a}} : \mathcal{D}(Q^{\mathbf{a}}) \to \ell^2_\pi$ is self-adjoint; this means

$$\mathcal{D}(Q^{\mathbf{a}*}) = \mathcal{D}(Q^{\mathbf{a}}) \quad \text{and} \quad Q^{\mathbf{a}} = Q^{\mathbf{a}*} \quad \text{on the domain } \mathcal{D}(Q^{\mathbf{a}*}).$$

We recall that

$$\mathcal{D}(Q^{\mathbf{a}*}) = \{g \in \ell^2_\pi : \exists g^* \in \ell^2_\pi \text{ such that } \langle Q^{\mathbf{a}} f, g\rangle_\pi = \langle f, g^*\rangle_\pi \ \forall f \in \mathcal{D}(Q^{\mathbf{a}})\},$$

and $Q^{\mathbf{a}*} g = g^*$ for all $g \in \mathcal{D}(Q^{\mathbf{a}*})$. This map is well-defined because when $g \in \mathcal{D}(Q^{\mathbf{a}*})$, the density of $\mathcal{D}(Q^{\mathbf{a}})$ implies that the element g^* is unique. Let us show the self-adjointness.

Since $\mathbf{1}_{\{j\}} \in \mathcal{D}(Q^{\mathbf{a}})$ $\forall j \in \mathbb{N}$, we can take $f = \mathbf{1}_{\{j\}}$ in $\langle Q^{\mathbf{a}} f, g\rangle_\pi = \langle f, g^*\rangle_\pi$ to obtain $\langle Q^{\mathbf{a}} \mathbf{1}_{\{j\}}, g\rangle_\pi = g^*(j)\pi_j = Q^* g(j)\pi_j$ $\forall j \in \mathbb{N}$. On the other hand, by defining $Q^{\mathbf{a}} g$ pointwise as in (5.58), we get $\langle Q^{\mathbf{a}} \mathbf{1}_{\{j\}}, g\rangle_\pi = \langle \mathbf{1}_{\{j\}}, Q^{\mathbf{a}} g\rangle_\pi = Q^{\mathbf{a}} g(j)\pi_j$. We have shown $Q^* g(j) = Q^{\mathbf{a}} g(j)$ $\forall j \in \mathbb{N}$. By definition, we have $Q^* g \in \ell^2_\pi$, then (5.59) implies $g \in \mathcal{D}(Q^{\mathbf{a}})$, so $\mathcal{D}(Q^{\mathbf{a}}) = \mathcal{D}(Q^{\mathbf{a}*})$ and $Q^{\mathbf{a}*} = Q^{\mathbf{a}}$. The claim has been proved.

Since $Q^{\mathbf{a}} : \mathcal{D}(Q^{\mathbf{a}}) \to \ell^2_\pi$ is self-adjoint, it has spectral decomposition

$$Q^{\mathbf{a}} = \int_{-\infty}^{\infty} \theta \, dE_\theta,$$

where $(E_\theta : \theta \in (-\infty, \infty))$ is a right-continuous increasing family of orthogonal projections, see Riesz and Nagy (1955, Sect. VIII.120).

On the other hand,

$$\forall f \in \mathcal{D}(Q^{\mathbf{a}}): \quad -\langle Q^{\mathbf{a}} f, f\rangle_\pi = \sum_{i \in \mathbb{N}} (\lambda_i + \mu_i)\big(f(i) - f(i+1)\big)^2 \pi_i + \mu_1 f_1^2 \pi_1.$$

In particular,

$$\forall f \in \mathcal{D}(Q^{\mathbf{a}}) \setminus \{0\}: \quad -\langle Q^{\mathbf{a}} f, f\rangle_\pi > 0. \tag{5.64}$$

Then 0 does not belong to the spectrum of $Q^{\mathbf{a}}$. We also have

$$\inf\{-\langle Q^{\mathbf{a}} f, f\rangle_\pi : \|f\|_\pi = 1\} = \lim_{N \to \infty} \inf\{-\langle Q^{\mathbf{a}} f, f\rangle_\pi : \|f\|_\pi = 1, f \in \ell^{2(N)}_\pi\}.$$

Since the spectrum of $-Q^{(N)}$ satisfies $\{a_{j,N} : j \in I_N\} \subset [\theta^*, \infty) \cap (0, \infty)$, we get

$$\inf\{-\langle Q^{\mathbf{a}} f, f\rangle_\pi : \|f\|_\pi = 1\} \ge \theta^*. \tag{5.65}$$

Then, the relation (5.60) is shown:

$$Q^{\mathbf{a}} = -\int_{\theta^*}^{\infty} \theta \, dE_\theta,$$

where from (5.64) and (5.65) the integral runs on the domain $[\theta^*, \infty) \cap (0, \infty)$. From the spectral decomposition, we get that

$$\forall t \geq 0: \quad e^{Q^{\mathbf{a}}t} := \int_{\theta^*}^{\infty} e^{-\theta t} \, dE_\theta$$

is a semigroup. By using the Dominated Convergence Theorem, it is shown that this semigroup is analytic. So, it verifies the backward and forward equations

$$\left(e^{Q^{\mathbf{a}}t}\right)' = Q^{\mathbf{a}} e^{Q^{\mathbf{a}}t} = e^{Q^{\mathbf{a}}t} Q^{\mathbf{a}},$$

and our condition $A = \infty$ guarantees that the unique solution to these equations is given by the minimal semigroup (P_t) (see the end of Sect. 5.1). We have proved (5.61), that is, the equality

$$P_t = \int_{\theta^*}^{\infty} e^{-\theta t} \, dE_\theta$$

is satisfied on the set of bounded functions in $\mathcal{D}(Q^{\mathbf{a}})$. Since

$$p_t(i, j) = (P_t \mathbf{1}_{\{j\}})(i) = \pi_i^{-1} \langle P_t \mathbf{1}_{\{j\}}, \mathbf{1}_{\{i\}} \rangle_\pi,$$

we obtain that (5.62) holds. Then,

$$p_t(1, 1) = \int_{\theta^*}^{\infty} e^{-\theta t} \, d\mathcal{E}_{1,1}(\theta).$$

Since $p_0(1, 1) = 1$, we get that $p_t(1, 1)$ is the Laplace transform of the distribution probability $\mathcal{E}_{1,1,\cdot}$. On the other hand, from (5.15) and by using $\psi_1(\theta) \equiv 1$ and $\pi_1 = 1$, we get

$$p_t(1, 1) = \int_{\theta^*}^{\infty} e^{-t\theta} \, d\Gamma(\theta),$$

where Γ is a probability distribution on $(0, \infty)$. Since the Laplace transform determines the distribution (see Sect. VII.6 in Feller 1971), we find $\Gamma = \mathcal{E}_{1,1,\cdot}$. Then (5.63) is satisfied.

From the spectral decomposition and relation (5.15), we get

$$p_t(i, i) = \int_{\theta^*}^{\infty} e^{-\theta t} \, d\mathcal{E}_{i,i}(\theta) = \pi_i \int_{\theta^*}^{\infty} e^{-t\theta} \left(\psi_i(\theta)\right)^2 d\Gamma(\theta).$$

Then, also by arguing that the Laplace transform determines the distribution, we get $\mathcal{E}_{i,i} \ll \mathcal{E}_{1,1}$ and $\frac{d\mathcal{E}_{i,i}}{d\mathcal{E}_{1,1}}(\theta) = \pi_i (\psi_i(\theta))^2$.

Let $i \neq j$. We have $\mathcal{E}_{i,j} \leq \frac{1}{2}(\mathcal{E}_{i,i} + \mathcal{E}_{j,j})$ because by polarization $\mathcal{E}_{i,i} + \mathcal{E}_{j,j} + 2\mathcal{E}_{i,j}$ is equal to the measure defined by $d\langle E_\theta(\mathbf{1}_{\{i\}} - \mathbf{1}_{\{j\}}), \mathbf{1}_{\{i\}} - \mathbf{1}_{\{j\}}\rangle_\pi$, which is a positive measure. Hence

$$\mathcal{E}_{i,j} \ll \mathcal{E}_{i,i} + \mathcal{E}_{j,j} \ll \mathcal{E}_{1,1}.$$

From

$$p_t(i,j) = \int_{\theta^*}^\infty e^{-\theta t}\, d\mathcal{E}_{i,j}(\theta) = \pi_j \int_{\theta^*}^\infty e^{-t\theta} \psi_i(\theta)\psi_j(\theta)\, d\Gamma(\theta),$$

we deduce

$$\mathrm{Support}\big(\mathcal{E}^+_{i,j}\big) = \big\{\theta \geq \theta^* : \psi_i(\theta)\psi_j(\theta) > 0\big\} \cap \mathrm{Support}(\Gamma) \quad \text{and}$$
$$\mathrm{Support}\big(\mathcal{E}^-_{i,j}\big) = \big\{\theta \geq \theta^* : \psi_i(\theta)\psi_j(\theta) < 0\big\} \cap \mathrm{Support}(\Gamma).$$

Finally, again by using the fact that the Laplace transform determines the distribution, we conclude $\frac{d\mathcal{E}_{i,j}}{d\mathcal{E}_{1,1}}(\theta) = \pi_j \psi_i(\theta)\psi_j(\theta)$. □

Remark 5.19 Assume $\theta^* > 0$. We recall that $\Gamma(\{\theta^*\}) > 0$ is a necessary and sufficient condition in order that the birth-and-death chain be R-positive. But $\Gamma(\{\theta^*\}) > 0$ is equivalent to the fact that θ^* is an eigenvalue of $Q^{\mathbf{a}} : \mathcal{D}(Q^{\mathbf{a}}) \to \ell^2_\pi$. Since $f = \psi(\theta)$ is the unique solution to $Q^{\mathbf{a}} f = -\theta f$ up to a multiplicative constant, θ^* is a simple eigenvalue and the associated eigenfunction is proportional to $\psi(\theta^*)$. Hence, the chain is R-positive if and only if $\psi(\theta^*)$ is an eigenfunction of $Q^{\mathbf{a}} : \mathcal{D}(Q^{\mathbf{a}}) \to \ell^2_\pi$, in particular, $\psi(\theta^*) \in \ell^2_\pi$.

Now, let us recall (5.32),

$$\forall j \in \mathbb{N}: \quad \nu_{\theta^*}(j) = c\psi_j\big(\theta^*\big)\pi_j = c < \mathbf{1}_{\{j\}}, \psi\big(\theta^*\big) >_\pi \quad \text{with } c = \nu_{\theta^*}(1). \tag{5.66}$$

In the next result, we require the existence of a gap after the bottom point of the spectrum.

Proposition 5.20 *Assume $\theta^* > 0$ and $\sum_{j\in\mathbb{N}} \pi_j < \infty$. If $Q^{\mathbf{a}}$ has a spectral gap after θ^*, that is,* $\mathrm{Support}(\Gamma) \cap [\theta^*, \theta^* + \epsilon) = \{\theta^*\}$ *for some $\epsilon > 0$, then the conditional evolution of every probability measure $\rho \in \mathcal{P}(\mathbb{N})$ that satisfies*

$$\sum_{n\in\mathbb{N}} \frac{\rho_j^2}{\pi_j} < \infty \tag{5.67}$$

converges to ν_{θ^*}, that is,

$$\forall j \in \mathbb{N}: \quad \mathbb{P}_\rho(Y_t = j | T > t) = \nu_{\theta^*}(j). \tag{5.68}$$

Proof Define $\rho/\pi(i)=\rho(i)/\pi_i\ \forall i\in I^{\mathbf{a}}$. From hypothesis (5.67), we have $\rho/\pi\in\ell^2_\pi$. Notice that $\rho(f)=\langle\rho/\pi,f\rangle_\pi\ \forall f\in\ell^2_\pi$. From the condition $\sum_{j\in\mathbb{N}}\pi_j<\infty$, we get $\mathbf{1}\in\ell^2_\pi$. Then $P_t\mathbf{1}\in\ell^2_\pi$ and $\int_0^\infty d\langle E_\theta f,\mathbf{1}\rangle_\pi=\langle f,\mathbf{1}\rangle_\pi$ is finite $\forall f\in\ell^2_\pi$.

Hence,

$$\mathbb{P}_\rho(Y_t=j|T>t)=\frac{\sum_{i\in I^{\mathbf{a}}}\rho(i)p_t(i,j)}{\sum_{i\in I^{\mathbf{a}}}\sum_{k\in I^{\mathbf{a}}}\rho(i)p_t(i,k)}=\frac{\rho(P_t\mathbf{1}_{\{j\}})}{\rho(P_t\mathbf{1})}=\frac{\langle\rho/\pi,P_t\mathbf{1}_{\{j\}}\rangle_\pi}{\langle\rho/\pi,P_t\mathbf{1}\rangle_\pi}.$$

Since $\theta^*\in\text{Support}(\Gamma)$, the gap hypothesis implies $\Gamma\{\theta^*\}>0$. Then θ^* is a simple eigenvalue and $\psi(\theta^*)$ is an eigenfunction, so necessarily $\psi(\theta^*)\in\ell^2_\pi$. Then, $\widehat{\psi}(\theta^*):=\psi(\theta^*)/\|\psi(\theta^*)\|_\pi$ is a normalized eigenfunction, and for all bounded function $h,g\in\mathcal{D}(Q^{\mathbf{a}})$, we have

$$\langle h,P_tg\rangle_\pi=e^{-\theta^*t}\langle g,\widehat{\psi}(\theta^*)\rangle_\pi\langle h,\widehat{\psi}(\theta^*)\rangle_\pi+\int_{\theta^*+\epsilon}^\infty e^{-\theta t}\,d\langle E_\theta h,g\rangle_\pi.$$

Hence,

$$\frac{\rho(P_t\mathbf{1}_{\{j\}})}{\rho(P_t\mathbf{1})}=\frac{e^{-\theta^*t}\langle\mathbf{1}_{\{j\}},\widehat{\psi}(\theta^*)\rangle_\pi\langle\rho/\pi,\widehat{\psi}(\theta^*)\rangle_\pi+\int_{\theta^*+\epsilon}^\infty e^{-\theta t}\,d\langle E_\theta\rho/\pi,\mathbf{1}_{\{j\}}\rangle_\pi}{e^{-\theta^*t}\langle\mathbf{1},\widehat{\psi}(\theta^*)\rangle_\pi\langle\rho/\pi,\widehat{\psi}(\theta^*)\rangle_\pi+\int_{\theta^*+\epsilon}^\infty e^{-\theta t}\,d\langle E_\theta\rho/\pi,\mathbf{1}\rangle_\pi}.$$

By the Dominated Convergence Theorem, we get

$$\lim_{t\to\infty}\int_{\theta^*+\epsilon}^\infty e^{-(\theta-\theta^*)t}\,d\langle E_\theta\rho/\pi,\mathbf{1}_{\{j\}}\rangle_\pi$$
$$=0=\lim_{t\to\infty}\int_{\theta^*+\epsilon}^\infty e^{-(\theta-\theta^*)t}\,d\langle E_\theta\rho/\pi,\mathbf{1}\rangle_\pi=0.$$

By using (5.66), we obtain

$$\mathbb{P}_\rho(Y_t=j|T>t)=\lim_{t\to\infty}\frac{\rho(P_t\mathbf{1}_{\{j\}})}{\rho(P_t\mathbf{1})}=\frac{\langle\mathbf{1}_{\{j\}},\widehat{\psi}(\theta^*)\rangle_\pi}{\langle\mathbf{1},\widehat{\psi}(\theta^*)\rangle_\pi}=c'\nu_{\theta^*}(j)$$

for some constant c'. So, necessarily $c'=1$ and the limit is ν_{θ^*}. Then, (5.68) holds. □

Now we turn to the conditions ensuring discreteness of the spectrum of the operator $Q^{\mathbf{a}}:\mathcal{D}(Q^{\mathbf{a}})\to\ell^2_\pi$.

The measure $\Gamma=\Gamma_{1,1}$ was constructed as the limit of spectral measures of the truncated matrices $Q^{(N)}$, and the spectral measures of these truncated matrices charge the roots $0<a_{1,N}<\cdots<a_{N-1,N}$ of the polynomial $\psi_N(\theta)$ when $N\ge 2$. By the interlacing property (5.10), the limit $\xi_k=\lim_{N\to\infty}a_{k,N}$ exists for $k\ge 1$. The sequence $(\xi_k:k\in\mathbb{N})$ is increasing, and we define

$$\xi_\infty:=\lim_{k\to\infty}\xi_k.$$

It is straightforward to show, see Van Doorn (1985), that Support(Γ) is discrete, that is, constituted of isolated points, only when

$$\xi_\infty = \infty. \tag{5.69}$$

In this case,

$$\text{Support}(\Gamma) = \{\xi_k : k \geq 1\}.$$

So, under condition (5.69), from (5.15) we get

$$p_t(i,j) = \pi_j \sum_{k \in \mathbb{N}} e^{-\xi_k t} \psi_i(\xi_k) \psi_j(\xi_k) \Gamma\big(\{\xi_k\}\big), \tag{5.70}$$

where $\Gamma(\{\xi_k\}) > 0$ and $\sum_{k \in \mathbb{N}} \Gamma(\{\xi_k\}) = 1$.

The first eigenfunction $\psi(\xi_1)$ is strictly positive and all other eigenfunctions $(\psi(\xi_k) : k \geq 2)$ do not have a defined sign. Since $\psi_1 \equiv 1$, this is written as

$$\forall k \geq 2, \exists i_k \quad \text{such that} \quad \psi_{i_k}(\xi_k) < 0.$$

This follows from the equality $\xi_1 = \sup\{\theta \geq 0 : \psi_k(\theta) > 0 \; \forall k \geq 1\}$, see (5.12).

Theorem 5.21 *The operator $Q^{\mathbf{a}} : \mathcal{D}(Q^{\mathbf{a}}) \to \ell^2_\pi$ has discrete pure point spectrum if and only if $\xi_\infty = \infty$. In this case, the spectrum consists of isolated simple eigenvalues $\{-\xi_k : k \geq 1\}$ with eigenfunctions $\{\psi(\xi_k) : k \geq 1\}$. The eigenfunction $\psi(\xi_1)$ is strictly positive and all other eigenfunctions $(\psi(\xi_n) : n \geq 2)$ do not have a defined sign.*

The relation (5.70) *holds, $p_t(i,j) = \pi_j \sum_{k \in \mathbb{N}} e^{-\xi_k t} \psi_i(\xi_k) \psi_j(\xi_k) \Gamma(\{\xi_k\})$, with*

$$\forall k \in \mathbb{N}: \quad \Gamma(\xi_k) = \big\| \psi(\xi_k) \big\|_\pi^{-2}.$$

Proof The first part is a simple consequence of Theorem 5.18. The set of eigenvalues is $\{\xi_k : k \in \mathbb{N}\}$, let $E_{\xi_k} - E_{\xi_k^-}$ be the eigenspace associated to $-\xi_k$. From (5.9) and the discussion after it, $f = \psi(\theta)$ is the unique solution to $Q^{\mathbf{a}} f = -\theta f$ up to a multiplicative constant. Hence, all the eigenvalues are simple and the associated eigenfunctions are proportional to $\psi(\xi_k)$, so necessarily $\psi(\xi_k) \in \ell^2_\pi$. Denote $\widetilde{\psi}(\xi_k) = \psi(\xi_k)/\|\psi(\xi_k)\|_\pi$, the normalized eigenfunction associated to $-\xi_k$. Since $Q^{\mathbf{a}} : \mathcal{D}(Q^{\mathbf{a}}) \to \ell^2_\pi$ is self-adjoint, the set of eigenfunctions $\{\widetilde{\psi}(\xi_k) : k \in \mathbb{N}\}$ is an orthonormal family on ℓ^2_π, and for all $g \in \mathcal{D}(Q^{\mathbf{a}})$,

$$Q^{\mathbf{a}} g = -\sum_{k \in \mathbb{N}} \xi_k \big\langle \widetilde{\psi}(\xi_k), g \big\rangle_\pi \widetilde{\psi}(\xi_k),$$

and so for all g bounded in $\mathcal{D}(Q^{\mathbf{a}})$ we find

$$P_t g = \sum_{k \in \mathbb{N}} e^{-\xi t} \big\langle \widetilde{\psi}(\xi_k), g \big\rangle_\pi \widetilde{\psi}(\xi_k).$$

Since $\mathbf{1}_{\{j\}} \in \ell^2_\pi$ for all $j \in I^{\mathbf{a}}$ and $\widetilde{\psi}_j(\xi_k) = \pi_j^{-1} \langle \mathbf{1}_{\{j\}}, \widetilde{\psi}(\xi_k) \rangle_\pi$, we obtain

$$p_t(i,j) = \pi_i^{-1} \langle P_t \mathbf{1}_{\{j\}}, \mathbf{1}_{\{i\}} \rangle_\pi = \pi_j \sum_{k \in \mathbb{N}} e^{-\xi_k t} \widetilde{\psi}_i(\xi_k) \widetilde{\psi}_j(\xi_k).$$

From

$$p_t(i,j) = \pi_j \sum_{k \in \mathbb{N}} e^{-\xi_k t} \psi_i(\xi_k) \psi_j(\xi_k) \Gamma(\{\xi_k\}),$$

we conclude $\Gamma(\xi_k) = \|\psi(\xi_k)\|_\pi^{-2}$. □

In the sequel, we will give some conditions for the $Q^{\mathbf{a}}$ to have a discrete spectrum on ℓ_π^2. These conditions and the result we give now follow closely Kreer (1994). Below we denote by $O(1/j)$ a quantity depending on j that satisfies $O(1/j) \to 0$ as $j \to \infty$.

Proposition 5.22 *The assumptions*

$$\lim_{j\to\infty} \mu_j = \infty = \lim_{j\to\infty} \lambda_j, \qquad \lim_{j\to\infty} \frac{\mu_j}{\lambda_j} = q \quad \textit{with } q \neq 1; \tag{5.71}$$

$$\frac{\lambda_j}{\lambda_{j+1}} = 1 + O\left(\frac{1}{j}\right), \qquad \frac{\mu_j}{\mu_{j+1}} = 1 + O\left(\frac{1}{j}\right), \tag{5.72}$$

imply that the operator $Q^{\mathbf{a}} : \mathcal{D}(Q^{\mathbf{a}}) \to \ell_\pi^2$ *has a discrete spectrum, and so the relations stated in Theorem* 5.21 *are satisfied.*

Proof Let $\mathcal{D}_\pi^{1/2}$ be the diagonal matrix $\mathcal{D}_\pi^{1/2}$ whose (j,j)-term is $\sqrt{\pi_j}$ (also considered in the finite case in Sect. 3.3). Define $R = \mathcal{D}_\pi^{1/2} Q^{\mathbf{a}} \mathcal{D}_\pi^{-1/2}$, which is a tri-diagonal symmetric matrix. The diagonal elements of R are the same as those of $Q^{\mathbf{a}}$, that is, $R(j,j) = -q_j$; and the nondiagonal terms are given by $R(j,j+1) = \sqrt{\lambda_j \mu_{j+1}} = R(j+1,j)$.

Let $\ell^2 = \{(f_n : n \in \mathbb{N}) \in \mathbb{C}^{\mathbb{N}} : \sum_{n\in\mathbb{N}} |f_n|^2 < \infty\}$. We have that $\mathcal{D}_\pi^{-1/2} : \ell^2 \to \ell_\pi^2$ is an isometry satisfying $R = \mathcal{D}_\pi^{1/2} Q^{\mathbf{a}} \mathcal{D}_\pi^{-1/2}$, and so the spectrum of $Q^{\mathbf{a}}$ on ℓ_π^2 is the same as the one of R on ℓ^2. Then, the spectrum of R is contained in $(-\infty, \theta^*]$.

Let $z > 0$, so it is not in the spectrum of R. We will find conditions ensuring that $(R - z\mathbb{I})^{-1}$ is a compact operator, that is, $(R - z\mathbb{I})^{-1} B_1$ is compact where B_1 denotes the unit ball in ℓ^2 and $\mathbb{I}$ denotes the identity operator. Let $\mathcal{D}$ be a diagonal matrix whose elements are the diagonal elements of R, so its ith element is $-q_i$ for $i \in \mathbb{N}$. Let $M = R - \mathcal{D}$ and $\widehat{\mathcal{D}} = \mathcal{D} - z\mathbb{I}$. Then $R - z\mathbb{I} = M + \widehat{\mathcal{D}} = (\mathbb{I} + M\widehat{\mathcal{D}}^{-1})\widehat{\mathcal{D}}$. The matrix $M\widehat{\mathcal{D}}^{-1}$ is a tri-diagonal matrix whose diagonal elements vanish and it satisfies

$$\left(M\widehat{\mathcal{D}}^{-1}\right)(i,i+1) = -\frac{R(i,i+1)}{q_{i+1}+z}, \qquad \left(M\widehat{\mathcal{D}}^{-1}\right)(i+1,i) = -\frac{R(i+1,i)}{q_i+z}.$$

Then, under the condition

$$\exists z > 0\ \exists \delta > 0\ \forall i \in \mathbb{N}: \quad \frac{R(i,i-1)}{q_{i-1}+z} < \frac{1}{2+\delta} \quad \text{and} \quad \frac{R(i,i+1)}{q_{i+1}+z} < \frac{1}{2+\delta}, \tag{5.73}$$

we have $\|M\widehat{\mathcal{D}}^{-1}\|_{\ell^2} < 1$. In this case, the Neumann expansion implies that $(\mathbb{I} + M\widehat{\mathcal{D}}^{-1})^{-1}$ is a bounded linear operator on ℓ^2.

Since by hypothesis $q_i = \lambda_i + \mu_i \to \infty$ as $i \to \infty$, the elements in $\widehat{\mathcal{D}}^{-1}(B_1)$ satisfy

$$\lim_{n\to\infty}\Big(\sup_{f\in\widehat{\mathcal{D}}^{-1}(B_1)} \sum_{i\ge n} |f_i|^2\Big) = 0,$$

and so $\widehat{\mathcal{D}}^{-1}(B_1)$ is relatively compact (see Dunford and Schwartz 1958, Exercise 3, Sect. IV.13). Hence $\widehat{\mathcal{D}}^{-1}$ is a compact operator acting on ℓ^2. Since $(\mathbb{I}+M\widehat{\mathcal{D}}^{-1})^{-1}$ is a bounded operator, we obtain that $(R - z\mathbb{I})^{-1} = \widehat{\mathcal{D}}^{-1}(\mathbb{I} + M\widehat{\mathcal{D}}^{-1})^{-1}$ is a compact operator. Therefore, R has a compact resolvent.

The restrictions $R(i, i+1)/(q_{i+1} + z) < 1/(2+\delta)$ and $R(i+1, i)/(q_i + z) < 1/(2+\delta)$ are respectively equivalent to

$$\forall i \in \mathbb{N}: \quad (2+\delta)\sqrt{\lambda_i\mu_{i+1}} < z + \min\{\lambda_{i+1} + \mu_{i+1}, \lambda_i + \mu_i\}.$$

These inequalities are equivalent to the following ones:

$$\forall i \in \mathbb{N}: \quad V_i(\delta) < \frac{z}{\mu_{i+1}} \quad \text{and} \quad V_i'(\delta) < \frac{z}{\mu_i},$$

where

$$V_i(\delta) = (2+\delta)\sqrt{\frac{\lambda_{i+1}}{\mu_{i+1}}}\left(\sqrt{\frac{\lambda_i}{\lambda_{i+1}}} - \frac{2}{2+\delta}\right) - \left(\sqrt{\frac{\lambda_{i+1}}{\mu_{i+1}}} - 1\right)^2$$

and

$$V_i'(\delta) = (2+\delta)\sqrt{\frac{\lambda_i}{\mu_i}}\left(\sqrt{\frac{\mu_{i+1}}{\mu_i}} - \frac{2}{2+\delta}\right) - \left(\sqrt{\frac{\lambda_i}{\mu_i}} - 1\right)^2.$$

Hence, conditions (5.71) and (5.72) imply that there exists $\delta > 0$ such that

$$\limsup_{i\to\infty} V_i(\delta) < 0 \quad \text{and} \quad \limsup_{i\to\infty} V_i'(\delta) < 0.$$

On the other hand, for all $\delta > 0$ and all $i_0 \in \mathbb{N}$, we can find $z > 0$ such that

$$\sup\{V_i(z,\delta) : i \le i_0\} < 0 \quad \text{and} \quad \sup\{V_i'(z,\delta) : i \le i_0\} < 0.$$

Hence, conditions (5.71) and (5.72) imply (5.73), and so R has a compact resolvent. We deduce from Theorem 6.29 of Sect. III.8 in Kato (1995) that R has a pure point spectrum that can only accumulate at 0. Then, $Q^{\mathbf{a}}$ has a discrete spectrum. □

5.8 The Process Conditioned to Hitting Barriers

We are assuming that $D=\infty$, so ∞ is a natural boundary. For $\theta=0$, the polynomial family $\psi(0)=(\psi_i(0): i\in\mathbb{N})$ is given by

$$\psi_i(0)=1+\sum_{j=1}^{i-1}\prod_{l=1}^{j}\frac{\mu_l}{\lambda_l}.$$

Fix $i\in\mathbb{N}$ and let $M>i$. From $Q^{\mathbf{a}}\psi(0)=0$, we have

$$\mathbb{E}_i\big(\psi_{Y_{t\wedge T_M\wedge T}}(0)\big)=\psi_i(0).$$

By taking $t\to\infty$ on the left-hand side and using $\mathbb{P}_i(T_M\wedge T<\infty)=1$ and $\psi_0(0)=0$, we get

$$\lim_{t\to\infty}\mathbb{E}_i\big(\psi_{Y_{t\wedge T_M\wedge T}}(0)\big)=\psi_M(0)\mathbb{P}_i(T_M<T).$$

Therefore,

$$\forall M>i:\quad \mathbb{P}_i(T_M<T)=\frac{\psi_i(0)}{\psi_M(0)}. \tag{5.74}$$

Proposition 5.23 *Assume $D=\infty$. Let $i_0,\ldots,i_k\in\mathbb{N}$, $s_0=0<\cdots<s_k$. Then, the following limit exists*

$$\mathbb{P}_{i_0}\big(Z^0_{s_1}=i_1,\ldots,Z^0_{s_k}=i_k\big):=\lim_{M\to\infty}\mathbb{P}_i(Y_{s_1}=i_1,\ldots,Y_{s_k}=i_k|T<T_M).$$

The process $Z^0=(Z^0_t:t\geq 0)$ is a Markov chain and its transition jump matrix $\widehat{Q}=(\widehat{q}(i,j):i,j\in\mathbb{N})$ is given by

$$\forall i,j\in\mathbb{N}:\quad \widehat{q}(i,j)=q(i,j)\frac{\psi_j(0)}{\psi_i(0)}.$$

Proof We take $s_k<M$. From the Markov property and (5.74), we get

$$\begin{aligned}\mathbb{P}_{i_0}(Y_{s_1}=i_1,\ldots,Y_{s_k}=i_k,s_k<T<T_M)&=\mathbb{P}_{i_0}(Y_{s_1}=i_1,\ldots,Y_{s_k}=i_k)\mathbb{P}_{i_k}(T<T_M)\\&=\mathbb{P}_i(Y_{s_1}=i_1,\ldots,Y_{s_k}=i_k)\frac{\psi_{i_k}(0)}{\psi_M(0)}.\end{aligned}$$

Note that $i_k\in\mathbb{N}$ implies $s_k<T$. Then, by using (5.74) again, we obtain

$$\begin{aligned}\mathbb{P}_{i_0}(Y_{s_1}=i_1,\ldots,Y_{s_k}=i_k|T<T_M)&=\mathbb{P}_{i_0}(Y_{s_1}=i_1,\ldots,Y_{s_k}=i_k)\frac{\psi_{i_k}(0)}{\psi_{i_0}(0)}\\&=\prod_{l=0}^{k-1}\widehat{q}(i_l,i_{l+1}).\end{aligned}$$

The result follows. □

5.9 Examples

Below we compute some of the quantities in three examples of birth-and-death chains. We recall that $\lambda_0 = 0 = \mu_0$. Below λ and μ represent two strictly positive constants.

5.9.1 The Random Walk

The random walk $\lambda_i = \lambda$, $\mu_i = \mu$. This case was treated in Cavender (1978); Seneta and Vere-Jones (1996). The system of equations associated to the QSD equality $\nu' Q^{\mathbf{a}} = -\theta \nu'$ is equivalent to

$$p\nu(i-1) + (1-p)\nu(i+1) = \gamma \nu(i), \quad i \geq 1, \tag{5.75}$$

where $\nu(0) = 0$ and

$$p = \frac{\lambda}{\lambda + \mu}, \qquad \gamma = 1 - \frac{\theta}{\lambda + \mu}.$$

In order for some γ to exist for which there exists a probability measure as a solution of (5.75), a necessary and sufficient condition is $p < 1/2$, which is equivalent to $\mu > \lambda$. In this case, the set of γ's for which there exists such a solution is the interval $[\gamma^*, 1)$ with $\gamma^* = 2\sqrt{p(1-p)}$. Then $\theta^* = (\lambda + \mu)(1 - 2\sqrt{p(1-p)}) = (\lambda + \mu) - 2\sqrt{\lambda\mu}$. For each $\gamma \in [\gamma^*, 1)$, or equivalently, for $\theta \in (0, \theta^*]$, the associated QSD (which, by abuse of notation, we denote by ν_γ instead of ν_θ) $\nu_\gamma = (\nu_\gamma(i) : i \in \mathbb{N})$ is

$$\nu_{\gamma^*}(i) = Ai\left(\sqrt{\frac{p}{1-p}}\right)^{i-1} = Ai\left(\sqrt{\frac{\lambda}{\mu}}\right)^{i-1}$$

and

$$\nu_\gamma(i) = B\left(\left(\frac{\gamma + \Delta}{2(1-p)}\right)^{i-1} - \left(\frac{\gamma - \Delta}{2(1-p)}\right)^{i-1}\right) \quad \text{for } \gamma \in (\gamma^*, 1),$$

where A and B are normalizing constants and

$$\Delta = \sqrt{\gamma^2 - 4p(1-p)} = \frac{1}{(\lambda + \mu)}\sqrt{(\lambda + \mu - \theta)^2 - 4\lambda\mu}.$$

We claim that the process Y^T is θ^*-transient. Indeed, we use (5.50) to show that $\overline{\pi}$ verifies

$$\overline{\pi}_i = \prod_{r=1}^{i-1}\left(\frac{\nu_{\gamma^*}(r+1)\sqrt{\mu}}{\nu_{\gamma^*}(r)\sqrt{\lambda}}\right)^2 = \prod_{r=1}^{i-1}\left(\frac{(r+1)}{r}\right)^2 = i^2.$$

Then

$$\overline{A} = \sum_{i \geq 1} \frac{\sqrt{i}}{i^2\sqrt{(i+1)}\sqrt{\lambda\mu}} < \infty,$$

and the θ^*-transience is shown.

5.9.2 The Linear Birth-and-Death Chain

Let $\lambda_i = i\lambda$ and $\mu_i = i\mu$ for $i \in \mathbb{N}$, with $\mu > \lambda$. This case was treated in Van Doorn (1991). Here $\theta^* = \mu - \lambda$ and the extremal QSD ν_{θ^*} verifies

$$\forall i \in \mathbb{N}: \quad \nu_{\theta^*}(i) = \left(\frac{\lambda}{\mu}\right)^{i-1}\left(1 - \frac{\lambda}{\mu}\right).$$

By using (5.48) and (5.50), we get

$$\forall i \in \mathbb{N}: \quad \overline{\lambda}_i = i\lambda \quad \text{and} \quad \overline{\pi}_i = i\left(\frac{\lambda}{\mu}\right)^{i-1}.$$

Hence, the process Y^T is θ^*-positive recurrent.

5.9.3 The $M/M/\infty$ Queue

In this case, $\lambda_i = \lambda$ and $\mu_i = i\mu$ for $i \in \mathbb{N}$. This example was treated in Martínez and Ycart (2001). Consider the generating function

$$g(s) = \sum_{i \geq 1} \nu_\theta(i) s^i, \qquad s \geq 0,$$

where ν_θ verifies (5.34). We find

$$g(s) = 1 + (1-s)^{\theta/\mu} e^{\lambda s/\mu}\big(-1 + L(\theta, s)\big),$$

where

$$L(\theta, s) = \frac{\lambda}{\mu}\int_0^s (1-x)^{-\theta/\mu} e^{-\lambda x/\mu}\, dx.$$

The function $L(\theta, 1)$ is smooth and strictly increasing in the interval of $\theta \in (0, \mu)$. Since $L(0, 1) = 1 - e^{-\lambda/\mu}$ and $L(0, \mu) = \infty$, there exists a unique $\theta_0 \in (0, \mu)$ such that $L(\theta_0, 1) = 1$. If $\theta > \theta_0$, we get that for s close enough to 1 the following inequality holds: $-1 + L(\theta, s) > 0$. From $g(1) = 1$, we get that g decreases close to $s = 1$. Therefore, in the domain $\theta > \theta_0$, the function g cannot be the generator function of some ν_θ, so $\theta^* \leq \theta_0 < \mu$.

On the other hand, in Martínez and Ycart (2001) it was also shown that the exponential rate to equilibrium of this chain satisfies $\alpha^*(0) = \mu$ (see (4.45)). Therefore, the inequality $\theta(\nu) \leq \alpha^*(0)$ proved for monotone process turns to be strict in this case. Indeed, $\theta^* = \sup\{\theta(\nu) : \nu \text{ QSD}\} < \mu = \alpha^*(0)$.

5.10 Structured Birth-and-Death Processes with Mutations

We consider a discrete model describing a structured population whose dynamics takes into account all reproduction and death events. Each individual is characterized by a trait which is a heritable quantitative parameter. During the reproduction process, mutations of the trait can occur, implying some variability in the trait space. In the general model, the individual reproduction and death rates, as well as the mutation distribution, depend on the trait of the individual and on the whole population. In particular, cooperation or competition between individuals in this population are taken into account.

In a general model (see Collet et al. 2011), the set of traits is a compact metric space. Here we will deal with a simplified model, so we assume that the set of traits is $\mathcal{A}$, a finite set.

At a given time, the structured population is described by an atomic measure on the trait space with integer weights, these weights are the numbers of individuals with the given trait. Then, the time evolution is an atomic measure-valued process on the trait space. The state space is thus the collection I of finite point measures on $\mathcal{A}$, which is a countable set. So, a configuration $x \in I$ is described by $(x_a : a \in \mathcal{A})$ with $x_a \in \mathbb{Z}_+ = \{0, 1, \dots\}$, where only a finite subset of elements $a \in \mathcal{A}$ satisfy $x_a > 0$. The finite set of present traits (i.e. traits of alive individuals) is denoted by $\{x\} := \{a \in \mathcal{A} : x_a > 0\}$ and called the support of x, its cardinality is denoted by $\#x$. We denote by $\|x\| = \sum_{a\in\{x\}} x_a$ the total number of individuals in x. The void configuration is denoted by $x = 0$, hence $\#0 = \|0\| = 0$, and we define $I^{\mathbf{a}} = I \setminus \{0\}$ the set of nonempty configurations.

The clonal birth rate, the mutation birth rate and the death rate of an individual with trait a in a population $x \in I$ are denoted respectively by $\lambda^{\mathbf{c}}_{x,a}$, $\lambda^{\mathbf{m}}_{x,a}$ and $\mu_{x,a}$. The total reproduction rate for an individual with trait $a \in \{x\}$ is equal to $\lambda_{x,a} = \lambda^{\mathbf{c}}_{x,a} + \lambda^{\mathbf{m}}_{x,a}$. We assume that $\lambda^{\mathbf{c}}_{x,a} > 0$ and $\mu_{x,a} > 0$ for all $x \in I^{\mathbf{a}}$ and $a \in \mathcal{A}$.

For each $a \in \mathcal{A}$, let κ_a be a probability measure on $\mathcal{A} \setminus \{a\}$, $\sum_{b\in\mathcal{A}\setminus\{a\}} \kappa_a(b) = 1$, that represents the probability distribution of the trait of the new mutated individual born from a. To simplify notations, we express the mutation part using location kernel $G(x, b)$ given by

$$\forall x \in I \ \forall b \in \mathcal{A}: \quad G(x, b) = \sum_{a\in\{x\}} x_a \lambda^{\mathbf{m}}_{x,a} \kappa_a(b).$$

Note that the ratio $G(x, b)/\sum_{c\in\mathcal{A}} G(x, c)$ is the probability that, given there is a mutation from x, the new trait is located at b.

Let $Y = (Y_t : t \geq 0)$ be a Markov process taking values on I and with càdlàg trajectories, defined by the following jump rates' matrix $Q = (q(x, x') : x, x' \in I)$:

$$\begin{aligned}
&q(x, x + \delta_a) = x_a \lambda^{\mathbf{c}}_{x,a}, && a \in \{x\} && \text{(clonal birth);}\\
&q(x, x - \delta_a) = x_a \mu_{x,a}, && a \in \{x\} && \text{(death);}\\
&q(x, x + \delta_b) = G(x, b), && b \notin \{x\} && \text{(mutated birth);}
\end{aligned}$$

and

$$-q(x,x) = \sum_{a\in\{x\}} \big(q(x, x+\delta_a) + q(x, x-\delta_a)\big) + \sum_{b\in\mathcal{A}\setminus\{x\}} q(x, x+\delta_b).$$

All other values of $q(x, x')$ vanish. As usual we put $q_x := -q(x,x)$ for $x \in I$. The trajectories of the process Y starting from x are such that after an exponential time of parameter q_x, the process jumps to $x + \delta_a$ for $a \in \{x\}$ with probability $q(x, x + \delta_a)/q_x$, or to $x - \delta_a$ for $a \in \{x\}$ with probability $q(x, x - \delta_a)/q_x$, or to a point $x + \delta_b$ for $b \in \mathcal{A} \setminus \{x\}$ with probability distribution $q(x, x + \delta_b)/q_x$. The process restarts independently at the new configuration.

Since $q_0 = 0$, 0 is an absorbing state, which is natural for a population dynamics. We denote by $T = \inf\{t > 0 : Y_t = 0\}$ the time of extinction. In what follows, we will assume that the process is a.s. extinct when starting from any initial configuration:

$$\forall x \in I\colon \quad \mathbb{P}_x(T < \infty) = 1.$$

So, in our setting, we assume that competition between individuals yields the discrete population to extinction with probability 1. We will exhibit some conditions ensuring the existence of a QSD.

Denote by

$$\forall k \in \mathbb{Z}_+\colon \quad I_k = \big\{x \in I : \|x\| = k\big\}$$

the sets of configurations with a fixed number of individuals. It is also useful to introduce the following parameters:

$$\underline{\mu} = \liminf_{k\to\infty}\Big(\inf_{x\in I_k}\inf_{a\in\{x\}} \mu_{x,a}\Big), \qquad \overline{\mu} = \limsup_{k\to\infty}\Big(\sup_{x\in I_k}\sup_{a\in\{x\}} \mu_{x,a}\Big),$$

$$\overline{\lambda} = \limsup_{k\to\infty}\Big(\sup_{x\in I_k}\sup_{a\in\{x\}} \lambda_{x,a}\Big).$$

Note that when $\overline{\lambda} < \infty$ we also have $\sup\lambda < \infty$ where $\sup\lambda := \sup_{x\in I}\sup_{a\in\{x\}} \lambda_{x,a}$. On the other hand, when $\underline{\mu} > 0$, we deduce that $\inf\mu > 0$ where $\inf\mu := \inf_{x\in I}\inf_{a\in\{x\}} \mu_{x,a}$.

Theorem 5.24 *Under the following assumption*

$$\overline{\lambda} < \underline{\mu} \le \overline{\mu} < \infty, \tag{5.76}$$

there exists a QSD.

Proof To show the existence of a QSD, we shall use Theorem 4.5, so we must prove that its hypotheses are fulfilled, and for this we must prove that the following two properties are satisfied:

$$\forall s > 0\ \forall \epsilon > 0\ \exists k_0 = k_0(\epsilon, s)\ \forall k \ge k_0\colon \quad \sup_{x\in I_k} \mathbb{P}_x(T < s) < \epsilon; \tag{5.77}$$

and the existence of an exponential moment

$$\exists\theta > 0: \quad \mathbb{E}_x\big(e^{\theta T}\big) < \infty. \tag{5.78}$$

Let us show (5.77). Let $(R_n : n \geq 1)$ be a sequence of independent random variables such that R_n is exponentially distributed with mean $(\overline{\mu} n)^{-1}$. From the definition of the process Y and $\overline{\mu}$, we have that for all $x \in I_k$, $\mathbb{P}_x(T < s) \leq \mathbb{P}(\sum_{n=1}^k R_n < s)$. On the other hand, it is known that $\sum_{n\geq 1}(\overline{\mu} n)^{-1} = \infty$ implies $\mathbb{P}(\sum_{n\geq 1} R_n = \infty) = 1$ (see Lemma (33) in Sect. 5.5 in Freedman 1971). Then, for all $s > 0$ and all $\epsilon > 0$ there exists $k_0 = k_0(\epsilon, s)$ such that for all $k \geq k_0$ we have $\mathbb{P}(\sum_{n=1}^k R_n \geq s) \geq 1 - \epsilon$. Then, (5.77) is satisfied.

The proof will be finished once we show that Y is exponentially killed, which is (5.78). Note that from the hypothesis of the theorem, we have $\sup \lambda < \infty$ and $\inf \mu > 0$. Let us introduce the following notation: for any integer $m \geq 1$,

$$\widehat{\lambda}_m = \sup_{0<\|x\|\leq m} \frac{1}{m} \sum_{a\in\{x\}} x_a \lambda_{x,a}$$

and

$$\underline{\mu}_m = \inf_{\|x\|=m} \inf_{a\in\{x\}} \mu_{x,a}.$$

Since $\sup \lambda < \infty$, we have $\sup_{m\geq 1} \widehat{\lambda}_m < \infty$. On the other hand, $\inf \mu > 0$ implies $\inf_{m\geq 1} \underline{\mu}_m > 0$. We claim that the hypothesis of Theorem 5.24 implies

$$\limsup_{m\to\infty} \widehat{\lambda}_m \leq \overline{\lambda} < \underline{\mu} = \liminf_{m\to\infty} \underline{\mu}_m. \tag{5.79}$$

Let us show this claim. By definition of $\overline{\lambda}$ and (5.76), we need only to prove the first inequality. By definition of $\overline{\lambda}$, for any $\epsilon > 0$, there exists an integer $K(\epsilon)$ such that for any $k > K(\epsilon)$ we have

$$\sup_{x\in I_k} \sup_{a\in\{x\}} \lambda_{x,a} \leq \overline{\lambda} + \epsilon.$$

Therefore, for $m \geq K(\epsilon)$,

$$\widehat{\lambda}_m \leq \sup_{0<\|x\|\leq K(\epsilon)} \frac{1}{m} \sum_{a\in\{x\}} x_a \lambda_{x,a} + \sup_{K(\epsilon)\leq\|x\|\leq m} \frac{1}{m} \sum_{a\in\{x\}} x_a \lambda_{x,a}.$$

This implies

$$\limsup_{m\to\infty} \widehat{\lambda}_m \leq \overline{\lambda} + \epsilon.$$

Since this holds for any $\epsilon > 0$ the claim follows.

To show the existence of exponential moment of Y, it is useful to introduce a birth-and-death process $X = (X_t : t \geq 0)$ on $\mathbb{Z}_+$ killed at 0, with individual birth rates $(\widehat{\lambda}_m : m \geq 1)$ and individual death rates $(\underline{\mu}_m : m \geq 1)$. The result will be shown once we prove that $\|Y\|$ is dominated by X and that X is exponentially killed. In fact, this ensures that $\|Y\|$ is exponentially killed, and so also Y.

Let us first prove that $\|Y\|$ is dominated by X. We introduce a coupling on the subset $\mathcal{J}$ of $I \times \mathbb{Z}_+$ defined by

$$\mathcal{J} = \big\{(x, m) \in I \times \mathbb{Z}_+ : \|x\| \le m\big\}.$$

The coupled process is defined by the nonzero rates

$$\begin{aligned}
J(x, m; x + \delta_a, m + 1) &= x_a \lambda^{\mathbf{c}}_{x,a}, \quad a \in \{x\},\\
J(x, m; x + \delta_b, m + 1) &= G(x, b), \quad b \notin \{x\},\\
J(x, m; x, m + 1) &= m\widehat{\lambda}_m - \sum_{a\in\{x\}} x_a \lambda_{x,a},\\
J(x, m; x - \delta_a, m - 1) &= \underline{\mu}_m x_a \mathbb{1}_{\|x\|=m}, \quad a \in \{x\},\\
J(x, m; x - \delta_a, m) &= x_a(\mu_{x,a} - \underline{\mu}_m \mathbb{1}_{\|x\|=m}), \quad a \in \{x\},\\
J(x, m; x, m - 1) &= m\underline{\mu}_m \mathbb{1}_{\|x\|<m},\\
J(0, m; 0, m + 1) &= m\widehat{\lambda}_m,\\
J(0, m; 0, m - 1) &= m\underline{\mu}_m.
\end{aligned}$$

It is easily checked that the marginals of this process have respectively the laws of Y and X. On the other hand, when the coupled process starts from $\mathcal{J}$, it remains in $\mathcal{J}$ forever, so the domination follows.

As noted, from the coupling it suffices to show that X is exponentially killed. We claim that starting from any k the time of first visit to $k-1$ has exponential moment, that is, there exists $\theta_k > 0$ such that $\mathbb{E}_k(e^{\theta T^X_{k-1}}) < \infty$ for all $\theta \le \theta_k$, where we put $T^X_l = \inf\{t > 0 : X_t = l\}$ for $l \in \mathbb{Z}_+$. This proves that T^X_0 has exponential moment, starting from any k because $\mathbb{E}_k(e^{\theta T^X_0}) = \prod_{j=k}^{1} \mathbb{E}_j(e^{\theta T^X_{j-1}})$ which is finite for all $0 < \theta \le \min\{\theta_j : j = 1, \ldots, k\}$.

The condition (5.79), namely $\limsup_{m\to\infty} \widehat{\lambda}_m < \liminf_{m\to\infty} \underline{\mu}_m$, implies there exists n_0 such that for $n \ge n_0$

$$0 < \widehat{\lambda}^*_n < \underline{\mu}_{*n}, \quad \text{where } \widehat{\lambda}^*_n = \sup_{l\ge n} \widehat{\lambda}_l,\ \underline{\mu}_{*n} = \inf_{l\ge n} \underline{\mu}_l.$$

From the results of Van Doorn (1991) on linear birth-and-death chain, we get that for the birth-and-death chain X (killed at any $n \ge n_0$) there exists $\theta_n \ge \underline{\mu}_{*n} - \widehat{\lambda}^*_n > 0$ such that

$$\mathbb{E}_{n+1}\big(e^{\theta_n T^X_n}\big) < \infty.$$

Therefore, the existence of exponential moment for $n \ge n_0$ is satisfied. The existence of exponential moment for T^X_0 follows from the solidarity property shown in Lemma 5.11. Thus, the result is proved. □

We observe that Theorem 5.24 also holds even when $\overline{\mu} < \infty$ in (5.76) is not verified. In fact, in Collet et al. (2011) it was shown under the unique condition $\overline{\lambda} < \underline{\mu}$ and also when the set of traits $\mathcal{A}$ is a compact metric space. Theorem 2.11 is a keystone to study the existence of a QSD in this more general case.

Chapter 6
Regular Diffusions on $[0, \infty)$

6.1 Hypotheses and Notations

The purpose of this chapter is to develop the theory of QSDs in the framework of one dimensional diffusions absorbed at 0. So here the space is $\mathcal{X} = \mathbb{R}_+$ and the trap $\partial\mathcal{X} = \{0\}$. Some of the applications of this theory can be found in models of ecology, economy (see Sect. 7.8 in the next chapter) and Markov mortality models (see Steinsaltz and Evans 2004).

In what follows, we assume (B_t) is a Brownian motion that starts at x, under $\mathbb{P}_x$. The main object of this chapter is the diffusion $X = (X_t : t \geq 0)$ given by the unique solution to the stochastic differential equation (SDE)

$$X_t = B_t - \int_0^t \alpha(X_s)\, ds. \tag{6.1}$$

In this chapter, we shall assume that $\alpha \in \mathcal{C}^1(\mathbb{R}_+)$, although most of the results hold for piecewise $\mathcal{C}^1$ functions. Under this condition, there exists a unique solution to the SDE (6.1), with initial condition $x > 0$, up to an explosion time.

The main interest of this chapter is to develop the theory of QSDs for the process X under the hypothesis that ∞ is a natural boundary (see Hypothesis **H** in (6.4)). After studying the finite interval case in Sect. 6.1.1, and some other preliminaries, we prove a ratio limit theorem for the unbounded interval in Sect. 6.2.2, Theorem 6.15. The main result of the chapter can be found in Sect. 6.2.4, Theorem 6.4, where the Yaglom limit is shown, as well as the existence of the Q-process. The rest of the chapter is devoted to the study of the h-process associated to X (Theorem 6.35) and the R-classification of X (Theorems 6.45 and 6.50). We finish with some examples in Sect. 6.7.

We denote by T_a the hitting time of a, that is, $T_a = \inf\{t > 0 : X_t = a\}$. In what follows, we denote $T = T_0$. Then the process X is well defined up to $\tau = T \wedge T_\infty$. We shall mainly consider the case where $\mathbb{P}_x(T_\infty < \infty) = 0$, which means that the process cannot explode to ∞ in a finite time. In the context of Markov processes, $\tau = T$ will be the explosion time of X. We are interested in the absorption to 0

P. Collet et al., *Quasi-Stationary Distributions*, Probability and Its Applications,
DOI 10.1007/978-3-642-33131-2_6, © Springer-Verlag Berlin Heidelberg 2013

and we define, as usual, $X_t = 0$ for $t > \tau$. Associated to X we consider the sub-Markovian semigroup given by

$$P_t f(x) = \mathbb{E}_x\big(f(X_t), t < T\big),$$

defined in principle for all bounded and measurable functions f. We denote by $p_t(x, y)$ the transition density of this semigroup. Under some extra conditions on α, this transition density can be computed using the Girsanov's Theorem by

$$p_t(x, y)\,dy = e^{-\int_x^y \alpha(\xi)\,d\xi}\mathbb{E}_x\big(e^{-\frac{1}{2}\int_0^t \alpha^2(B_s)-\alpha'(B_s)\,ds}, B_t \in dy, t < T\big).$$

The generator of this semigroup depends on the behavior of α near ∞, but is given by

$$\mathcal{L} = \frac{1}{2}\partial_{xx} - \alpha\partial_x \tag{6.2}$$

when applied to smooth functions with compact support contained in $(0,\infty)$. Unless specified, we shall denote by $\mathcal{L}$ the second order differential operator given by (6.2) and by $\mathcal{L}^*$ its formal adjoint given by

$$\mathcal{L}^* u = \frac{1}{2}\partial_{xx}u + \partial_x(\alpha u).$$

The functions given by $\gamma(x) = 2\int_0^x \alpha(z)\,dz$ and

$$\Lambda(x) = \int_0^x e^{\gamma(z)}\,dz$$

will be relevant for our study. The importance of Λ lies on the fact that it satisfies $(\frac{1}{2}\partial_{xx} - \alpha\partial_x)\Lambda = 0$, $\Lambda(x) > 0$, for $x > 0$, and $\Lambda(0) = 0$. This implies that $\Lambda(X)$ is a nonnegative local martingale. Λ is called the scale function of X. The other important piece of information is the measure defined as

$$d\mu(y) := e^{-\gamma(y)}\,dy.$$

We notice that this measure is not necessarily finite and μ is half the speed measure of X. The unbounded operator $\mathcal{L}$ is formally self adjoint in $L^2(d\mu)$, which here means that for all $f, g \in C_0^2((0,\infty))$ we have

$$\int \mathcal{L}(f)g\,d\mu = \int f\mathcal{L}(g)\,d\mu.$$

On the other hand, we point out that $\mathcal{L}^*$ acts naturally on $L^2(d\Lambda)$ and is formally self adjoint there, that is, for all $f, g \in \mathcal{C}_0^2((0,\infty))$ we have[1]

$$\int \mathcal{L}^*(f) g \, d\Lambda = \int f \mathcal{L}^*(g) \, d\Lambda.$$

Let us introduce some of the elements we need for the spectral study of $\mathcal{L}$. Consider the eigenfunction φ_λ, the solution of $\mathcal{L}^*\varphi_\lambda = -\lambda\varphi_\lambda$, with the boundary conditions $\varphi_\lambda(0) = 0$, $\varphi_\lambda'(0) = 1$. Also, ψ_λ denotes the solution of $\mathcal{L}\psi_\lambda = -\lambda\psi_\lambda$, with the boundary conditions $\psi_\lambda(0) = 0$, $\psi_\lambda'(0) = 1$. Since α is regular at 0, the $\mathcal{C}^2$ functions with compact support contained in $[0,\infty)$, vanishing at 0, belong to the domain of $\mathcal{L}$. This is the main reason why we impose the boundary condition $\psi_\lambda(0) = 0$. The other condition $\psi_\lambda'(0) = 1$ is imposed for a normalization of the eigenfunctions (see Coddington and Levinson 1955, p. 226). The spectral properties of $\mathcal{L}^*$ and $\mathcal{L}$ are related by the equality $\varphi_\lambda = e^{-\gamma}\psi_\lambda$. We point out that ψ_λ and ψ_λ' satisfy the equalities

$$\begin{aligned}\psi_\lambda'(x) &= e^{\gamma(x)}\left(1 - 2\lambda \int_0^x \psi_\lambda(z) e^{-\gamma(z)} \, dz\right) = e^{\gamma(x)}\left(1 - 2\lambda \int_0^x \varphi_\lambda(z) \, dz\right),\\ \psi_\lambda(x) &= \int_0^x e^{\gamma(y)}\left(1 - 2\lambda \int_0^y \varphi_\lambda(z) \, dz\right) dy.\end{aligned} \tag{6.3}$$

Then $\psi_\lambda(x)$, $\psi_\lambda'(x)$, $\psi_\lambda''(x)$ are continuous functions of (x,λ).

We shall use the theory developed in Coddington and Levinson (1955). For that purpose, we introduce the operator $\mathcal{A}u = \frac{1}{2}\partial_{xx}u + \frac{1}{2}(\partial_x\alpha - \alpha^2)u$. We denote by $u(x,\lambda)$ the solution to the problem $\mathcal{A}u = -\lambda u$ with $u(0) = 0$ and $u'(0) = 1$. These functions are related to the eigenfunctions of $\mathcal{L}$ by the equality

$$\psi_\lambda(x) = u(x,\lambda) e^{\frac{1}{2}\gamma(x)}.$$

Notice that $\mathcal{A}$ is an unbounded self adjoint operator on $L^2(dx)$.

The main hypothesis of this chapter, denoted by **H**, is

$$\mathbf{H}: \quad \begin{cases} J1 = \int_0^\infty e^{\gamma(x)} (\int_0^x e^{-\gamma(z)} \, dz) \, dx = \infty, \\ J2 = \int_0^\infty e^{-\gamma(x)} (\int_0^x e^{\gamma(z)} \, dz) \, dx = \infty. \end{cases} \tag{6.4}$$

According to Feller's classification (see Feller 1952), this means that ∞ is a natural boundary for the process. Under Hypothesis **H**, we have (see Azencott 1974)

[1] Given a measure ρ, the integral $\int_a^b f(x) \, d\rho(x)$ is interpreted as the integral over the closed interval $[a,b]$, unless a or b are $\pm\infty$ in which case they are excluded from the integral. When it is necessary to exclude, for example, the point a, we shall use the notation $\int_{(a,b]} f(x) \, d\rho(x)$.

for any $s > 0$

$$\lim_{x\to\infty} \mathbb{P}_x(s < T) = 1,$$
$$\lim_{M\to\infty} \mathbb{P}_x(T_M < s) = 0, \tag{6.5}$$

which means that the process can neither explode from, nor explode to, infinity in finite time. On the other hand, Itô's formula gives

$$\begin{aligned} \Lambda(x) &= \mathbb{E}_x\big(\Lambda(X_{T\wedge T_M\wedge s})\big) \\ &= \mathbb{E}_x\big(\Lambda(M), T_M < T\wedge s\big) + \mathbb{E}_x\big(\Lambda(X_s), s < T_M \wedge T\big) \\ &\geq \mathbb{E}_x\big(\Lambda(X_s), s < T_M \wedge T\big). \end{aligned} \tag{6.6}$$

Then the Monotone Convergence Theorem implies that $\mathbb{E}_x(\Lambda(X_s), s < T) \leq \Lambda(x)$. Similarly, we have for $x \in (0, M)$

$$\mathbb{P}_x(T_M < T) = \frac{\Lambda(x)}{\Lambda(M)}. \tag{6.7}$$

In particular, we have, using (6.5),

$$\mathbb{P}_x(T = \infty) = \frac{\Lambda(x)}{\Lambda(\infty)}. \tag{6.8}$$

We introduce the second hypothesis we use in most of our results

$$\textbf{H1}: \quad \forall x > 0 \quad \mathbb{P}_x(T = \infty) = 0. \tag{6.9}$$

This hypothesis is equivalent to $\Lambda(\infty) = \infty$ (recall that we assume the drift is regular near 0).

Under Hypotheses **H**, the operator $\mathcal{A}$ is of the limit point type at infinity. This means that for every (or equivalently, for some) complex value ζ the ODE $\mathcal{A}u = \zeta u$ has a solution which is not square integrable (see Coddington and Levinson 1955, Theorem 2.1).

To see that $\mathcal{A}$ is of the limit point type, it is enough to consider $\zeta = 0$. The function $u_0 = \Lambda e^{-\gamma/2}$ satisfies $\mathcal{A}u_0 = 0$ and is not in $L^2(dx)$ as the following inequality shows

$$\int_0^\infty e^{-\gamma(x)} \Lambda^2(x)\, dx \geq \Lambda(z) \int_z^\infty e^{-\gamma(x)} \Lambda(x)\, dx = \infty.$$

Here we have used that Λ is increasing and positive on $(0, \infty)$ and $J2 = \infty$.

The study of general one-dimensional diffusions can be reduced, under suitable conditions on the diffusion coefficient, to the previous setting. In fact, consider the solution of

$$dY_t = \sigma(Y_t)dB_t - h(Y_t)\, dt, \qquad Y_0 = y,$$

where we assume $\sigma > 0$, $\sigma \in \mathcal{C}^1(\mathbb{R})$, $\int_0^\infty \frac{1}{\sigma(u)}\,du = \infty$. Take $F(y) = \int_0^y \frac{1}{\sigma(u)}\,du$, then $F \in \mathcal{C}^2(\mathbb{R})$ and F^{-1} exists. By Itô's formula, we obtain

$$F(Y_t) = F(y) + B_t - \int_0^t \alpha\big(F(Y_s)\big)\,ds,$$

where $\alpha(x) = (\frac{h}{\sigma} + \frac{1}{2}\sigma') \circ F^{-1}(x)$. If we consider the process $X_t = F(Y_t)$, $X_0 = x = F(y)$ then

$$dX_t = dB_t - \alpha(X_t)\,dt.$$

Since the hitting time of 0 is the same for both processes, $T_0^X = T^Y$, we get

$$\mathbb{P}\big(X \in A, t < T_0^X, X_0 = x\big) = \mathbb{P}\big(Y \in F^{-1}(A), t < T_0^Y, Y_0 = y\big).$$

Therefore, for our purposes it is enough to study the semigroup associated with the process (X_t).

The multidimensional analogues of the results we are going to consider (see mainly Theorem 6.26) have been studied by Pinsky (see Pinsky 1985) in bounded regions. The methods he used are based upon the theory of Stroock and Varadhan about the martingale problem. We remark that for unbounded regions, as occurs in our case where the domain is $(0, \infty)$, the Hypotheses 2 and 3 of Pinsky do not hold, in general. This is due to the fact that the spectrum of $\mathcal{L}$ is not necessarily discrete. This is the case, for instance, if the drift α is a positive constant.

6.1.1 QSDs on a Bounded Interval

In this section, we consider the diffusion X living inside the interval $(0, b)$ and the process killed at 0 and b. In this situation, both ends 0 and b are regular points for the diffusion X. We will see that the killed semigroup

$$P_t^{(b)} f(x) = \mathbb{E}_x\big(f(X_t), t < T \wedge T_b\big)$$

has a discrete spectrum.

Using Girsanov's Theorem, we conclude that $P_t^{(b)}$ has a kernel with respect to the Lebesgue measure given by (see Theorem 3.2.7 of Royer 1999 for a similar calculation)

$$p^{(b)}(t, x, y)\,dy = e^{\frac{1}{2}(\gamma(x) - \gamma(y))}\mathbb{E}_x\big(e^{-\int_0^t \wp(B_s)\,ds}, B_t \in dy, t < T \wedge T_b\big),$$

where $\wp = \frac{1}{2}(\alpha^2 - \alpha')$. Thus, $P_t^{(b)}$ has a symmetric kernel with respect to μ given by

$$r_t^{(b)}(x, y) = e^{\frac{1}{2}(\gamma(x) + \gamma(y))}\mathbb{E}_x\big(e^{-\int_0^t \wp(B_s)\,ds}, t < T \wedge T_b | B_t = y\big)\frac{e^{-\frac{(y-x)^2}{2t}}}{\sqrt{2\pi t}}.$$

We point out that for each $t > 0$ and $x \in [0, b]$ the function $r_t^{(b)}(x, y)$ is well defined dy-a.s. If we define $C = C(b) = \max\{-\frac{1}{2}(\alpha^2(x) - \alpha'(x)) : 0 \le x \le b\}$, we obtain the following Gaussian bound for $r^{(b)}$

$$r_t^{(b)}(x, y) \le e^{\frac{1}{2}(\gamma(x)+\gamma(y)+Ct)} \frac{e^{-\frac{(y-x)^2}{2t}}}{\sqrt{2\pi t}},$$

and the following uniform bound

$$r_t^{(b)}(x, y) \le \kappa(t, b) = \max_{x,y\in[0,b]^2} e^{\frac{1}{2}(\gamma(x)+\gamma(y)+Ct)} \frac{e^{-\frac{(y-x)^2}{2t}}}{\sqrt{2\pi t}}. \tag{6.10}$$

We shall show later that $r^{(b)}$ has a smooth version.

On the other hand, in $L^2([0, b], dx)$ there is a complete orthonormal basis $\theta_{n,b}$ of eigenfunctions for the operator $\mathcal{A}$ with Dirichlet boundary conditions at $x = 0$ and $x = b$ (see Coddington and Levinson 1955, Chap. 7). Thus, there exists a sequence of numbers $r_{n,b} > 0$ and eigenvalues $\lambda_{n,b}$ such that $\theta_{n,b}(x) = r_{n,b} u(x, \lambda_{n,b})$. The completeness theorem says that

$$\int_0^b |f(x)|^2 \, dx = \sum_{n=1}^{\infty} |r_{n,b}|^2 \left| \int_0^b f(x) u(x, \lambda_{n,b}) \, dx \right|^2. \tag{6.11}$$

The mass distribution Γ_b associated to $[0, b]$ is given by the distribution concentrated on $\{\lambda_{n,b} : n \ge 1\}$ with weight $|r_{n,b}|^2$ at $\lambda_{n,b}$, thus

$$\int_0^b |f(x)|^2 \, dx = \int |g(\lambda)|^2 \Gamma_b(d\lambda)$$

where

$$g(\lambda) = \int_0^b f(x) u(x, \lambda_{n,b}) \, dx.$$

We notice that each $u(\bullet, \lambda_{n,b})$ is zero at $x = 0$ and $x = b$, which means that $u(\bullet, \lambda_{n,b})$ changes sign on $\mathbb{R}_+$ (being a solution of $\mathcal{A}u = -\lambda_{n,b} u$ it cannot satisfy $u'(b) = 0$ because in this case it is the zero function). This shows that the support of Γ_b is contained in $J = \{\lambda : u(\bullet, \lambda)$ changes sign in $\mathbb{R}_+\}$. The basic oscillatory theorem for second-order linear ODE (see Coddington and Levinson 1955, Theorem 1.1) allows us to conclude that J is an interval. Since $u(\bullet, 0)$ is nonnegative (actually, positive in $(0, \infty)$), we conclude that $J \subset (0, \infty)$. We also have the following important property.

Lemma 6.1 *Each $\lambda_{n,b}$ is simple and the only possible accumulation point of the set $\{\lambda_{n,b} : n \ge 1\}$ is ∞.*

Proof Assuming that $\lambda_{n,b} = \lambda_{k,b}$, for $n \neq k$, we conclude that $u(\bullet, \lambda_{n,b})$ and $u(\bullet, \lambda_{k,b})$ are equal because they satisfy the same second order differential equation with the same boundary conditions at 0. Since they are orthogonal, we arrive at a contradiction.

The function of λ given by $u(b, \lambda)$ is analytic and not constant, which implies that there cannot exist a sequence (λ_m) which accumulates on a finite point $\lambda \in \mathbb{C}$ for which $u(b, \lambda_m) = 0$. □

This result shows that we can assume without loss of generality that

$$0 < \lambda_{1,b} < \lambda_{2,b} < \cdots < \lambda_{n,b} < \lambda_{n+1,b} < \cdots.$$

Given the relation between the eigenfunctions of $\mathcal{A}$ and $\mathcal{L}$, we can translate the previous discussion as follows. The linear operator Ψ_b defined as

$$g(\lambda) = \Psi_b(h)(\lambda) = \int_0^b h(x)\psi_\lambda(x)\,d\mu(x)$$

is an isomorphism between $L^2([0, b], d\mu)$ and $L^2(d\Gamma_b)$. Its inverse is given by $\Psi_b^{-1}(g)(x) = \int g(\lambda)\psi_\lambda(x)\,d\Gamma_b(\lambda)$. This is obtained by replacing $h = fe^{1/2\gamma}$ and $d\mu = e^{-\gamma}\,dx$. We also get

$$\int_0^b |h(x)|^2\,d\mu(x) = \int |g(\lambda)|^2\,d\Gamma_b(\lambda).$$

Consider $\ell \in C_0^2((0, b))$ and take $h = \mathcal{L}\ell$. Then integration by parts shows that

$$\int \mathcal{L}\ell(x)\psi_\lambda(x)\,d\mu(x) = \int \ell(x)\mathcal{L}\psi_\lambda(x)\,d\mu(x) = -\lambda \int \ell(x)\psi_\lambda(x)\,d\mu(x).$$

This equality for any ℓ implies that $\Psi_b\mathcal{L} = \mathcal{K}\Psi_b$ on $C_0^2((0, b))$, where $\mathcal{K}$ is the (unbounded) operator of multiplication by $-\lambda$ in $L^2(d\Gamma_b)$. Then $\bar{\mathcal{L}}^b$, the minimal closure of $\mathcal{L}^b$, has the domain given by the functions $h \in L^2([0, b], d\mu)$ such that $-\lambda\Psi_b(h)(\lambda) \in L^2(d\Gamma_b)$. This minimal closure satisfies $\bar{\mathcal{L}}^b = \Psi_b^{-1}\mathcal{K}\Psi_b$. Also the resolvent of $\bar{\mathcal{L}}^b$ can be obtained from this isomorphism Ψ_b. For $\zeta \in \mathbb{C}$ such that $\Re\zeta > 0$, we get

$$\Psi_b\big(\zeta I - \bar{\mathcal{L}}^b\big)^{-1}(h)(\lambda) = \frac{1}{\zeta + \lambda}\Psi_b(h)(\lambda).$$

In particular, for every $n \geq 1$ and $h \in L^2([0, b], d\mu)$, we have

$$\begin{aligned}\int_0^\infty \big|\big(I - 1/n\bar{\mathcal{L}}^b\big)^{-1}h(x)\big|^2\,d\mu(x) &= \int_0^\infty (1 + \lambda/n)^{-2}|g(\lambda)|^2\,d\Gamma_b(\lambda)\\ &\leq \int_0^\infty |g(\lambda)|^2\,d\Gamma_b(\lambda) = \int_0^\infty |h(x)|^2\,d\mu(x).\end{aligned}$$

This implies that $\bar{\mathcal{L}}^b$ is the generator of a uniquely determined contraction semigroup $(T_t^{(b)})$ of class $\mathcal{C}_0$ in $L^2([0,b], d\mu)$ (see, for example, Corollary 7.1, Sect. IX in Yosida 1980), and moreover,

$$\Psi_b T_t^{(b)}(h)(\lambda) = e^{-\lambda t}\Psi_b(h)(\lambda).$$

This gives the formula

$$T_t^{(b)}h(x) = \int e^{-\lambda t}\psi_\lambda(x)\int_0^b h(y)\psi_\lambda(y)\,d\mu(y)\,d\Gamma_b(\lambda)$$

$$= \sum_n |r_{n,b}|^2 e^{-\lambda_{n,b}t}\int_0^b h(y)\psi_{\lambda_{n,b}}(y)\,d\mu(y)\psi_{\lambda_{n,b}}(x).$$

According to Theorem 4.1, Chap. 7 in Coddington and Levinson (1955), this series converges uniformly in $x \in [0,b]$, $t \in \mathbb{R}_+$. We now prove that this semigroup is the one associated to the killed process $(X_{t\wedge T\wedge T_b})$.

Proposition 6.2 *For all t, all $x \in [0,b]$ and all $h \in L^2([0,b], d\mu)$,*

$$\mathbb{E}_x\big(h(X_t), t < T \wedge T_b\big) = T_t^{(b)}h(x).$$

We also have the representation (that holds for all $t > 0$ and almost surely in x, y)

$$r_t^{(b)}(x,y) = \sum_n |r_{n,b}|^2 e^{-\lambda_{n,b}t}\psi_{\lambda_{n,b}}(x)\psi_{\lambda_{n,b}}(y), \tag{6.12}$$

where the series converges uniformly on $[\epsilon, \infty) \times [0,b]^2$, for every $\epsilon > 0$. This representation gives a continuous version of the transition kernel $r^{(b)}$ on $(0,\infty) \times [0,b]^2$.

Proof Consider a compactly supported function $g(\lambda)$ and define

$$h(x) = \int g(\lambda)\psi_\lambda(x)\,d\Gamma_b(\lambda) = \sum_n |r_{n,b}|^2 g(\lambda_{n,b})\psi_{\lambda_{n,b}}(x),$$

$$u(t,x) = T_t^{(b)}h(x) = \sum_n |r_{n,b}|^2 e^{-\lambda_{n,b}t} g(\lambda_{n,b})\psi_{\lambda_{n,b}}(x).$$

Recall that both series are actually a finite sum. It is straightforward to prove that $u \in \mathcal{C}^{1,2}$ solves the differential problem

$$\partial_t u = \mathcal{L}u,$$
$$u(t,0) = u(t,b) = 0,$$
$$u(0,x) = h(x).$$

Applying Itô's formula to the function $(s, x) \to u(t-s, x)$, we obtain for $x \le b, s \le t$

$$u(t,x) = \mathbb{E}_x\big(u(t - s \wedge T \wedge T_b, X_{s \wedge T \wedge T_b})\big) = \mathbb{E}_x\big(u(t-s, X_s), s < T \wedge T_b\big).$$

If we take $s = t$, we conclude that

$$T_t^{(b)} h(x) = u(t,x) = \mathbb{E}_x\big(h(X_t), t < T \wedge T_b\big) = P_t^{(b)} h(x).$$

A density argument shows simultaneously that $(P_t^{(b)})$ is a contraction semigroup on $L^2([0,b], d\mu)$ and that $(T_t^{(b)}) = (P_t^{(b)})$.

Consider now a nonnegative continuous function supported on $[0, b]$ and compute

$$\begin{aligned}
&\int \mathbb{E}_x\big(h(X_t), t < T \wedge T_b\big) h(x)\, d\mu(x) \\
&= \iint h(x) h(y) r_t^{(b)}(x,y)\, d\mu(y)\, d\mu(x) \\
&= \int T_t^{(b)} h(x) h(x)\, d\mu(x) \\
&= \sum_n |r_{n,b}|^2 e^{-\lambda_{n,b} t} \left(\int h(y) \psi_{\lambda_{n,b}}(y)\, d\mu(y) \right)^2.
\end{aligned}$$

For $t > 0$ the density $r_t^{(b)}(x, y)$ is bounded (see (6.10)), which gives the inequality

$$\sum_{n=1}^{l} |r_{n,b}|^2 e^{-\lambda_{n,b} t} \left(\int h(y) \psi_{\lambda_{n,b}}(y)\, d\mu(y) \right)^2 \le \kappa \left(\int h(y)\, d\mu(y) \right)^2,$$

for all $l \ge 1$, all h ($\kappa = \kappa(t, b)$ is given in (6.10)). If we now take a sequence (h_m) converging to the Dirac mass at $z \in [0, b]$, we conclude that

$$\sum_{n=1}^{l} |r_{n,b}|^2 e^{-\lambda_{n,b} t} \big(\psi_{\lambda_{n,b}}(z)\big)^2 \le \kappa,$$

and therefore, the series

$$\sum_n |r_{n,b}|^2 e^{-\lambda_{n,b} t} \big(\psi_{\lambda_{n,b}}(z)\big)^2$$

is convergent and bounded by κ. The Cauchy–Schwarz Inequality shows that the series

$$\zeta(t, x, y) = \sum_n |r_{n,b}|^2 e^{-\lambda_{n,b} t} \psi_{\lambda_{n,b}}(y) \psi_{\lambda_{n,b}}(x)$$

is also absolutely convergent for any $x, y \in [0, b]$ and bounded by κ. Now, we show the series is uniformly convergent in t, x, y.

In fact, the equality $\psi'_\lambda(x) = e^{\gamma(x)}(1 - 2\lambda \int_0^x \psi_\lambda(z)e^{-\gamma(z)}\,dz)$ implies for $\lambda \geq 0$ that $|\psi'_\lambda(x)| \leq e^{\gamma(x)}(1 + 2\lambda \int_0^x |\psi_\lambda(z)|e^{-\gamma(z)}\,dz)$. Hence, there exists a constant $K = K(b)$ such that, for all $\lambda \geq 0, x \in [0, b]$,

$$\left|\psi'_\lambda(x)\right| \leq K\left(1 + 2\lambda \int_0^b \left|\psi_\lambda(z)\right| dz\right). \tag{6.13}$$

We also get

$$\left|\psi_\lambda(x)\right| \leq Kb\left(1 + 2\lambda \int_0^b \left|\psi_\lambda(z)\right| dz\right).$$

Using the inequality $(a+b)^2 \leq 2(a^2+b^2)$, we obtain the following uniform domination, for $t \geq t_0 > 0, x \in [0, b]$,

$$\begin{aligned}\zeta(t,x,x) &= \sum_n |r_{n,b}|^2 e^{-\lambda_{n,b}t}\left(\psi_{\lambda_{n,b}}(x)\right)^2 \\ &\leq 2K^2b^2 \sum_n |r_{n,b}|^2 e^{-\lambda_{n,b}t}\left(1 + 4\lambda_{n,b}^2\left(\int_0^b \left|\psi_{\lambda_{n,b}}(z)\right| dz\right)^2\right).\end{aligned}$$

Let us define

$$G(t_0) = \max_{\lambda\geq 0}\left\{\left(1+\lambda^2\right)e^{-\lambda t_0}\right\} < \infty, \qquad H(t_0) = 4\max_{\lambda \geq 0}\left\{\lambda^2 e^{-\lambda t_0/2}\right\} < \infty.$$

Hence,

$$\begin{aligned}\zeta(t,x,x) &\leq 2K^2b^2G(t_0)\sum_n \frac{|r_{n,b}|^2}{1+\lambda_{n,b}^2} \\ &\quad + 2K^2b^2H(t_0)\int_0^b\int_0^b \sum_n |r_{n,b}|^2 e^{-\lambda_{n,b}t_0/2}\left|\psi_{\lambda_{n,b}}(z)\right|\left|\psi_{\lambda_{n,b}}(z')\right| dz\,dz' \\ &\leq 2K^2b^2\left[G(t_0)\sum_n \frac{|r_{n,b}|^2}{1+\lambda_{n,b}^2} + H(t_0)b^2\kappa(t_0/2, b)\right] < \infty.\end{aligned}$$

This last quantity is finite because, according to Coddington and Levinson (1955), the measure Γ_b satisfies (where M is independent of b)

$$\sum_n \frac{1}{\lambda_{n,b}^2+1}|r_{n,b}|^2 \leq M < \infty.$$

This argument and Cauchy–Schwarz Inequality show the uniform convergence of the series defining ζ on $[t_0, \infty) \times [0, b]^2$. Therefore, this function is continuous in the three variables on $(0, \infty) \times [0, b]^2$.

We shall prove that for all $t > 0$ $r_t^{(b)}(x, y) = \zeta(t, x, y)\,dx\,dy$-a.s. For that purpose, we consider two continuous functions f and h. We get from the Dominated Convergence Theorem

$$\begin{aligned}
&\iint f(x)h(y)\zeta(t, x, y)\,d\mu(x)\,d\mu(y) \\
&= \sum_n |r_{n,b}|^2 e^{-\lambda_{n,b}t} \int h(y)\psi_{\lambda_{n,b}}(y)\,d\mu(y) \int \psi_{\lambda_{n,b}}(x)\,d\mu(x) \\
&= \iint f(x)h(y)r_t^{(b)}(x, y)\,d\mu(x)\,d\mu(y),
\end{aligned}$$

implying that $\zeta(t, x, y) = r_t^{(b)}(x, y)$ holds $d\mu(x)\,d\mu(y)$-a.s. □

In what follows, we use the continuous version of $r^{(b)}$ given by the series (6.12).

Proposition 6.3

(i) $r^{(b)} \in \mathcal{C}^{1,2}((0, \infty), [0, b]^2)$ *and*

$$\begin{aligned}
\partial_t r_t^{(b)}(x, y) &= \mathcal{L}_x r_t^{(b)}(x, y) = \mathcal{L}_y r_t^{(b)}(x, y), \\
r_t^{(b)}(x, 0) &= r_t^{(b)}(x, b) = 0, \\
r_0^{(b)}(x, \bullet)\,d\mu &= \delta_x(\bullet).
\end{aligned}$$

On the other hand, $r_t^{(b)}(x, x)$, *as a function of* t, *is decreasing for every fixed* x.

(ii) (*Harnack*) *Consider* $0 < \delta < 1$ *and* $x_0 > 0$ *for which* $3\delta \le x_0$. *Then, there exists a constant* $C = C(x_0)$ *which depends only on the coefficients of* $\mathcal{L}$ *and not on* b, *such that for all* $5\delta^2 < t$, $|x - x_0| < \delta$, $b > x_0 + 3\delta$ *and* $0 \le y \le b$,

$$r_t^{(b)}(x, y) \le C r_{t+\delta^2}^{(b)}(x_0, y).$$

Moreover, since $r_t^{(b)}(\bullet, \bullet)$ *is symmetric, we have the following extension. Let* $0 < \delta < 1$ *and* $x_0 > 0$, $y_0 > 0$ *for which* $3\delta \le (x_0 \wedge y_0)$. *Then, there exists a constant* $C = C(x_0, y_0)$ *which again depends only on the coefficients of* $\mathcal{L}$ *and not on* b, *such that for all* $5\delta^2 < t$, $|x - x_0| < \delta$, $|y - y_0| < \delta$ *and every* $b > x_0 \vee y_0 + 3\delta$,

$$r_t^{(b)}(x, y) \le C r_{t+2\delta^2}^{(b)}(x_0, y_0).$$

(iii) (*Uniform Hölder continuity*) *Consider* $0 < \delta < 1$ *and again* $x_0 > 0$, $y_0 > 0$ *with* $3\delta \le (x_0 \wedge y_0)$. *Take also* $t_0 > 5\delta^2$ *and define*

$$\begin{aligned}
A_1 &= \big\{(t, x, y) : t_0 - \delta^2/4 \le t \le t_0, \big\|(x, y) - (x_0, y_0)\big\| \le \delta/2\big\}, \\
A_2 &= \big\{(t, x, y) : t_0 - \delta^2 \le t \le t_0, \big\|(x, y) - (x_0, y_0)\big\| \le \delta\big\}.
\end{aligned}$$

Then, there exist $C = C(x_0, y_0)$, $\beta = \beta(x_0, y_0)$, independent of b, such that for all $b \ge x_0 \vee y_0 + 3\delta$ and all $\xi = (t, x, y)$, $\zeta = (s, u, v) \in A_1$ we have

$$\begin{aligned}&\left|r_t^{(b)}(x,y) - r_s^{(b)}(u,v)\right|\\&\quad \le C\delta^{-\beta}\left(\left\|(x,y)-(u,v)\right\| + |t-s|^{\frac{1}{2}}\right)^{\beta} \sup\left\{r_{t'}^{(b)}\left(x',y'\right) : \left(t',x',y'\right) \in A_2\right\}.\end{aligned}$$

Proof We recall that $r_t^{(b)}(x,y) = \sum_n |r_{n,b}|^2 e^{-\lambda_{n,b}t}\psi_{\lambda_{n,b}}(x)\psi_{\lambda_{n,b}}(y)$. We shall differentiate this series term by term. For that purpose, we use the bound $|\psi_\lambda'(x)| \le K(1 + 2\lambda\int_0^b |\psi_\lambda(z)|\,dz)$, see (6.13). Hence we get the domination, for $t \ge t_0 > 0$,

$$\begin{aligned}&\sum_n |r_{n,b}|^2 e^{-\lambda_{n,b}t}\left|\psi_{\lambda_{n,b}}(y)\right|\left|\psi_{\lambda_{n,b}}'(x)\right|\\&\quad \le K\sum_n |r_{n,b}|^2 e^{-\lambda_{n,b}t}\left|\psi_{\lambda_{n,b}}(y)\right|\left(1 + 2\lambda_{n,b}\int_0^b \left|\psi_{\lambda_{n,b}}(z)\right| dz\right)\\&\quad \le K\Bigg(\sum_n |r_{n,b}|^2 e^{-\lambda_{n,b}t}\left|\psi_{\lambda_{n,b}}(y)\right|\\&\qquad + 2\int_0^b \sum_n |r_{n,b}|^2 e^{-\lambda_{n,b}t}\lambda_{n,b}\left|\psi_{\lambda_{n,b}}(y)\right|\left|\psi_{\lambda_{n,b}}(z)\right| dz\Bigg)\\&\quad \le K\left[\kappa(t_0/2)\right]^{\frac{1}{2}}\left[\sum_n |r_{n,b}|^2 e^{-\lambda_{n,b}t_0}\right]^{\frac{1}{2}} + 2Kb\sup_{\lambda\ge 0}\left\{\lambda e^{-\lambda t_0/2}\right\}\kappa(t_0/2) < \infty.\end{aligned}$$

This last term is finite again because the measure Γ_b satisfies

$$\sum_n \frac{1}{\lambda_{n,b}^2 + 1}|r_{n,b}|^2 \le M < \infty.$$

This allows us to differentiate term by term to get

$$\partial_x r_t^{(b)}(x,y) = \sum_n |r_{n,b}|^2 e^{-\lambda_{n,b}t}\partial_x\psi_{\lambda_{n,b}}(x)\psi_{\lambda_{n,b}}(y).$$

For the second derivative with respect to x, we notice that $\psi_\lambda'' = 2\alpha\psi_\lambda' - 2\lambda\psi_\lambda$.

Similar arguments show that $r^{(b)} \in \mathcal{C}^{1,2}((0,\infty),[0,b]^2)$ and $\partial_t r_t^{(b)}(x,y) = \mathcal{L}_x r_t^{(b)}(x,y) = \mathcal{L}_y r_t^{(b)}(x,y)$. The boundary conditions

$$r_t^{(b)}(x,0) = r_t^{(b)}(x,b) = 0$$

are obtained immediately. For all continuous function h and all $x \in (0,b)$, we have, as $t \to 0$,

$$\int h(y) r_t^{(b)}(x,y)\,d\mu(y) = \mathbb{E}_x\left(h(X_t), t < T \wedge T_b\right) \to h(x),$$

since the trajectories of X are continuous. This shows part (i).

Parts (ii) and (iii) are deduced from the standard Harnack's Inequality for a nonnegative solution of a parabolic differential equation. The reader may consult Karatzas and Shreve (1988, Theorem 1.1 and Lemma 4.1), where we have taken $R = 2\delta$ and $\theta = 5/4$. □

These two propositions are the basis to show the main result of this section. We denote by $\underline{\lambda}_b = \lambda_{1,b}$ and $\overline{\psi} = r_{1,b}\psi_{\lambda_{1,b}}$. The function $\overline{\psi}$ is the unique solution to $\mathcal{L}\psi = -\underline{\lambda}_b\psi$, $\psi(0) = \psi(b) = 0$, $\int_0^b \psi^2\, d\mu = 1$, $\psi'(0) > 0$.

Theorem 6.4

(i) *Uniformly in* x, y *the following holds*:

$$\lim_{t\to\infty} e^{\underline{\lambda}_b t} r_t^{(b)}(x, y) = \overline{\psi}(x)\overline{\psi}(y).$$

In particular, $\overline{\psi}$ *is nonnegative in* $(0, b)$, *which implies it is positive there.*

(ii) *For all* $h \in L^2([0, b], d\mu)$ *and uniformly in* x,

$$\lim_{t\to\infty} e^{\underline{\lambda}_b t}\mathbb{E}_x\big(h(X_t), t < T \wedge T_b\big) = \overline{\psi}(x)\int h(y)\overline{\psi}(y)\, d\mu(y).$$

In particular, by taking $h \equiv 1$, *we get*

$$\lim_{t\to\infty} e^{\underline{\lambda}_b t}\mathbb{P}_x(t < T \wedge T_b) = \overline{\psi}(x)\int \overline{\psi}(y)\, d\mu(y).$$

(iii) *The Yaglom limit holds*: *for all bounded and measurable functions* h *and uniformly in* x *on compact sets in* $(0, b)$, *we have the limit*

$$\lim_{t\to\infty} \mathbb{E}_x\big(h(X_t)|t < T \wedge T_b\big) = \int h(y)\, d\nu(y),$$

where ν *is the QSD given by the density* $\frac{\overline{\psi}(z)e^{-\gamma(z)}}{\int \overline{\psi}(y)e^{-\gamma(y)}\, dy}$. *Recall that* $\varphi_{\lambda_{1,b}}(z) = \psi_{\lambda_{1,b}}(z)e^{-\gamma(z)}$ *which yields that the Yaglom limit is the probability measure whose density (with respect to the Lebesgue measure) is proportional to* $\varphi_{\lambda_{1,b}}$.

(iv) *Let* ξ *be any probability measure supported on* $(0, b)$. *Then for all* $h \in L^2([0, b], d\mu)$,

$$\lim_{t\to\infty} \mathbb{E}_\xi\big(h(X_t)|t < T \wedge T_b\big) = \int h(y)\, d\nu(y).$$

(v) *The following ratio limits hold*

$$\lim_{t\to\infty} \frac{r_{t+s}^{(b)}(x, y)}{r_t^{(b)}(z, w)} = e^{-s\underline{\lambda}_b}\frac{\overline{\psi}(x)\overline{\psi}(y)}{\overline{\psi}(z)\overline{\psi}(w)}$$

uniformly on compacts in $[0, \infty) \times (0, b)^4$ *and*

$$\lim_{t\to\infty} \frac{\mathbb{P}_x(t < T \wedge T_b)}{\mathbb{P}_y(t < T \wedge T_b)} = \frac{\overline{\psi}(x)}{\overline{\psi}(y)}$$

uniformly on compacts in $(0, b)^2$.

Proof The only point that requires some detail is (iv). The result is clear when ξ has a density with respect to μ that belongs to $L^2(d\mu)$. Nevertheless, this is the general case since

$$\mathbb{E}_\xi\big(h(X_{t+1}) | t + 1 < T \wedge T_b\big) = \mathbb{E}_\omega\big(h(X_t) | t < T \wedge T_b\big),$$

where $d\omega(x) = C^{-1} \int r_1^{(b)}(z, x)\, d\xi(z)\, d\mu(x)$, where the normalizing constant $C = \iint r_1^{(b)}(z, y)\, d\xi(z)\, d\mu(y)$. It is straightforward to show that the density of ω with respect to μ is bounded. □

6.2 QSDs on $[0, \infty)$

We consider now the diffusion X on $[0, \infty)$ killed at 0. Under **H**, we have that ∞ is a natural boundary for the process. Then, according to Coddington and Levinson (1955, Theorem 3.1), there exists a distribution function $\Gamma(\lambda)$ over $\mathbb{R}$ such that

$$g(\lambda) = \mathcal{J} f(\lambda) = \int_0^\infty f(x) u(x, \lambda)\, dx \tag{6.14}$$

is an isomorphism from $L^2(dx)$ to $L^2(d\Gamma)$ and its inverse is given by

$$f(x) = \mathcal{J}^{-1} g(x) = \int_{\mathbb{R}} g(\lambda) u(x, \lambda)\, d\Gamma(\lambda). \tag{6.15}$$

Equality (6.14) is interpreted in the $L^2(d\Gamma)$ sense as follows:

$$\lim_{b\to\infty} \int_0^\infty \left| g(\lambda) - \int_0^b f(x) u(x, \lambda)\, dx \right|^2 d\Gamma(\lambda) = 0.$$

Similar interpretation holds for (6.15). We also know that the measure Γ satisfies the integrability condition

$$\int_{\mathbb{R}} \frac{1}{1 + \lambda^2}\, d\Gamma(\lambda) < \infty. \tag{6.16}$$

The measure Γ is an accumulation point of Γ_b, as $b \to \infty$. Actually, in the limit point case, Γ is the limit of Γ_b as $b \to \infty$. We also know that the support of Γ

is included in the adherence of the interval $J = \{\lambda : u(\bullet, \lambda) \text{ changes sign in } \mathbb{R}_+\} \subset \mathbb{R}_+$.

As in the finite interval case, the linear operator Ψ defined as

$$g(\lambda) = \Psi(h)(\lambda) = \int_0^\infty h(x)\psi_\lambda(x)\,d\mu(x)$$

is an isomorphism between $L^2(d\mu)$ and $L^2(d\Gamma)$, whose inverse is given by

$$\Psi^{-1}(g)(x) = \int_0^\infty g(\lambda)\psi_\lambda(x)\,d\Gamma(\lambda).$$

This is obtained again by replacing $h = fe^{1/2\gamma}$ and $d\mu = e^{-\gamma}\,dx$. We also get

$$\int_0^\infty |h(x)|^2\,d\mu(x) = \int_0^\infty |g(\lambda)|^2\,d\Gamma(\lambda).$$

Consider $\ell \in \mathcal{C}_0^2((0,\infty))$ and take $h = \mathcal{L}\ell$, then integration by parts shows that

$$\int_0^\infty \mathcal{L}\ell(x)\psi_\lambda(x)\,d\mu(x) = \int_0^\infty \ell(x)\mathcal{L}\psi_\lambda(x)\,d\mu(x) = -\lambda\int_0^\infty \ell(x)\psi_\lambda(x)\,d\mu(x).$$

And then $\Psi\mathcal{L} = \mathcal{K}\Psi$ holds on $\mathcal{C}_0^2$. Let $\bar{\mathcal{L}}$ be the minimal closure of $\mathcal{L}$. It is easy to prove that if $h \in L^2(d\mu)$ and h is in the domain of $\bar{\mathcal{L}}$ then $-\lambda\Psi(h)(\lambda) \in L^2(d\Gamma)$ and this minimal closure satisfies $\bar{\mathcal{L}} = \Psi^{-1}\bar{\mathcal{K}}\Psi$, where $\bar{\mathcal{K}}$ is the minimal closure of $(\mathcal{K}, \Psi(\mathcal{C}_0^2))$. It is easy to show that $\bar{\mathcal{K}}(g)(\lambda) = -\lambda g(\lambda)$. For $\ell \in \mathbb{C}$ such that $\Re\ell > 0$, we get $\Psi(\ell I - \bar{\mathcal{L}})^{-1}(h)(\lambda) = \frac{1}{\ell+\lambda}\Psi(h)(\lambda)$, and also for every $n \geq 1$ and $h \in L^2(d\mu)$ we have

$$\begin{aligned}\int_0^\infty \left|\left(I - \frac{1}{n}\bar{\mathcal{L}}\right)^{-1} h(x)\right|^2 d\mu(x) &= \int_0^\infty \left(1 + \frac{\lambda}{n}\right)^{-2} |g(\lambda)|^2\,d\Gamma(\lambda) \\ &\leq \int_0^\infty |g(\lambda)|^2\,d\Gamma(\lambda) = \int_0^\infty |h(x)|^2\,d\mu(x).\end{aligned}$$

This implies again that $\bar{\mathcal{L}}$ is the generator of a uniquely determined contraction semigroup (T_t) of class C_0 in $L^2(d\mu)$, and moreover $\Psi T_t(h)(\lambda) = e^{-\lambda t}\Psi(h)(\lambda)$.

This gives the formula

$$T_t h(x) = \int_0^\infty e^{-\lambda t}\psi_\lambda(x)\int_0^\infty h(y)\psi_\lambda(y)\,d\mu(y)\,d\Gamma(\lambda).$$

We prove this semigroup is the one associated to the killed process $(X_{t\wedge T})$.

Proposition 6.5 *The semigroup (P_t) has an extension to $L^2(d\mu)$ and that extension is (T_t).*

Proof We use the semigroup $(T_t^{(b)})$ associated to $\mathcal{L}$ with Dirichlet boundary conditions at $x = 0$ and $x = b$, which is given by

$$T_t^{(b)} h(x) = \int_0^\infty e^{-\lambda t} \psi_\lambda(x) \int_0^\infty h(y)\psi_\lambda(y)\, d\mu(y)\, d\Gamma_b(\lambda).$$

Here we assume that $h \in L^2([0, b], d\mu)$. Using what we have proved for the finite interval case, we have for all t, b, all $0 \le x \le b$ and all $h \in \mathcal{C}_0((0, b))$

$$\mathbb{E}_x\big(h(X_t), t < T \wedge T_b\big) = \int_0^\infty e^{-\lambda t} \psi_\lambda(x) \int_0^\infty h(y)\psi_\lambda(y)\, d\mu(y)\, d\Gamma_b(\lambda). \quad (6.17)$$

We divide the integral on the right side in two pieces:

$$I_{n,b} = \int_{[0,n]} e^{-\lambda t} \psi_\lambda(x) \int_0^\infty h(y)\psi_\lambda(y)\, d\mu(y)\, d\Gamma_b(\lambda),$$
$$J_{n,b} = \int_{(n,\infty)} e^{-\lambda t} \psi_\lambda(x) \int_0^\infty h(y)\psi_\lambda(y)\, d\mu(y)\, d\Gamma_b(\lambda).$$

The first integral converges pointwise (see Coddington and Levinson 1955) in t, x, when $b \to \infty$, to

$$I_n = \int_{[0,n]} e^{-\lambda t} \psi_\lambda(x) \int_0^\infty h(y)\psi_\lambda(y)\, d\mu(y)\, d\Gamma(\lambda).$$

On the other hand, the Dominated Convergence Theorem shows that the left-hand side of (6.17) converges pointwise in t, x, when $b \to \infty$, to

$$\mathbb{E}_x\big(h(X_t), t < T\big).$$

Notice that under **H** we have $T \wedge T_b \uparrow T$. This argument also shows that $J_{n,b}$ also converges pointwise in t, x, when $b \to \infty$, to a certain limit J_n. So far we have the equality

$$\mathbb{E}_x\big(h(X_t), t < T\big) = I_n + J_n.$$

The isometry between $L^2([0, b])$ and $L^2(d\Gamma_b)$ gives that

$$\begin{aligned}
\int_0^b J_{n,b}^2(z)\, d\mu(z) &= \int_{(n,\infty)} e^{-2\lambda t} \left(\int h(y)\psi_\lambda(y)\, d\mu(y) \right)^2 d\Gamma_b(\lambda) \\
&\le e^{-2nt} \int_0^\infty \left(\int h(y)\psi_\lambda(y)\, d\mu(y) \right)^2 d\Gamma_b(\lambda) \\
&= e^{-2nt} \int h^2(z)\, d\mu(z).
\end{aligned}$$

Therefore, from Fatou's Lemma, we conclude the inequality

$$\int J_n^2(z)\,d\mu(z) \le e^{-2nt}\int h^2(z)\,d\mu(z),$$

showing that J_n converges to 0 in $L^2(d\mu)$. Finally, I_n converges, in the same sense, to $T_t h$, which shows that

$$P_t h(x) = \mathbb{E}_x\big(h(X_t), t < T\big) = T_t h(x)$$

holds in the sense of $L^2(d\mu)$. Hence, (P_t) has an extension to $L^2(d\mu)$ and that extension coincides with (T_t). □

We denote by $\underline{\lambda}$ the infimum of the closed support for Γ. Clearly, we have that $\underline{\lambda} \ge 0$.

Definition 6.6 We define the quantity

$$\lambda^+ = \sup\big\{\lambda : \psi_\lambda(\bullet) \ge 0\big\}.$$

Since $\psi_0 = \Lambda$ is nonnegative, we conclude that $\lambda+ \ge 0$. In Mandl (1961), the following result is proved.

Lemma 6.7

$$\underline{\lambda} = \lambda^+.$$

In particular, $\psi_{\underline{\lambda}}(\bullet) \ge 0$.

Proof We know that $\lambda^+ \ge 0$ and $\psi_{\lambda^+} \ge 0$. On the other hand, the support of Γ is included in $\overline{\{\lambda : \psi_\lambda \text{ changes sign}\}}$, which allows us to conclude that $\lambda^+ \le \underline{\lambda}$. Then, it is enough to prove that $\psi_{\underline{\lambda}}$ has constant sign.

Assume that $\psi_{\underline{\lambda}}$ changes sign. Then there are $0 < x_0 < x_1 < x_2 < x_3 < \infty$ such that $\psi_{\underline{\lambda}} > 0$ on $I = (x_0, x_1)$ and $\psi_{\underline{\lambda}} < 0$ on $J = (x_2, x_3)$. By a continuity argument, we can find $\epsilon > 0$ such that

$$\begin{aligned} &\forall \underline{\lambda} \le \lambda \le \underline{\lambda} + \epsilon \quad \psi_\lambda(x) > 0 \quad \text{for } x \in (x_0, x_1), \\ &\forall \underline{\lambda} \le \lambda \le \underline{\lambda} + \epsilon \quad \psi_\lambda(x) < 0 \quad \text{for } x \in (x_2, x_3), \\ &\Gamma\big([\underline{\lambda}, \underline{\lambda} + \epsilon/2]\big) > 0. \end{aligned}$$

Take two nonnegative continuous functions f, h, not identically 0, whose supports are included in (x_0, x_1) and (x_2, x_3), respectively. Then,

$$\int f(x) P_t h(x)\,d\mu(x) = \int e^{-\lambda t}\int f(x)\psi_\lambda(x)\,d\mu(x)\int h(y)\psi_\lambda(y)\,d\mu(y)\,d\Gamma(\lambda).$$

We divide the integral in λ in two pieces:

$$I = \int_{\underline{\lambda}}^{\underline{\lambda}+\epsilon} e^{-\lambda t} \int f(x)\psi_\lambda(x)\,d\mu(x) \int h(y)\psi_\lambda(y)\,d\mu(y)\,d\Gamma(\lambda),$$

$$J = \int_{(\underline{\lambda}+\epsilon,\infty)} e^{-\lambda t} \int f(x)\psi_\lambda(x)\,d\mu(x) \int h(y)\psi_\lambda(y)\,d\mu(y)\,d\Gamma(\lambda).$$

Notice that

$$\begin{aligned} I &\le \int_{\underline{\lambda}}^{\underline{\lambda}+\epsilon/2} e^{-\lambda t} \int f(x)\psi_\lambda(x)\,d\mu(x) \int h(y)\psi_\lambda(y)\,d\mu(y)\,d\Gamma(\lambda) \\ &\le e^{-(\underline{\lambda}+\epsilon/2)t} \int_{\underline{\lambda}}^{\underline{\lambda}+\epsilon/2} \int f(x)\psi_\lambda(x)\,d\mu(x) \int h(y)\psi_\lambda(y)\,d\mu(y)\,d\Gamma(\lambda) \\ &\le -C_1 e^{-(\underline{\lambda}+\epsilon/2)t}, \end{aligned}$$

for some positive C_1 that depends on f, h, ϵ, but not on t. On the other hand,

$$\begin{aligned} J^2 &\le \int_{\underline{\lambda}+\epsilon}^{\infty} e^{-\lambda t}\left(\int f(x)\psi_\lambda(x)\,d\mu(x)\right)^2 d\Gamma(\lambda) \\ &\quad \times \int_{\underline{\lambda}+\epsilon}^{\infty} e^{-\lambda t}\left(\int h(y)\psi_\lambda(y)\,d\mu(y)\right)^2 d\Gamma(\lambda) \\ &\le e^{-2(\underline{\lambda}+\epsilon)t} \int f^2\,d\mu \int h^2\,d\mu. \end{aligned}$$

Thus, $J \le C_2 e^{-(\underline{\lambda}+\epsilon)t}$, for some finite constant that does not depend on t. Therefore, for large t, we get $\int f(x) P_t h(x)\,d\mu(x) < 0$, which is not possible, and the result is proved. □

Notice that $\psi_{\underline{\lambda}}(x) > 0$ for $x > 0$, otherwise $\psi_{\underline{\lambda}}$ must change its sign in $(0,\infty)$.

6.2.1 Existence and Regularity of a Kernel with Respect to μ

We shall prove that the semigroup (P_t) has a regular kernel with respect to μ. The starting point is the following lemma.

Lemma 6.8

(i) *For all $t > 0, x, y$, as b increases to ∞, we have*

$$r_t^{(b)}(x,y) \uparrow r_t(x,y).$$

This limit kernel is finite and symmetric. For all $x > 0$, the function $r_\bullet(x, x)$ is decreasing; for all $0 < t$, $x \in \mathbb{R}_+$, the function $r_t(x, \bullet)$ belongs to $L^2(d\mu)$ and for all $h \in L^2(d\mu)$

$$P_t h(x) = \int h(y) r_t(x, y)\, d\mu(y).$$

Also $\int (\int h(y) r_t(x, y)\, d\mu(y))^2\, d\mu(x) \le \int h^2(y)\, d\mu(y)$.

(ii) *(Harnack) Consider $0 < \delta < 1$ and $x_0 > 0$ for which $3\delta \le x_0$. Then, there exists a constant $C = C(x_0)$ which depends only on the coefficients of $\mathcal{L}$ such that for all $5\delta^2 < t$, $|x - x_0| < \delta$, $y \ge 0$,*

$$r_t(x, y) \le C r_{t+\delta^2}(x_0, y).$$

Furthermore, since $r_t(\bullet, \bullet)$ is symmetric, we have the following extension. Let $0 < \delta < 1$ and $x_0 > 0$, $y_0 > 0$ for which $3\delta \le (x_0 \wedge y_0)$. Then, there exists a constant $C = C(x_0, y_0)$ which again depends only on the coefficients of $\mathcal{L}$, such that for all $5\delta^2 < t$, $|x - x_0| < \delta$, $|y - y_0| < \delta$,

$$r_t(x, y) \le C r_{t+2\delta^2}(x_0, y_0).$$

(iii) *(Hölder continuity) Consider $0 < \delta < 1$ and again $x_0 > 0$, $y_0 > 0$ with $3\delta \le (x_0 \wedge y_0)$. Take also $t_0 > 5\delta^2$ and define*

$$A_1 = \{(t, x, y) : t_0 - \delta^2/4 \le t \le t_0, \|(x, y) - (x_0, y_0)\| \le \delta/2\},$$
$$A_2 = \{(t, x, y) : t_0 - 2\delta^2 \le t \le t_0, \|(x, y) - (x_0, y_0)\| \le \delta\}.$$

Then, there exists $C = C(x_0, y_0)$, $\beta = \beta(x_0, y_0)$ such that we have

$$|r_t(x, y) - r_s(u, v)| \le C\delta^{-\beta}\left(\|(x, y) - (u, v)\| + |t - s|^{\frac{1}{2}}\right)^\beta r_{t_0 - 2\delta^2}(x_0, x_0),$$

for all $\xi = (t, x, y)$, $\chi = (s, u, v) \in A_1$. In particular, r is continuous on $(0, \infty) \times \mathbb{R}_+^2$.

Proof Let h be a nonnegative and compactly supported measurable function. Then, $\mathbb{E}_x(h(X_t), t < T \wedge T_b)$ is monotone increasing in b, proving that $r_t^{(b)}(x, y)$ is monotone increasing for all $t > 0$, x, y and all $b > x \vee y$. Thus, by the Monotone Convergence Theorem, we have, for all x, t,

$$\int h(y) r_t^{(b)}(x, y)\, d\mu(y) = \mathbb{E}_x\big(h(X_t), t < T \wedge T_b\big) \uparrow \mathbb{E}_x\big(h(X_t), t < T\big).$$

This proves that the limit kernel $r_t(x, y)$ exists for $t > 0$, x, y and it is finite $d\mu(y)$-a.s. (the set of measure 0 where it is infinite depends, in principle, on t, x). Moreover, r is a kernel for the semigroup (P_t).

Part (ii) follows at once from Proposition 6.3. This also shows that $r_t(x, y)$ is finite for all $t > 0$, x, y and part (i) is also proved.

For part (iii) we obtain, as in the cited proposition, a bound that depends on

$$\sup_{(t',x',y')\in A_2} \{r_{t'}(x',y')\}.$$

Now, we show how to replace this bound by $r_{t_0-2\delta^2}(x_0,x_0)$. So let us start with $(t',x',y')\in A_2$ and we move from x' to x_0. We use Harnack's Inequality with a step of size δ. Then we have

$$r_{t'}(x',y') \le C(x_0)^p r_{t'+p\delta^2}(x_0,y'),$$

with $p=0$ when $x'=x_0$, otherwise $p=1$. Similarly, if we move from y' to y_0 then

$$r_{t'}(x',y') \le C^p(x_0)C^i(y_0) r_{t'+(p+i)\delta^2}(x_0,y_0).$$

Now we can move from y_0 to x_0 with a step size of δ each time and then we have the existence of $\bar{C}=\bar{C}(x_0,y_0)$ and an integer $n=n(x_0,y_0)$ (n reflects the number of times Harnack's Inequality is used to move from y_0 to x_0) such that for all $(t',x',y')\in A_2$

$$r_{t'}(x',y') \le \bar{C} r_{t'+(n+p+i)\delta^2}(x_0,x_0).$$

The result is obtained by noticing that $r_\bullet(x_0,x_0)$ is decreasing. □

Lemma 6.9

(i) *For each $x,t>0$, the function $\lambda \to e^{-\lambda t}\psi_\lambda(x)$ belongs to $L^p(d\Gamma)$, for all $1\le p\le 2$. Also for any $a>0$, the function $\lambda\to e^{-\lambda t}\int_0^a |\psi_\lambda(x)|\,dx$ belongs to $L^p(d\Gamma)$, for all $1\le p\le 2$.*

(ii) *For all $t>0$, $x,y\in\mathbb{R}_+$,*

$$r_t(x,y) = \int e^{-\lambda t}\psi_\lambda(x)\psi_\lambda(y)\,d\Gamma(\lambda). \tag{6.18}$$

(iii) *$r\in\mathcal{C}^{1,2}((0,\infty),\mathbb{R}_+^2)$ and*

$$\partial_t r_t(x,y) = \mathcal{L}_x r_t(x,y) = \mathcal{L}_y r_t(x,y),$$
$$r_t(x,0)=0,$$
$$\forall h\in\mathcal{C}_0\big((0,\infty)\big)\quad \lim_{t\to 0}\int h(y) r_t(x,y)\,d\mu(y) = h(x).$$

(iv) *For all $t>0$, $x>0$, we have*

$$\partial_y r_t(x,y) = \int e^{-\lambda t}\psi_\lambda(x)\psi'_\lambda(y)\,d\Gamma(\lambda),$$
$$\partial_y r_t(x,0) = \partial_y r_t(x,0^+) = \int e^{-\lambda t}\psi_\lambda(x)\,d\Gamma(\lambda).$$

Proof (i) Let us first prove $e^{-\lambda t}\psi_\lambda(x) \in L^2(d\Gamma)$, for all $t > 0, x \in \mathbb{R}_+$. For that purpose, fix $t > 0, x > 0$ (for $x = 0$ the property is obvious). Consider a nonnegative continuous function h with support contained in $(0, \infty)$ and such that $h(x) > 0$. Then, for any $n \geq 0$,

$$\iint h(y)h(z)r_t(z, y)\,d\mu(y)\,d\mu(z) = \int e^{-\lambda t}\left(\int \psi_\lambda(z)h(z)\,d\mu(z)\right)^2 d\Gamma(\lambda)$$
$$\geq \int_0^n e^{-\lambda t}\left(\int \psi_\lambda(z)h(z)\,d\mu(z)\right)^2 d\Gamma(\lambda).$$

We now take a sequence of such functions h converging to the Dirac mass at x to conclude the inequality

$$r_t(x, x) \geq \int_0^n e^{-\lambda t}\psi_\lambda^2(x)\,d\Gamma(\lambda).$$

The Monotone Convergence Theorem then gives

$$r_t(x, x) \geq \int e^{-\lambda t}\psi_\lambda^2(x)\,d\Gamma(\lambda),$$

proving the case $p = 2$. Since $e^{-\lambda t}$ belongs to $L^p(d\Gamma)$, for all $1 \leq p$, Hölder's Inequality implies that $e^{-\lambda t}\psi_\lambda(x) \in L^p(d\Gamma)$ for $1 \leq p \leq 2$.

For the second part, consider first a small $b > 0$. We notice that for all $\lambda \geq 0$ and $0 \leq x \leq b$, we have an upper bound of the type

$$\left|\psi_\lambda'(x)\right| \leq A(b)\left(1 + 2\lambda\int_0^x \left|\psi_\lambda(y)\right| dy\right),$$

where $A(b) = e^{2b\max\{|\alpha(z)|:z\in[0,b]\}}$ depends only on b (see (6.13)). Hence we deduce for all $0 \leq x \leq b$

$$\int_0^x \left|\psi_\lambda(u)\right| du \leq \int_0^x A(b)\left(1 + 2\lambda\int_0^u \left|\psi_\lambda(y)\right| dy\right) u\,du$$
$$\leq A(b)\frac{b^2}{2} + 2bA(b)\lambda\int_0^x\int_0^u \left|\psi_\lambda(y)\right| dy\,du,$$

and Gronwall's Inequality then shows

$$\int_0^b \left|\psi_\lambda(y)\right| dy \leq A(b)\frac{b^2}{2}e^{2b^2A(b)\lambda}.$$

We choose now b such that $2b^2A(b) < t$ and we obtain that $e^{-\lambda t}\int_0^b |\psi_\lambda(x)|\,dx \in L^p(d\Gamma)$. If $a > b$, we split the integral $\int_0^a |\psi_\lambda(x)|\,dx$ into two pieces and we need to show that

$$e^{-\lambda t}\int_b^a \left|\psi_\lambda(x)\right| dx \in L^p(d\Gamma).$$

We only show the case $p = 2$ (the other cases are similar). We have

$$\begin{aligned}
&\int e^{-2\lambda t}\left(\int_b^a |\psi_\lambda(x)|\,dx\right)^2 d\Gamma(\lambda) \\
&\quad \le (a-b)\int e^{-2\lambda t}\int_b^a |\psi_\lambda(x)|^2\,dx\,d\Gamma(\lambda) \\
&\quad \le (a-b)\int_b^a r_{2t}(x,x)\,dx \le C(a,b,t)(a-b)^2 r_{2t}(a,a),
\end{aligned}$$

where in the last inequality we have used Harnack's Inequality (which gives a finite $C(a,b,t)$) and the fact that $r_\bullet(a,a)$ is a decreasing function.

(ii) We denote by $\zeta(t,x,y) = \int e^{-\lambda t}\psi_\lambda(x)\psi_\lambda(y)\,d\Gamma(\lambda)$. We shall prove that $\zeta = r$. For that purpose, fix $t > 0$, and take two nonnegative continuous functions h and l whose supports are compact and contained in $(0,\infty)$. Let us denote by $\mathcal{K} \subset (0,\infty)$ a compact set containing the supports of h and l. Let us fix $x_0 \in \mathcal{K}$. Then, using Harnack's Inequality and the monotonicity of $r_\bullet(x_0,x_0)$, we conclude the existence of a constant C that depends only on $\mathcal{K}, x_0, t$ such that for all $x \in \mathcal{K}$

$$r_t(x,x) \le C r_t(x_0,x_0).$$

Hence,

$$\begin{aligned}
&\int e^{-\lambda t}\int h(x)|\psi_\lambda(x)|\,d\mu(x)\int l(y)|\psi_\lambda(y)|\,d\mu(y)\,d\Gamma(\lambda) \\
&\quad \le \iint h(x)l(y)\left(\int e^{-\lambda t}|\psi_\lambda(x)|^2\,d\Gamma(\lambda)\right)^{1/2} \\
&\qquad \times \left(\int e^{-\lambda t}|\psi_\lambda(y)|^2\,d\Gamma(\lambda)\right)^{1/2} d\mu(x)\,d\mu(y) \\
&\quad \le \iint h(x)l(y)\big(r_t(x,x)\big)^{1/2}\big(r_t(y,y)\big)^{1/2}\,d\mu(x)\,d\mu(y) \\
&\quad \le C r_t(x_0,x_0)\iint h(x)l(y)\,d\mu(x)\,d\mu(y).
\end{aligned}$$

An application of Fubini's Theorem gives

$$\begin{aligned}
\iint h(x)l(y)r_t(x,y)\,d\mu(x)\,d\mu(y) &= \int h(x)P_t l(x)\,d\mu(x) \\
&= \iint h(x)l(y)\zeta(t,x,y)\,d\mu(x)\,d\mu(y),
\end{aligned}$$

proving that (6.18) holds $d\mu(x) \otimes d\mu(y)$-a.s. We shall now prove that ζ is regular. This is done by application of the Dominated Convergence Theorem, together with the following inequality for all $\lambda \geq 0$ and $0 \leq x \leq z$,

$$|\psi_\lambda'(x)| \leq K\left(1 + 2\lambda \int_0^z |\psi_\lambda(y)|\, dy\right).$$

Thus, we can differentiate under the integral defining ζ to obtain

$$\partial_x \zeta(t, x, y) = \int e^{-\lambda t} \partial_x \psi_\lambda(x) \psi_\lambda(y)\, d\Gamma(\lambda).$$

Similarly, we get $\zeta \in \mathcal{C}^{1,2}((0,\infty) \times \mathbb{R}_+^2)$, and ζ is the solution to the problem in part (iii).

Given that r is Hölder continuous and coincides almost surely with ζ, we conclude that $r = \zeta$ on $(0,\infty) \times \mathbb{R}_+^2$, and the result is proved.

(iv) The result follows from the Dominated Convergence Theorem. □

Remark 6.10 Representation 6.18 is similar to the representation of Karlin–McGregor in the context of birth-and-death processes; see (5.15).

On the other hand, using the representation

$$r_t^{(b)}(x, y) = e^{\frac{1}{2}(\gamma(x)+\gamma(y))} \mathbb{E}_x\left(e^{-\int_0^t \wp(B_s)\, ds}, t < T \wedge T_b | B_t = y\right) \frac{1}{\sqrt{2\pi t}} e^{-\frac{(y-x)^2}{2t}}$$

where $\wp = \frac{1}{2}(\alpha^2 - \alpha')$, and the Monotone Convergence Theorem, we obtain

$$r_t(x, y) = e^{\frac{1}{2}(\gamma(x)+\gamma(y))} \mathbb{E}_x\left(e^{-\int_0^t \wp(B_s)\, ds}, t < T | B_t = y\right) \frac{1}{\sqrt{2\pi t}} e^{-\frac{(y-x)^2}{2t}}.$$

We need the following result to control the behavior of ψ_λ near $\underline{\lambda}$ (see Lemmas 6.48 and 6.49 for further details).

Lemma 6.11 *Assume that ψ_{λ_0} is positive on the interval $(0, x_0]$. Then there exists $\epsilon > 0$ such that ψ_β is positive in the same interval, for all $\beta \in [\lambda_0 - \epsilon, \lambda_0 + \epsilon]$.*

Also, if ψ_{λ_0} is nonnegative on $[0, x_0]$ and $\beta < \lambda_0$ then $\psi_{\lambda_0}(x) < \psi_\beta(x)$ for all $x \in (0, x_0]$.

Proof The proof is based on the continuity of $\psi_\lambda(x)$, $\psi_\lambda'(x)$ in both arguments. Consider $a > 0$ small enough such that $\psi_{\lambda_0}'(x) > 1/2$ for all $x \in [0, a]$. By continuity, this inequality holds for all β close to λ_0. Thus, for all these values of β and all $x \in [0, a]$, we have

$$\psi_\beta(x) \geq \frac{x}{2}.$$

To finish the first part, consider $c = \inf\{\psi_{\lambda_0}(x) : x \in [a, x_0]\} > 0$. Then, again by continuity, for all β close to λ, we have $\inf\{\psi_\beta(x) : x \in [a, x_0]\} > c/2 > 0$, and the lemma is proved.

Let us prove now the second part of this lemma. Clearly, ψ_{λ_0} is positive on $(0,x_0)$, otherwise if it has a zero in that interval then it is negative in some part of it (it cannot happen that $\psi_{\lambda_0}(a)=\psi'_{\lambda_0}(a)=0$). Now, consider the *Wronskian* $W(x)=\psi_\beta(x)\psi'_{\lambda_0}(x)-\psi'_\beta(x)\psi_{\lambda_0}(x)$. On the one hand, $W(0)=0$, and on the other hand,

$$W'(x)=-2\alpha(x)W(x)-2(\lambda_0-\beta)\psi_\beta(x)\psi_{\lambda_0}(x),$$

or equivalently,

$$\left(We^\gamma\right)'=-2e^\gamma(\lambda_0-\beta)\psi_\beta\psi_{\lambda_0}.$$

This shows that while $\psi_\beta,\psi_{\lambda_0}$ are positive on $(0,a)$ the function We^γ is strictly increasing there and then W is positive on $(0,a)$. This happens at least in a small neighborhood of $x=0$, that is, for some small $a>0$. Since

$$\left(\frac{\psi_{\lambda_0}}{\psi_\beta}\right)'=\frac{W}{\psi_\beta^2},$$

the ratio $\frac{\psi_{\lambda_0}}{\psi_\beta}$ is strictly increasing in the same interval $[0,a]$. Since, this ratio is 1 at $x=0$ (by l'Hôpital's Rule), we conclude the strict inequality

$$\psi_{\lambda_0}(x)<\psi_\beta(x),$$

for all $x\in(0,a]$. This shows that ψ_β is positive on $(0,x_0]$, and then again $\psi_{\lambda_0}<\psi_\beta$ in this interval. □

Corollary 6.12 *The sequence $(\lambda_{1,n})_n$ is strictly decreasing and its limit is $\underline{\lambda}$. Also, we have the lower bound for the mass that Γ puts at $\underline{\lambda}$*

$$\Gamma(\{\underline{\lambda}\})\geq\left(\int\psi^2_{\underline{\lambda}}(x)\,d\mu(x)\right)^{-1}. \tag{6.19}$$

Proof We recall that $\lambda_{1,n}$ is the bottom of the spectrum of $\mathcal{L}$ on the interval $[0,n]$ with Dirichlet boundary conditions at $x=0$ and n. This means in particular that $\psi_{\lambda_{1,n}}$ is positive on $(0,n)$ and $\psi_{\lambda_{1,n}}(n)=0$. This implies that $\psi_{\lambda_{1,n}}$ is negative for all values $x\in(n,n+\epsilon)$ for some small $\epsilon>0$. On the other hand, the function $\psi_{\lambda_{1,n+1}}$ is nonnegative on $[0,n+1]$. Then, according to Lemma 6.11, we have $\lambda_{1,n+1}\leq\lambda_{1,n}$, but since $\psi_{\lambda_{1,n+1}}\neq\psi_{\lambda_{1,n}}$, we deduce that $\lambda_{1,n+1}<\lambda_{1,n}$. The same argument shows that $\underline{\lambda}<\lambda_{1,n}$ for all n. Again Lemma 6.11 shows that $0\leq\psi_{\lambda_{1,n}}(x)\leq\psi_{\underline{\lambda}}(x)$ for $x\in[0,n]$.

The convergence of (Γ_n) to Γ and the fact that $\underline{\lambda}$ is the minimum of the closed support of Γ imply for all $\epsilon>0$ such that $\underline{\lambda}+\epsilon$ is a point of continuity for Γ

$$0<\Gamma(\underline{\lambda}+\epsilon)-\Gamma(-1)=\lim_{n\to\infty}\Gamma_n(\underline{\lambda}+\epsilon)-\Gamma_n(-1),$$

where -1 plays no role and it is a point to the left of the support of Γ, which then is clearly a point of continuity for that measure. Since the support of Γ_n is contained on $[\lambda_{1,n},\infty)$, we conclude that $\lambda_{1,n}\leq\underline{\lambda}+\epsilon$ for all n large enough. We conclude the convergence $\lambda_{1,n}\downarrow\underline{\lambda}$.

On the other hand, if n is large enough such that $\lambda_{1,n} < \underline{\lambda} + \epsilon$, we get (see (6.11))

$$\Gamma_n(\underline{\lambda} + \epsilon) - \Gamma_n(-1) = \sum_{k:\lambda_{k,n} \le \underline{\lambda}+\epsilon} |r_{k,n}|^2 \ge |r_{1,n}|^2 = \left(\int_0^n \psi_{\lambda_{1,n}}(x)^2 \, d\mu(x) \right)^{-1}.$$

The conclusion is that $\Gamma(\{\underline{\lambda}\}) \ge \limsup_n (\int_0^n \psi_{\lambda_{1,n}}(x)^2 \, d\mu(x))^{-1}$. Finally, since $\psi^2_{\lambda_{1,n}} \mathbb{1}_{[0,n]} \uparrow \psi^2_{\underline{\lambda}}$, we get the lower bound

$$\Gamma(\{\underline{\lambda}\}) \ge \left(\int \psi^2_{\underline{\lambda}}(x) \, d\mu(x) \right)^{-1}. \qquad \square$$

We end this section with the next proposition that will be useful for the asymptotic analysis of $\mathbb{P}_x(t < T)$.

Proposition 6.13 *We have the following exact formulas for the distribution and density of T:*

$$\begin{aligned}
\mathbb{P}_x(t < T) &= \frac{\Lambda(x)}{\Lambda(\infty)} + \int_{\underline{\lambda}}^{\infty} \frac{e^{-\lambda t}}{2\lambda} \psi_\lambda(x) \, d\Gamma(\lambda), \\
\frac{\partial}{\partial t}\mathbb{P}_x(t < T) &= -\frac{1}{2}\frac{\partial}{\partial y} r_t(x, 0) = -\frac{1}{2} \int_{\underline{\lambda}}^{\infty} e^{-\lambda t} \psi_\lambda(x) \, d\Gamma(\lambda).
\end{aligned} \tag{6.20}$$

Proof We shall prove first $\frac{\partial}{\partial t}\mathbb{P}_x(t < T) = -\frac{1}{2}\frac{\partial}{\partial y} r_t(x, 0)$. For that purpose, take $0 < h < 1$ and

$$\mathbb{P}_x(t < T) - \mathbb{P}_x(t + h < T) = \mathbb{E}_x\big(\mathbb{P}_{X_t}(T \le h), t < T\big).$$

Consider now $0 < \epsilon < 1/2$ fixed. We have $\mathbb{P}_y(T \le h) \le \mathbb{P}_\epsilon(T \le h)$ for all $y \ge \epsilon$, and we shall prove $\mathbb{P}_\epsilon(T \le h) = o(h)$. This probability is split into two terms:

$$\mathbb{P}_\epsilon(T \le h) = \mathbb{P}_\epsilon(T \le h, T_1 \le h) + \mathbb{P}_\epsilon(T \le h < T_1). \tag{6.21}$$

For the first term, consider $\beta = \max\{-\alpha(z) : z \in [0, 1]\}$ and the process $dY_t = dB_t + \beta \, dt$ with $Y_0 = \epsilon$. Since $X_t \le Y_t$ for all $t \le T \wedge T_1$, we obtain

$$\mathbb{P}_\epsilon(T \le h, T_1 \le h) = \mathbb{P}_\epsilon(T_1 < T \le h) \le \mathbb{P}_\epsilon\big(T_1^Y \le h\big).$$

Since Y is a Brownian motion with constant drift, is easy to estimate $\mathbb{P}_\epsilon(T_1^Y \le h) = o(h)$.

For the second term in (6.21), we use Girsanov's Theorem

$$\begin{aligned}
\mathbb{P}_\epsilon(T \le h < T_1) &= \mathbb{E}_\epsilon\big(e^{\frac{1}{2}[\gamma(\epsilon) - \int_0^{T_0^B} (\alpha^2(B_s) - \alpha'(B_s)) \, ds]}, T_0^B \le h < T_1^B\big) \\
&\le K\mathbb{P}_\epsilon\big(T_0^B \le h\big) = o(h),
\end{aligned}$$

where $K = e^{\frac{1}{2}[\gamma(\epsilon) + h \max\{|\alpha^2(z)| + |\alpha'(z)| : z \in [0,1]\}]}$.

So far we have proved that

$$\mathbb{P}_x(t<T)-\mathbb{P}_x(t+h<T)=\mathbb{E}_x\big(\mathbb{P}_{X_t}(T\le h),\,X_t\le\epsilon,\,t<T\big)+o(h).$$

Using the same technique, we can prove

$$\mathbb{P}_z(T\le h,\,T_1\le h)=o(h),$$

uniformly for $z\in[0,\epsilon]$. Again, using Girsanov's Theorem, we obtain for $z\in[0,\epsilon]$

$$\begin{aligned}\mathbb{P}_z(T\le h<T_1)&=e^{\frac{1}{2}\gamma(z)}\mathbb{P}_z\big(T_0^B\le h<T_1^B\big)+o(h)\\&=e^{\frac{1}{2}\gamma(z)}\mathbb{P}_z\big(T_0^B\le h\big)+o(h)=e^{\frac{1}{2}\gamma(z)}G(z/\sqrt{h})+o(h),\end{aligned}$$

where $G(w)=2\int_w^\infty\frac{1}{\sqrt{2\pi}}e^{-\frac{u^2}{2}}\,du$.

From this discussion, we obtain

$$\begin{aligned}\mathbb{P}_x(t<T)-\mathbb{P}_x(t+h<T)&=\mathbb{E}_x\big(e^{\frac{1}{2}\gamma(X_t)}G(X_t/\sqrt{h}),\,X_t\le\epsilon,\,t<T\big)+o(h)\\&=\int_0^\epsilon r_t(x,z)G(z/\sqrt{h})e^{-\frac{1}{2}\gamma(z)}\,dz+o(h)\\&=h\int_0^{\epsilon/\sqrt{h}}\frac{r_t(x,v\sqrt{h})}{v\sqrt{h}}vG(v)e^{-\frac{1}{2}\gamma(v\sqrt{h})}\,dv.\end{aligned}$$

Now, we use the Dominated Convergence Theorem. From Lemma 6.9, we get, as h goes to zero, the pointwise convergence

$$\frac{r_t(x,v\sqrt{h})}{v\sqrt{h}}e^{-\frac{1}{2}\gamma(v\sqrt{h})}\to\frac{\partial}{\partial y}r_t(x,0).$$

Since $v\sqrt{h}\le\epsilon$, we need only to dominate $\frac{r_t(x,z)}{z}$ for $0<z<\epsilon$, and this is done as follows:

$$\left|\frac{r_t(x,z)}{z}\right|\le\max\left\{\left|\frac{\partial}{\partial y}r_t(x,y)\right|:y\in[0,\epsilon]\right\}.$$

On the other hand, from Lemma 6.9, we get

$$\left|\frac{\partial}{\partial y}r_t(x,y)\right|\le\int_{\underline{\lambda}}^\infty e^{-\lambda t}\big|\psi_\lambda(x)\big|\big|\psi_\lambda'(y)\big|\,d\Gamma(\lambda).$$

The last ingredient we need is the inequality

$$\big|\psi_\lambda'(y)\big|\le e^{\gamma(y)}\left(1+2\lambda\int_0^y\big|\psi_\lambda(z)\big|e^{-\gamma(z)}\,dz\right)\le A\left(1+2\lambda\int_0^\epsilon\big|\psi_\lambda(z)\big|\,dz\right),$$

where A is a constant that depends on ϵ. The integrability condition of Lemma 6.9 gives

$$\lim_{h\to 0} \frac{\mathbb{P}_x(t<T)-\mathbb{P}_x(t+h<T)}{h} = \frac{\partial}{\partial y} r_t(x,0) \int_0^\infty v G(v)\, dv = \frac{1}{2}\frac{\partial}{\partial y} r_t(x,0)$$
$$= \frac{1}{2}\int_{\underline{\lambda}}^\infty e^{-\lambda t}\psi_\lambda(x)\, d\Gamma(\lambda).$$

The result for the tail distribution of T is obtained by Fubini's Theorem. This is simple in the case $\underline{\lambda} > 0$. Now, we provide the proof in the case $\underline{\lambda} = 0$. For that purpose, consider $\epsilon > 0$ small enough such that $\psi_\lambda(x) > 0$ for all $\lambda \in [0, \epsilon]$. Integrating the formula for the density of T, we get

$$\begin{aligned}\mathbb{P}_x(t<T)-\mathbb{P}_x(T=\infty) &= \frac{1}{2}\int_t^\infty \int_0^\infty e^{-\lambda s}\psi_\lambda(x)\, d\Gamma(\lambda)\, ds \\ &= \frac{1}{2}\int_t^\infty \int_0^\epsilon e^{-\lambda s}\psi_\lambda(x)\, d\Gamma(\lambda)\, ds \\ &\quad + \frac{1}{2}\int_t^\infty \int_\epsilon^\infty e^{-\lambda s}\psi_\lambda(x)\, d\Gamma(\lambda)\, ds.\end{aligned}$$

For the first term, we can use Tonelli's Theorem (the integrand and the measures are nonnegative) and for the second term we can use Fubini's Theorem (see the integrability at Lemma 6.9) to obtain that

$$\mathbb{P}_x(t<T)-\mathbb{P}_x(T=\infty) = \int_0^\epsilon \frac{e^{-\lambda t}}{2\lambda}\psi_\lambda(x)\, d\Gamma(\lambda) + \int_\epsilon^\infty \frac{e^{-\lambda t}}{2\lambda}\psi_\lambda(x)\, d\Gamma(\lambda),$$

and the second term is finite due to the integrability condition at the mentioned lemma. The left-hand side is obviously finite, which implies that the first integral in the right-hand side is finite. In this first integral, the function being integrated is nonnegative, which implies it is integrable. Finally, since $\mathbb{P}_x(T=\infty) = \frac{\Lambda(x)}{\Lambda(\infty)}$ (see (6.8)), we conclude

$$\mathbb{P}_x(t<T) = \frac{\Lambda(x)}{\Lambda(\infty)} + \int_0^\infty \frac{e^{-\lambda t}}{2\lambda}\psi_\lambda(x)\, d\Gamma(\lambda).$$ □

Remark 6.14 A straightforward consequence of this result is that Γ puts no mass at $\lambda = 0$, which is not trivial only when $\underline{\lambda} = 0$.

6.2.2 *A Ratio Limit for the Diffusion on* $(0, \infty)$

In this section, we prove a general ratio limit theorem, which is essentially the same proved in Mandl (1961) plus a pointwise version.

Theorem 6.15 *The following ratio limit holds*:

$$\lim_{t\to\infty} \frac{r_t(x, y)}{r_t(x', y')} = \frac{\psi_{\underline{\lambda}}(x)\psi_{\underline{\lambda}}(y)}{\psi_{\underline{\lambda}}(x')\psi_{\underline{\lambda}}(y')}, \tag{6.22}$$

uniformly in x, x', y, y' *on compact sets of* $(0, \infty)$.

Moreover, if ξ *is a probability measure with bounded support contained in* $(0, \infty)$ *and* A, B *are two bounded measurable sets in* $\mathbb{R}_+$ *of positive Lebesgue measure, then*

$$\lim_{t\to\infty} \frac{\mathbb{E}_\xi(X_t \in A, t < T)}{\mathbb{E}_\xi(X_t \in B, t < T)} = \frac{\int_A \psi_{\underline{\lambda}}(z)\, d\mu(z)}{\int_B \psi_{\underline{\lambda}}(z)\, d\mu(z)} = \frac{\int_A \varphi_{\underline{\lambda}}(z)\, dz}{\int_B \varphi_{\underline{\lambda}}(z)\, dz}. \tag{6.23}$$

Proof We start proving (6.22). Using the same ideas as in the proof of Lemma 6.7, we get this ratio is equivalent to the ratio

$$\frac{\int_{\underline{\lambda}}^{\underline{\lambda}+\epsilon} e^{-\lambda t}\psi_\lambda(x)\psi_\lambda(y)\, d\Gamma(\lambda)}{\int_{\underline{\lambda}}^{\underline{\lambda}+\epsilon} e^{-\lambda t}\psi_\lambda(x')\psi_\lambda(y')\, d\Gamma(\lambda)}$$

for every fixed $\epsilon > 0$. Consider $a > 0$ such that $x, x', y, y' \in (0, a]$. We can assume that ϵ is small enough such that for all $\lambda \in [\underline{\lambda}, \underline{\lambda} + \epsilon]$ and all $x \in (0, a]$ we have $\psi_\lambda(x) > 0$ (see Lemma 6.11, where $\lambda_0 = \underline{\lambda}$).

Now, from the Intermediate Value Theorem, we have

$$\int_{\underline{\lambda}}^{\underline{\lambda}+\epsilon} e^{-\lambda t}\psi_\lambda(x)\psi_\lambda(y)\, d\Gamma(\lambda) = \psi_{\lambda(t,\epsilon)}(x)\psi_{\lambda(t,\epsilon)}(y) \int_{\underline{\lambda}}^{\underline{\lambda}+\epsilon} e^{-\lambda t}\, d\Gamma(\lambda),$$

for some $\lambda(t, \epsilon) \in [\underline{\lambda}, \underline{\lambda} + \epsilon]$ ($\lambda(t, \epsilon)$ may also depend on x, y). A simple argument using lim sup and lim inf shows that (6.22) holds by letting first $t \to \infty$ and then $\epsilon \to 0$.

The second assertion (6.23) is obtained by dividing numerator and denominator by $r_t(x_0, x_0)$, for some fixed x_0, and then using the Dominated Convergence Theorem. The domination is obtained from Harnack's Inequality (jointly with the fact that $r_\bullet(x_0, x_0)$ is decreasing). The last equality in (6.22) comes from the relation between ψ_λ and φ_λ. □

6.2.3 *The Case* $\Lambda(\infty) < \infty$

We study in this section the simplest case where $\Lambda(\infty) < \infty$. According to (6.8), this happens if and only if $\mathbb{P}_x(T = \infty) > 0$ for all $x > 0$. This implies that Λ is bounded, and also

$$\lim_{t\to\infty} \mathbb{P}_x(t < T) = \mathbb{P}_x(T = \infty) > 0.$$

On the other hand, from Hypothesis **H**, we have $\int_0^\infty e^{-\gamma(x)} \Lambda(x)\,dx = \infty$. Since Λ is nonnegative, increasing and bounded, we deduce that $\int_0^\infty e^{-\gamma(x)}\,dx = \infty$, that is, the measure μ is not finite.

Theorem 6.16 *Assume hypotheses* **H** *and* $\Lambda(\infty) < \infty$ *hold.*

(i) *For all* $x > 0$, *we have the existence of the following exponential rate* $\eta =: \lim_{t\to\infty} -\frac{\log(\mathbb{P}_x(t<T))}{t} = 0$.
(ii) *Let* ξ *be a probability measure with compact support contained in* $(0, \infty)$ *and* A *a bounded measurable set, then* $\lim_{t\to\infty} \mathbb{P}_\xi(X_t \in A | t < T) = 0$. *In particular, for all* $x > 0$,

$$\lim_{t\to\infty} \mathbb{P}_x(X_t \in A | t < T) = 0.$$

(iii) *For all* $x, y > 0$ *and* $s > 0$, $\lim_{t\to\infty} \frac{\mathbb{P}_x(t+s<T)}{\mathbb{P}_y(t<T)} = \frac{\Lambda(x)}{\Lambda(y)}$.
(iv) *For all* $s \geq 0$, *all* $x > 0$ *and* $A \in \mathcal{B}(\mathcal{C}([0, s]))$, *we have*

$$\lim_{t\to\infty} \mathbb{P}_x(X \in A | t < T) = \mathbb{P}_x(Z \in A),$$

where Z *is the following* h*-transform of* X

$$\mathbb{P}_x(Z \in A) = \mathbb{E}_x\left(\frac{\Lambda(X_s)}{\Lambda(x)}, X \in A, s < T\right).$$

The transition density of Z *is given by* $\mathbb{P}_x(Z_s \in dy) = \frac{\Lambda(y)}{\Lambda(x)} \mathbb{P}_x(X_s \in dy, s < T)$ *for* $x > 0$, $y > 0$ *and* $s \geq 0$, *and* Z *is the unique solution to the SDE*

$$dZ_t = dB_t + \left(\frac{\Lambda'(Z_t)}{\Lambda(Z_t)} - \alpha(Z_t)\right) dt, \qquad Z_0 = x.$$

Proof Parts (i) and (iii) follow immediately from the fact that $\mathbb{P}_x(T = \infty) = \frac{\Lambda(x)}{\Lambda(\infty)} > 0$. Part (iv) follows from the Markov property and the Dominated Convergence Theorem.

Now, we prove part (ii). We consider two different cases. If $\underline{\lambda} > 0$ then for $t \geq 2$

$$\begin{aligned}
&\mathbb{P}_\xi(X_t \in A, t < T) \\
&= \int_{\underline{\lambda}}^\infty e^{-\lambda t} \int \psi_\lambda(x)\,d\xi(x) \int_A \psi_\lambda(y)\,d\mu(y)\,d\Gamma(\lambda) \\
&\leq e^{-\underline{\lambda} t/2} \iint_A \left[\int_{\underline{\lambda}}^\infty e^{-\lambda t/2} |\psi_\lambda(x)||\psi_\lambda(y)|\,d\Gamma(\lambda)\right] d\mu(y)\,d\xi(x) \\
&\leq e^{-\underline{\lambda} t/2} \int \left[r_{t/2}(x, x)\right]^{1/2} d\xi(x) \int_A \left[r_{t/2}(y, y)\right]^{1/2} d\mu(y) \\
&\leq C e^{-\underline{\lambda} t/2} r_{t/2}(x_0, x_0) \leq C e^{-\underline{\lambda} t/2} r_1(x_0, x_0).
\end{aligned}$$

Here again we have used Harnack's Inequality and the fact that $r_\bullet(x_0, x_0)$ is a decreasing function. The constant C depends on x_0, A and the support of ξ. This shows that $\mathbb{P}_\xi(X_t \in A, t < T)$ tends to 0.

On the other hand, $\mathbb{P}_\xi(t < T) \geq \int \mathbb{P}_x(T = \infty)\, d\xi(x) = \int \frac{\Lambda(x)}{\Lambda(\infty)}\, d\xi(x) > 0$, and the result holds in this case.

It remains to consider the case where $\underline{\lambda} = 0$. According to (6.23), we have for any bounded and measurable set B

$$\limsup_{t\to\infty} \frac{\mathbb{P}_\xi(X_t \in A, t < T)}{\mathbb{P}_\xi(t < T)} \leq \lim_{t\to\infty} \frac{\mathbb{P}_\xi(X_t \in A, t < T)}{\mathbb{P}_\xi(X_t \in B, t < T)} = \frac{\int_A \psi_0(x)\, d\mu(x)}{\int_B \psi_0(x)\, d\mu(x)}.$$

Now it is enough to make $B \uparrow \mathbb{R}_+$ to get

$$\limsup_{t\to\infty} \frac{\mathbb{P}_\xi(X_t \in A, t < T)}{\mathbb{P}_\xi(t < T)} \leq \frac{\int_A \Lambda(x)\, d\mu(x)}{\int \Lambda(x)\, d\mu(x)} = 0.$$

□

When $\Lambda(\infty) < \infty$, the limit conditional distribution $\mathbb{P}_\xi(X_t \in \bullet | t < T)$ is the Dirac mass at ∞ and $\eta = 0$. Nevertheless, $\underline{\lambda}$ may be (strictly) positive, as it is shown by the next example.

Example 6.17 Assume that α is constant. The distribution of X_t on $\{t < T\}$ is explicitly known and the spectrum of $\mathcal{L}$ is also known. Theorem 6.16 is quite simple to prove in this case (see Knight 1969 and Williams 1970 in the case $\alpha = 0$ and Martínez and San Martín 1994 for the case $\alpha > 0$). We have $\underline{\lambda} = \alpha^2/2$ independently of the sign of α, and **H** always holds. When $\alpha < 0$, that is, the drift is positive, the probability of eventually hitting 0 is smaller than 1, and in this case $\Lambda(\infty) < \infty$. When $\alpha \leq 0$, we have $\eta = 0$ and the limit distribution of Theorem 6.16 corresponds to a three dimensional Bessel process $BES(3, |\alpha|)$ (see Rogers and Pitman 1981). In this case, according to the cited theorem, one has $\lim_{t\to\infty} \mathbb{P}_x(X_t > b | t < T) = 1$, for all finite b.

6.2.4 The Case $\Lambda(\infty) = \infty$

Through this section, we shall assume that **H1** holds, that is, $\Lambda(\infty) = \infty$ and the process X is absorbed with probability one. We start with the following result which gives some basic relations among the eigenfunctions ψ_λ and φ_λ.

Lemma 6.18 *The following statements are equivalent for* $\lambda > 0$:

(i) ψ_λ *(or equivalently,* φ_λ*) is nonnegative*;
(ii) ψ_λ *is strictly increasing*;
(iii) φ_λ *is positive on* $(0, \infty)$ *and integrable*.

Moreover, if any of these conditions hold then

$$\lim_{M\to\infty}\frac{\psi_\lambda(M)}{\Lambda(M)}=0; \tag{6.24}$$

$$\int_0^\infty \varphi_\lambda(x)\,dx=\frac{1}{2\lambda}; \tag{6.25}$$

$$\mathbb{E}_x\big(\psi_\lambda(X_t),t<T\big)=e^{-\lambda t}\psi_\lambda(x)\quad \textit{for all } x>0,t\ge 0. \tag{6.26}$$

Proof Let us start with the representation (6.3):

$$\psi_\lambda'(x)=e^{\gamma(x)}\bigg(1-2\lambda\int_0^x\psi_\lambda(z)e^{-\gamma(z)}\,dz\bigg)=e^{\gamma(x)}\bigg(1-2\lambda\int_0^x\varphi_\lambda(z)\,dz\bigg),$$

$$\psi_\lambda(x)=\int_0^x e^{\gamma(y)}\bigg(1-2\lambda\int_0^y\varphi_\lambda(z)\,dz\bigg)\,dy.$$

In what follows, we denote by $\psi=\psi_\lambda$ and $\varphi=\varphi_\lambda$.

Let us prove (i) $\Rightarrow$ (ii). If we assume that $\psi'(x_0)<0$, then for all $x\ge x_0$ we have

$$\psi'(x)=e^{\gamma(x)-\gamma(x_0)}e^{\gamma(x_0)}\bigg(1-2\lambda\int_0^x\varphi(y)\,dy\bigg)\le e^{\gamma(x)-\gamma(x_0)}\psi'(x_0).$$

Thus, $\psi(x)\le\psi(x_0)+e^{-\gamma(x_0)}\psi'(x_0)(\Lambda(x)-\Lambda(x_0))$, which is eventually negative because $\Lambda(\infty)=\infty$. We deduce that ψ is increasing. Let us show that ψ is strictly increasing. Since $\psi(0)=0$ and $\psi'(0)=1$, we deduce that $\psi(z)>0$ for any $z>0$. Now, assume ψ is constant on some interval $[x,y]$. Hence, for $z\in(x,y)$, we get $\frac{1}{2}\psi''(z)-\alpha(z)\psi'(z)=0\neq-\lambda\psi(z)$, which is a contradiction, and (ii) is proved.

(ii) $\Rightarrow$ (iii). From (6.3), we get for all $x>0$ that $1-2\lambda\int_0^x\varphi(y)\,dy\ge 0$ and so $\int_0^\infty\varphi(y)\,dy\le\frac{1}{2\lambda}$. The function ψ is positive on $(0,\infty)$, and therefore φ is also positive there.

(iii) $\Rightarrow$ (i) is trivial.

Let us now show that $\lim_{M\to\infty}\frac{\psi(M)}{\Lambda(M)}$ exists. From Itô's formula, we get

$$\begin{aligned}\mathbb{E}_x\big(\psi(X_{s\wedge T_M\wedge T})\big)&=\psi(x)-\lambda\mathbb{E}_x\int_0^{s\wedge T_M\wedge T}\psi(X_t)\,dt\\&=\psi(M)\mathbb{P}_x(T_M<T\wedge s)+\mathbb{E}_x\big(\psi(X_s),s<T_M\wedge T\big).\end{aligned}$$

On the other hand, since ψ is increasing, we obtain

$$\big|\mathbb{E}_x\big(\psi(X_s),s<T_M\wedge T\big)\big|\le\psi(M)\mathbb{P}_x(s<T),$$

which converges to 0 as $s\to\infty$. Hence, from the Monotone Convergence Theorem, we deduce

$$\psi(x)-\lambda\mathbb{E}_x\int_0^{T_M\wedge T}\psi(X_t)\,dt=\psi(M)\mathbb{P}_x(T_M<T)=\psi(M)\frac{\Lambda(x)}{\Lambda(M)}.$$

Therefore, the following limit exists

$$\lim_{M\to\infty} \frac{\psi(M)}{\Lambda(M)} = \frac{\psi(x) - \lambda \mathbb{E}_x \int_0^T \psi(X_t)\,dt}{\Lambda(x)}.$$

Now, if $\lim_{M\to\infty} \frac{\psi(M)}{\Lambda(M)} > 0$, we obtain from **H** that $\int_0^\infty \varphi(x)\,dx = \int_0^\infty \psi(x) \times e^{-\gamma(x)}\,dx = \infty$, which contradicts (iii). Thus, $\lim_{M\to\infty} \frac{\psi(M)}{\Lambda(M)} = 0$.

On the other hand, if $\int_0^\infty \varphi(z)\,dz < \frac{1}{2\lambda}$, we would obtain from (6.3), $\liminf_{x\to\infty} \frac{\psi(x)}{\Lambda(x)} > 0$, which is a contradiction, and so (6.25) is proved.

Finally, we prove (6.26). For that purpose, consider the function $e^{\lambda t}\psi(x)$ and apply Itô's formula at the time $t \wedge T_M \wedge T$ to get

$$e^{\lambda t}\mathbb{E}_x\big(\psi(X_t), t < T_M \wedge T\big) + \psi(M)\mathbb{E}_x\big(e^{\lambda T_M}, T_M < t \wedge T\big) = \psi(x).$$

The first term on the left-hand side converges monotonically to (here we use that $\psi \geq 0$)

$$e^{\lambda t}\mathbb{E}_x\big(\psi(X_t), t < T\big).$$

Then, the result will follow as soon as we show that the second term tends to 0. This term is bounded by

$$e^{\lambda t}\psi(M)\mathbb{P}_x(T_M < T) = \frac{\psi(M)\Lambda(x)}{\Lambda(M)},$$

and the conclusion follows from (6.24). □

Corollary 6.19 *Assume* $\underline{\lambda} > 0$ *then*

(i) *For all* $\lambda \in (0, \underline{\lambda}]$, *we have* $\varphi_\lambda \in L^1(dx)$, *or equivalently,* $\psi_\lambda \in L^1(d\mu)$ *and*

$$\int \psi_\lambda(x)\,d\mu(x) = \frac{1}{2\lambda}.$$

Moreover, the measure μ *is finite and* $\mathbf{1} \in L^2(d\mu)$.

(ii) *The following limit exists in* $L^2(d\Gamma)$

$$b(\lambda) = \lim_{a\to\infty} \int_0^a \psi_\lambda(x)\,d\mu(x).$$

Moreover, $b(\lambda) = \frac{1}{2\lambda}\,d\Gamma$*-a.e.*

Proof (i) According to Lemmas 6.7 and 6.11, the function ψ_λ is nonnegative for all $\lambda \leq \underline{\lambda}$. Now, Lemma 6.18 shows that $\varphi_\lambda \in L^1(dx)$ for all $0 < \lambda \leq \underline{\lambda}$. Since $\psi_\lambda = \varphi_\lambda e^\gamma$, we conclude that $\psi_\lambda \in L^1(d\mu)$ for the same values of λ. This lemma also shows that

$$\int \psi_\lambda(x)\,d\mu(x) = \frac{1}{2\lambda}.$$

The function $\psi_{\underline{\lambda}}$ is nonnegative and strictly increasing (see again Lemma 6.18) and γ is regular on $[0, \infty)$, thus the measure μ is finite, and in particular $\mathbf{1} \in L^2(d\mu)$.

(ii) We know that $b(\lambda) = \int \psi_\lambda(x)\, d\mu(x)$ is well defined and belongs to $L^2(d\Gamma)$ (since $\mathbf{1} \in L^2(d\mu)$). Moreover, we have

$$\mathbb{P}_x(t < T) = P_t\mathbf{1}(x) = \int_{\underline{\lambda}}^{\infty} e^{-\lambda t}\psi_\lambda(x) b(\lambda)\, d\Gamma(\lambda).$$

According to Proposition 6.13, this integral is equal to $\int_{\underline{\lambda}}^{\infty} e^{-\lambda t}\psi_\lambda(x)\frac{1}{2\lambda}\, d\Gamma(\lambda)$ (recall that $\Lambda(\infty) = \infty$). Since $\Psi P_t = e^{-\lambda t}\Psi$ and Ψ is an isometry, we conclude $b(\lambda) = \frac{1}{2\lambda}$ for almost all λ. □

Lemma 6.20 *Assume ξ is a probability distribution with compact support contained in $(0, \infty)$. Take any sequence (t_n) converging to ∞, for which $\mathbb{P}_\xi(X_{t_n} \in \bullet | t_n < T)$ converges weakly in $\overline{\mathbb{R}}_+$ to a probability measure ν, and define $\theta = \nu(\{\infty\})$. Then, either $\theta = 1$, in which case $\nu = \delta_\infty$, or $\theta < 1$, in which case necessarily $\varphi_{\underline{\lambda}} \in L^1(dx)$ and*

$$d\nu(x) = \theta\, d\delta_\infty(x) + \frac{(1-\theta)}{\int \varphi_{\underline{\lambda}}(z)\, dz}\varphi_{\underline{\lambda}}(x)\, dx.$$

Proof Notice first that by a compactness argument such sequence and limit always exist in the compact set $\overline{\mathbb{R}}_+$. Also, if $\theta = 1$, we have $\nu = \delta_\infty$. So in the rest of the proof, we assume that $\theta < 1$. This is equivalent to the existence of a finite z_0 for which

$$\lim_{n\to\infty} \mathbb{P}_\xi(X_{t_n} \le z_0 | t_n < T) = \nu\big([0, z_0]\big) > 0.$$

Using Theorem 6.15, we conclude that for all $z > 0$ we have

$$\lim_{n\to\infty} \mathbb{P}_\xi(X_{t_n} \le z | t_n < T) = \nu\big([0, z]\big) = \nu\big([0, z_0]\big)\frac{\int_0^z \varphi_{\underline{\lambda}}(x)\, dx}{\int_0^{z_0} \varphi_{\underline{\lambda}}(x)\, dx}.$$

We deduce that $\varphi_{\underline{\lambda}} \in L^1(dx)$ and

$$\lim_{n\to\infty} \mathbb{P}_\xi(X_{t_n} \le z | t_n < T) = \nu\big([0, \infty)\big)\frac{\int_0^z \varphi_{\underline{\lambda}}(x)\, dx}{\int_0^\infty \varphi_{\underline{\lambda}}(x)\, dx} = \frac{(1-\theta)\int_0^z \varphi_{\underline{\lambda}}(x)\, dx}{\int_0^\infty \varphi_{\underline{\lambda}}(x)\, dx}.$$ □

Corollary 6.21

(i) $\varphi_{\underline{\lambda}} \in L^1(dx) \Leftrightarrow \underline{\lambda} > 0$;
(ii) *If $\varphi_{\underline{\lambda}} \notin L^1(dx)$, the Yaglom limit exists and is given by δ_∞, that is, for all ξ probability measure with bounded support contained in $(0, \infty)$ and all $z \ge 0$*

$$\lim_{t\to\infty} \mathbb{P}_\xi(X_t \ge z | t < T) = 1;$$

(iii) *If* $\varphi_{\underline{\lambda}} \in L^1(dx)$ *then for all probability measures* ξ *with bounded support contained in* $(0,\infty)$ *and all* $z \geq 0$

$$\liminf_{t\to\infty} \mathbb{P}_\xi(X_t \geq z | t < T) \geq \frac{\int_z^\infty \varphi_{\underline{\lambda}}(x)\,dx}{\int_0^\infty \varphi_{\underline{\lambda}}(x)\,dx}.$$

Proof (i) Assume that $\varphi_{\underline{\lambda}} \notin L^1(dx)$. Then $\underline{\lambda}$ cannot be positive because of Lemma 6.18 and the fact $\varphi_{\underline{\lambda}} = e^{-\gamma}\psi_{\underline{\lambda}}$ is nonnegative (see Lemma 6.7). On the other hand, if $\varphi_{\underline{\lambda}} \in L^1(dx)$ then $\underline{\lambda}$ cannot be 0 because $\varphi_0 = e^{-\gamma}\Lambda \notin L^1(dx)$ (see Hypothesis **H**).

(ii) According to the previous lemma, if (t_n) is a sequence for which $\mathbb{P}_\xi(X_{t_n} \in \bullet | t_n < T)$ converges weakly to a probability measure ν on $\overline{\mathbb{R}}_+$, then necessarily $\nu(\{\infty\}) = 1$, otherwise $\varphi_{\underline{\lambda}}$ is integrable. From this we have the convergence to δ_∞.

(iii) Consider a fixed probability measure ξ with bounded support contained in $(0,\infty)$ and $z \geq 0$. Take a sequence (t_n) converging to ∞ such that

$$\lim_{n\to\infty} \mathbb{P}_\xi(X_{t_n} \geq z | t_n < T) = \liminf_{t\to\infty} \mathbb{P}_\xi(X_t \geq z | t < T).$$

Take a further subsequence (t_{n_k}) and a probability measure ν on $\overline{\mathbb{R}}_+$ such that $\mathbb{P}_\xi(X_{t_{n_k}} \in \bullet | t_{n_k} < T)$ converges weakly to ν. According to the previous lemma, we get that for some $\theta \in [0,1]$ we have

$$d\nu(x) = \theta\, d\delta_\infty(x) + \frac{(1-\theta)}{\int \varphi_{\underline{\lambda}}(z)\,dz}\varphi_{\underline{\lambda}}(x)\,dx.$$

Then we deduce

$$\begin{aligned}\liminf_{t\to\infty} \mathbb{P}_\xi(X_t \geq z | t < T) &= \lim_{k\to\infty} \mathbb{P}_\xi(X_{t_{n_k}} \geq z | t_{n_k} < T) \\ &= \theta + \frac{(1-\theta)}{\int \varphi_{\underline{\lambda}}(x)\,dx}\int_z^\infty \varphi_{\underline{\lambda}}(x)\,dx \geq \frac{\int_z^\infty \varphi_{\underline{\lambda}}(x)\,dx}{\int \varphi_{\underline{\lambda}}(x)\,dx}. \quad \square\end{aligned}$$

Remark 6.22 The main result of this section is to prove that in (iii) the limit actually exists and coincides with the probability measure with density (with respect to the Lebesgue measure)

$$\frac{\varphi_{\underline{\lambda}}(x)}{\int \varphi_{\underline{\lambda}}(y)\,dy}.$$

Corollary 6.23 *For any* $x > 0$, *there exists a constant* $L(x) > 0$ *such that for all* $s \geq 0$

$$\mathbb{P}_x(X_s \geq x | s < T) \geq L(x).$$

Proof From the previous corollary, we obtain

$$\liminf_{s\to\infty}\mathbb{P}_x(X_s\ge x|s<T)\ge\begin{cases}1 & \text{if } \varphi_{\underline{\lambda}}\notin L^1(dy),\\ \frac{\int_x^\infty \varphi_{\underline{\lambda}}(y)\,dy}{\int_0^\infty \varphi_{\underline{\lambda}}(y)\,dy} & \text{if } \varphi_{\underline{\lambda}}\in L^1(dy).\end{cases}$$

From Girsanov's Theorem, we get, with $h=\alpha^2-\alpha'$,

$$\begin{aligned}\mathbb{P}_x(X_s\ge x|s<T)&\ge\mathbb{P}_x(X_s\ge x, s<T)\\ &\ge e^{\frac{\gamma(x)}{2}}\mathbb{E}_x\big(e^{-\frac{1}{2}\int_0^s h(B_r)\,dr-\frac{\gamma(B_s)}{2}}, s<T\wedge T_{x+1}\big)\\ &\ge e^{C(x)+D(x)s}\mathbb{P}_x(s<T\wedge T_{x+1}),\end{aligned}$$

where $2C(x)=\gamma(x)-\max\{\gamma(y):y\in[0,x+1]\}$ and $2D(x)=-\max\{h(y):y\in[0,x+1]\}$. The last term is computed under the Brownian law, and the result follows. □

Theorem 6.24
For all $x>0$, the following limit exists and is independent of x

$$0\le\eta=\lim_{t\to\infty}-\frac{\log(\mathbb{P}_x(t<T))}{t}.$$

Proof Consider a fixed $x>0$. The existence of the limit is guaranteed by the subadditive theorem. Indeed, the strong Markov property and the previous corollary yield

$$\begin{aligned}\mathbb{P}_x(t+s<T)&=\mathbb{E}_x\big(\mathbb{P}_{X_s}(t<T), s<T\big)\\ &\ge\mathbb{P}_x(t<T)\mathbb{P}_x(s<T)\mathbb{P}_x(X_s\ge x|s<T)\\ &\ge L(x)\mathbb{P}_x(t<T)\mathbb{P}_x(s<T).\end{aligned}$$

Here we have used that $\mathbb{P}_y(s<T)\ge\mathbb{P}_x(s<T)$ whenever $y\ge x$. Hence, the function $t\to a(t)=-\log(\mathbb{P}_x(t<T))-\log(L(x))$ is subadditive, and therefore the limit

$$\eta(x)=\lim_{t\to\infty}-\frac{\log(\mathbb{P}_x(t<T))}{t}$$

exists. We shall prove that η does not depend on x.

Clearly, $\eta(x)\ge\eta(y)$ if $x\le y$. On the other hand, Harnack's Inequality implies that for $t\ge 1$ there exists a constant $C=C(x,y)$ such that for all z

$$r_t(y,z)\le Cr_{t+1}(x,z).$$

Integrating this inequality yields $\mathbb{P}_y(t<T)\le C\mathbb{P}_x(t+1<T)$. Then $\eta(y)\ge\eta(x)$, proving the result. □

Corollary 6.25 *We have the inequality* $\eta \le \underline{\lambda}$. *Moreover, if* $\varphi_{\underline{\lambda}} \notin L^1(dx)$ *then* $\eta = 0 = \underline{\lambda}$.

Proof For any bounded measurable set B of positive Lebesgue measure and a fixed $x > 0$, we have $\mathbb{P}_x(t < T) \ge \mathbb{P}_x(X_t \in B, t < T)$. This last quantity is given by

$$\int_{\underline{\lambda}}^{\infty} e^{-\lambda t} \psi_\lambda(x) \int_B \psi_\lambda(y)\, d\mu(y)\, d\Gamma(\lambda).$$

As we have done before, this integral is asymptotically equivalent to

$$\int_{\underline{\lambda}}^{\underline{\lambda}+\epsilon} e^{-\lambda t} \psi_\lambda(x) \int_B \psi_\lambda(y)\, d\mu(y)\, d\Gamma(\lambda),$$

for all $\epsilon > 0$ small enough. We recall that we can assume $\psi_\lambda(x) \ge a$, $\int_B \psi_\lambda(y)\, d\mu(y) \ge a$, for some $a > 0$ and all $\lambda \in [\underline{\lambda}, \underline{\lambda} + \epsilon]$. Thus

$$\int_{\underline{\lambda}}^{\underline{\lambda}+\epsilon} e^{-\lambda t} \psi_\lambda(x) \int_B \psi_\lambda(y)\, d\mu(y)\, d\Gamma(\lambda) \ge a^2 e^{-(\underline{\lambda}+\epsilon)t} \Gamma\big([\underline{\lambda}, \underline{\lambda} + \epsilon]\big)$$

we conclude that $\eta \le \underline{\lambda} + \epsilon$, from where the first part follows.

The second part holds because $\underline{\lambda} = 0$, when $\varphi_{\underline{\lambda}} \notin L^1(dx)$. □

Now we can state the main theorem of this chapter. We point out that the existence of the Yaglom limit and that of a QSD for killed one-dimensional diffusion processes have been proved by various authors, following the pioneering work by Mandl (1961) (see, e.g., Collet et al. 1995a; Martínez and San Martín 1994; Steinsaltz and Evans 2007; Kolb and Steinsaltz 2012 and references therein).

Theorem 6.26 *Assume Hypotheses* **H** *and* **H1** *hold.*

(i) (*Yaglom limit*) *For any* $x > 0$, *the family* $(\mathbb{P}_x(X_t \in \bullet | t < T))_t$ *converges weakly, as* $t \to \infty$, *to*

$$\begin{cases} \delta_\infty & \text{if } \varphi_{\underline{\lambda}} \notin L^1(dx), \\ \nu & \text{if } \varphi_{\underline{\lambda}} \in L^1(dx), \end{cases}$$

where in the second case the measure ν *has a density proportional to* $\varphi_{\underline{\lambda}}$. *We notice that the classification for the limit is the same as*

$$\begin{cases} \delta_\infty & \text{if } \underline{\lambda} = 0, \\ \nu & \text{if } \underline{\lambda} > 0. \end{cases}$$

(ii) *We have* $\Gamma(\{\underline{\lambda}\}) = (\int \psi_{\underline{\lambda}}^2(z)\, d\mu(z))^{-1}$ *and for all* $x, y > 0$

$$\lim_{t\to\infty} e^{\underline{\lambda} t} r_t(x, y) = \psi_{\underline{\lambda}}(x) \psi_{\underline{\lambda}}(y) \Gamma\big(\{\underline{\lambda}\}\big), \tag{6.27}$$

which is 0 *in the case* $\underline{\lambda} = 0$ *because* $\Gamma(\{\underline{\lambda}\}) = 0$ *(Remark* 6.14*). Also, for all* $x > 0$*, it holds*

$$\lim_{t \to \infty} e^{\underline{\lambda} t} \mathbb{P}_x(t < T) = \begin{cases} \frac{\Gamma(\{\underline{\lambda}\})}{2\underline{\lambda}} \psi_{\underline{\lambda}}(x) & \text{if } \underline{\lambda} > 0, \\ 0 & \text{otherwise}. \end{cases} \tag{6.28}$$

(iii) *For all* $s \geq 0$ *and all* $x > 0$*, we have* $\lim_{t \to \infty} \frac{\mathbb{P}_x(t+s<T)}{\mathbb{P}_x(t<T)} = e^{-\underline{\lambda} s}$*, proving that* $\eta = \underline{\lambda}$.

(iv) *For all* $x > 0$ *and all* $t > 0$*, we have the bound*

$$\frac{\mathbb{P}_x(t < T)}{\mathbb{P}_1(t < T)} \leq \frac{\Lambda(x)}{\Lambda(1)} + 1, \tag{6.29}$$

and the limit

$$\lim_{t \to \infty} \frac{\mathbb{P}_x(t < T)}{\mathbb{P}_1(t < T)} = \frac{\psi_{\underline{\lambda}}(x)}{\psi_{\underline{\lambda}}(1)}. \tag{6.30}$$

(v) *For all* $s \geq 0$*, all* $x > 0$ *and* $A \in \mathcal{B}(\mathcal{C}([0, s]))$*, we have*

$$\lim_{t \to \infty} \mathbb{P}_x(X \in A | t < T) = \mathbb{P}_x(Z \in A).$$

The distribution of Z *is an* h*-transform of* X*, whose transition density is given by*

$$\mathbb{P}_x(Z_s \in dy) = e^{\underline{\lambda} s} \frac{\psi_{\underline{\lambda}}(y)}{\psi_{\underline{\lambda}}(x)} \mathbb{P}_x(X_s \in dy, s < T)$$

for $x > 0$*,* $y > 0$ *and* $s \geq 0$*, and* Z *is the unique solution to the SDE*

$$dZ_t = dB_t + \left(\frac{\psi'_{\underline{\lambda}}(Z_t)}{\psi_{\underline{\lambda}}(Z_t)} - \alpha(Z_t) \right) dt, \qquad Z_0 = x. \tag{6.31}$$

Remark 6.27 There are examples where $\Lambda(\infty) = \infty$, Hypothesis **H** is verified, the condition $\mu(\mathbb{R}_+) < \infty$, but still $\underline{\lambda} = \eta = 0$. It suffices to take $\alpha(x) = \frac{1}{1+x}$.

Proof (i) We shall assume first that $\varphi_{\underline{\lambda}} \in L^1(dx)$, or equivalently, that $\underline{\lambda} > 0$. In this situation, $\psi_{\underline{\lambda}} \in L^1(d\mu)$ and $\int \psi_{\underline{\lambda}}(z) \, d\mu(z) = \frac{1}{2\underline{\lambda}}$. Also the measure μ is finite and $\mathbf{1} \in L^2(d\mu)$.

Recall that

$$\mathbb{P}_x(t < T) = P_t \mathbf{1}(x) = \int_{\underline{\lambda}}^{\infty} \frac{e^{-\lambda t}}{2\lambda} \psi_\lambda(x) \, d\Gamma(\lambda).$$

Using the same ideas as in the proof of Theorem 6.15, we obtain that this integral, as $t \to \infty$, is equivalent to

$$\frac{\psi_{\underline{\lambda}}(x)}{2\underline{\lambda}} \int_{\underline{\lambda}}^{\underline{\lambda}+\epsilon} e^{-\lambda t} \, d\Gamma(\lambda)$$

for all $\epsilon > 0$, sufficiently small. On the other hand, if B is a bounded measurable set, we get that $\mathbb{P}_x(X_t \in B, t < T)$ is equivalent to

$$\int_B \psi_{\underline{\lambda}}(z)\,d\mu(z)\psi_{\underline{\lambda}}(x)\int_{\underline{\lambda}}^{\underline{\lambda}+\epsilon} e^{-\lambda t}\,d\Gamma(\lambda)$$

also for all $\epsilon > 0$ small enough. We conclude that

$$\lim_{t\to\infty}\mathbb{P}_x(X_t \in B|t < T) = 2\underline{\lambda}\int_B \psi_{\underline{\lambda}}(z)\,d\mu(z) = \frac{\int_B \psi_{\underline{\lambda}}(z)\,d\mu(z)}{\int \psi_{\underline{\lambda}}(z)\,d\mu(z)}.$$

This shows that the set of probability measures $(\mathbb{P}_x(X_t \in \bullet|t < T))_t$ is tight in $\mathbb{R}_+$ and, by Lemma 6.20, we obtain the desired weak convergence of this family to the probability measure ν on $\mathbb{R}_+$ whose density with respect to μ is

$$\frac{\psi_{\underline{\lambda}}(z)}{\int \psi_{\underline{\lambda}}(y)\,d\mu(y)},$$

and part (i) is proved.

(ii) We start with the formula $r_t(x,y) = \int_{\underline{\lambda}}^{\infty} e^{-\lambda t}\psi_\lambda(x)\psi_\lambda(y)\,d\Gamma(\lambda)$ to get

$$r_t(x,y) = e^{-\underline{\lambda}t}\psi_{\underline{\lambda}}(x)\psi_{\underline{\lambda}}(y)\Gamma\big(\{\underline{\lambda}\}\big) + \int_{(\underline{\lambda},\infty)} e^{-\lambda t}\psi_\lambda(x)\psi_\lambda(y)\,d\Gamma(\lambda).$$

So we need to prove that $\int_{(\underline{\lambda},\infty)} e^{(\underline{\lambda}-\lambda)t}\psi_\lambda(x)\psi_\lambda(y)\,d\Gamma(\lambda)$ converges to 0, which we do by the Dominated Convergence Theorem. This integral may be divided into two pieces. The first one is

$$\int_{(\underline{\lambda},\underline{\lambda}+\epsilon)} e^{(\underline{\lambda}-\lambda)t}\psi_\lambda(x)\psi_\lambda(y)\,d\Gamma(\lambda),$$

for some fixed $\epsilon > 0$. To show that this integral converges to zero, it is enough to bound the integrand on the interval $[\underline{\lambda}, \underline{\lambda}+\epsilon]$ by a constant. For the other term, that is, the integral over $[\underline{\lambda}+\epsilon,\infty)$, we assume $t > 1$ and we use the inequality

$$\left|e^{(\underline{\lambda}-\lambda)t}\psi_\lambda(x)\psi_\lambda(y)\right| \le e^{-\lambda\beta}\left|\psi_\lambda(x)\psi_\lambda(y)\right|$$

where $\beta = \frac{\epsilon}{\underline{\lambda}+\epsilon}$. This upper bound is integrable by Lemma 6.9, proving (6.27).

Formula (6.20) and the fact that $\Lambda(\infty)=\infty$ give

$$\mathbb{P}_x(t < T) = \int_{\underline{\lambda}}^{\infty}\frac{e^{-\lambda t}}{2\lambda}\psi_\lambda(x)\,d\Gamma(\lambda).$$

Then, (6.28) is shown using the same technique as before.

The only thing left to prove is $\Gamma(\{\underline{\lambda}\}) = (\int \psi_{\underline{\lambda}}^2(z)\,d\mu(z))^{-1}$. Formula (6.19) is exactly the inequality $\Gamma(\{\underline{\lambda}\}) \ge (\int \psi_{\underline{\lambda}}^2(x)\,d\mu(x))^{-1}$, and we show now the opposite one. For that purpose, we use equality (6.26)

$$\mathbb{E}_x\big(\psi_\lambda(X_t), t < T\big) = e^{-\lambda t}\psi_\lambda(x),$$

valid for $\lambda \in (0, \underline{\lambda}]$. This equality shows that

$$\begin{aligned}
&e^{-\underline{\lambda} t}\left(\int_0^n \psi_{\underline{\lambda}}^2(x)\,d\mu(x)\right)^2 \Gamma\big(\{\underline{\lambda}\}\big)\\
&\quad\le \int_{\underline{\lambda}}^\infty e^{-\lambda t}\left(\int_0^n \psi_\lambda(x)\psi_{\underline{\lambda}}(x)\,d\mu(x)\right)^2 d\Gamma(\lambda)\\
&\quad= \int_0^n \psi_{\underline{\lambda}}(z)\mathbb{E}_z\big(\psi_{\underline{\lambda}}(X_t), X_t \le n, t < T\big)\,d\mu(z)\\
&\quad\le \int_0^n \psi_{\underline{\lambda}}(z)\mathbb{E}_z\big(\psi_{\underline{\lambda}}(X_t), t < T\big)\,d\mu(z)\\
&\quad= e^{-\underline{\lambda} t}\int_0^n \psi_{\underline{\lambda}}^2(z)\,d\mu(z).
\end{aligned}$$

Hence we deduce

$$\Gamma\big(\{\underline{\lambda}\}\big) \le \left(\int_0^n \psi_{\underline{\lambda}}^2(z)\,d\mu(z)\right)^{-1},$$

from where the desired inequality follows.

(iii) Notice that

$$\frac{\mathbb{P}_x(t+s<T)}{\mathbb{P}_x(t<T)} = \frac{\int_{\underline{\lambda}}^\infty e^{-\lambda(t+s)}\frac{\psi_\lambda(x)}{2\lambda}\,d\Gamma(\lambda)}{\int_{\underline{\lambda}}^\infty e^{-\lambda(t)}\frac{\psi_\lambda(x)}{2\lambda}\,d\Gamma(\lambda)}.$$

The proof then is similar to that of part (i).

(iv) The same type of arguments show that (6.30) holds. We shall now prove the domination (6.29). The inequality is obvious if $x \le 1$. So, assume that $x > 1$ and consider

$$\begin{aligned}
\mathbb{P}_1(t<T) &\ge \mathbb{P}_1(T_x < T, t + T_x < T) = \mathbb{P}_1(T_x < T)\mathbb{P}_x(t<T)\\
&= \frac{\Lambda(1)}{\Lambda(x)}\mathbb{P}_x(t<T).
\end{aligned}$$

Here we have used (6.7), and the desired inequality is proved.

Part (v) follows from (iii), (iv), the strong Markov property and the Dominated Convergence Theorem. Indeed, conditioning at time s, we get

$$\mathbb{P}_x(X \in A \mid t < T) = \frac{\mathbb{P}_x(t-s<T)}{\mathbb{P}_x(t<T)}\mathbb{E}_x\left(\frac{\mathbb{P}_{X_s}(t-s<T)}{\mathbb{P}_x(t-s<T)}, X \in A, s < T\right).$$

The domination follows from (6.29) and the fact that (see (6.6))

$$\mathbb{E}_x\big(\Lambda(X_s), s<T\big) \le \Lambda(x). \tag{6.32}$$

We now prove the results under the hypothesis $\varphi_{\underline{\lambda}} \notin L^1(dx)$, or equivalently, $\underline{\lambda}=0$. In this case, notice that $\psi_0 = \Lambda$ and $\varphi_0 = \Lambda e^{-\gamma}$.

Part (i) was proved in Corollary 6.21, and part (ii) is straightforward in this case. As for part (iii), notice that

$$\begin{aligned}1 \ge \frac{\mathbb{P}_x(t+s<T)}{\mathbb{P}_x(t<T)} &\ge \frac{\mathbb{E}_x(\mathbb{P}_{X_t}(s<T), t<T)}{\mathbb{P}_x(t<T)}\\ &\ge \mathbb{P}_x(X_t \ge z | t<T)\mathbb{P}_z(s<T)\end{aligned}$$

holds for all $z>0$. From part (i), we deduce that the lower bound converges to $\mathbb{P}_z(s<T)$, as $t\to\infty$ for all $z>0$. Hence we conclude

$$1 \ge \limsup_{t\to\infty} \frac{\mathbb{P}_x(t+s<T)}{\mathbb{P}_x(t<T)} \ge \liminf_{t\to\infty} \frac{\mathbb{P}_x(t+s<T)}{\mathbb{P}_x(t<T)} \ge \lim_{z\to\infty} \mathbb{P}_z(s<T) = 1.$$

(iv) We notice that domination (6.29) holds independently of the value of $\underline{\lambda}$.

Let us prove the limit result in (6.30). Since $v_t(x) = \frac{\mathbb{P}_x(t<T)}{\mathbb{P}_1(t<T)}$ is increasing in x and the family of functions $(v_t)_{t>0}$ is bounded by the continuous function $\frac{\Lambda(x)}{\Lambda(1)}+1$, we can apply the Helly's Theorem (see, for example, Coddington and Levinson 1955, p. 233) to get the existence of a subsequence $t_n\to\infty$, such that v_{t_n} converges to a increasing nonnegative function w (w may depend on the subsequence (t_n)). Obviously, $w(0)=0$, $w(1)=1$ and $w(x)\le \frac{\Lambda(x)}{\Lambda(1)}+1$.

Let us show $w(y)>0$ for $y\in(0,1)$. Indeed, consider $y<x$ and compute

$$\begin{aligned}v_t(y) = \frac{\mathbb{P}_y(t<T)}{\mathbb{P}_1(t<T)} &\ge \frac{\mathbb{P}_y(t+T_x<T, T_x<T)}{\mathbb{P}_1(t<T)} \ge \frac{\mathbb{P}_y(T_x<T)\mathbb{P}_x(t<T)}{\mathbb{P}_1(t<T)}\\ &= \frac{\Lambda(y)}{\Lambda(x)} v_t(x).\end{aligned}$$

From this inequality, we get

$$w(x) \le \frac{\Lambda(x)}{\Lambda(y)} w(y)$$

for any $x\ge y$. From $w(1)=1$, we conclude $w(y)>0$. Let us now show the continuity of w for $x>0$. Since w is increasing, it is enough to prove that $w(x+)\le w(x-)$. We have $w(x+h)\le \frac{\Lambda(x+h)}{\Lambda(x-h)} w(x-h)$ for $h>0$, and the continuity of Λ implies $w(x+)\le w(x-)$, as we wanted.

The Markov property shows that

$$\frac{\mathbb{P}_x(t_n+s<T)}{\mathbb{P}_x(t_n<T)}\frac{\mathbb{P}_x(t_n<T)}{\mathbb{P}_1(t_n<T)} = \mathbb{E}_x\big(v_{t_n}(X_s), s<T\big).$$

The left side converges to $w(x)$, while the right side, due to the Dominated Convergence Theorem (see (6.29) and (6.32)), converges to $\mathbb{E}_x(w(X_s), s < T)$. Thus, for all $s \ge 0$ and all $x \ge 0$

$$w(x) = \mathbb{E}_x\big(w(X_s), s < T\big). \tag{6.33}$$

The goal is to prove that $w(x) = \frac{\Lambda(x)}{\Lambda(1)}$, from where we have a unique accumulation point for $(v_t)_t$ and therefore the convergence. Consider the function $G(x) = \frac{\Lambda(x)}{\Lambda(1)} - w(x)$. We know this function is continuous on $(0,\infty)$ and it is nonnegative on $[1,\infty)$. Let us prove it is nonnegative on $\mathbb{R}_+$. In fact, if there exists $x_0 < 1$ for which $G(x_0) < 0$ then

$$\min\big\{G(x) : x \in (0,1)\big\}\mathbb{P}_{x_0}(s < T) \le \mathbb{E}_{x_0}\big(G(X_s), s < T\big) \le G(x_0) < 0.$$

Letting s converge to ∞, and since $\mathbb{P}_{x_0}(T = \infty) = 0$, we arrive at a contradiction. Therefore, $G \ge 0$.

From $G(1) = 0$, $G \ge 0$ and $\mathbb{E}_1(G(X_s), s < T) \le G(1) = 0$, we deduce $G = 0$ with respect to the measure $\mathbb{P}_1(\cdot, s < T)$-a.e. for any $s > 0$. Given that G is continuous, to achieve the result, i.e. $G \equiv 0$, it is enough to prove that for any fixed interval $[a,b]$, $\mathbb{P}_1(X_s \in [a,b], s < T)$ is strictly positive for some (every) $s > 0$. Now $\mathbb{P}_1(X_s \in [a,b], s < T) \ge \mathbb{P}_1(X_s \in [a,b], s < T, s < T_M)$ and this quantity is positive by the Cameron–Martin formula and the fact that it is such for the Brownian motion.

Hence, we have for all $x > 0$ the equality $w(x) = \frac{\Lambda(x)}{\Lambda(1)}$, implying in particular that w is continuous at 0, and it is independent of the subsequence (t_n). This finishes the proof of (iii).

Finally, part (v) is proved in the same way it was proved under the hypothesis $\underline{\lambda} > 0$. □

Remark 6.28 Notice that in parts (i), (ii), (iii) we can replace the initial distribution δ_x by any distribution ξ with bounded support contained in $(0,\infty)$. Also the results hold when the initial distribution is given by $h\,d\mu$ with $h \in L^2(d\mu)$.

In general, the convergence in part (i) depends on the initial distribution. For example, in the case $\underline{\lambda} > 0$, each of the functions φ_λ, for $\lambda \in (0, \underline{\lambda}]$ is nonnegative and integrable. We shall see in the next section that $\frac{\varphi_\lambda(x)\,dx}{\int \varphi_\lambda(z)\,dz}$ is a QSD, that is, for any measurable set A and any $t \ge 0$ (see Definition 2.1 and Theorem 2.2),

$$\mathbb{P}_{\varphi_\lambda\,dx}(X_t \in A, t < T) = e^{-\lambda t}\int_A \varphi_\lambda(y)\,dy.$$

In particular,

$$\mathbb{P}_{\varphi_\lambda\,dx}(X_t \in A \mid t < T) = \frac{\int_A \varphi_\lambda(y)\,dy}{\int \varphi_\lambda(z)\,dz}.$$

Hence, the asymptotic distribution is not the one given in (i).

The following result gives some extra information on the asymptotic behavior of the exit probability.

Proposition 6.29

$$\limsup_{t\to\infty} e^{\underline{\lambda} t}\mathbb{P}_x(t < T) \le \frac{\psi_{\underline{\lambda}}(x)}{2\underline{\lambda}}\left(\sup_{a>0}\left\{\psi_{\underline{\lambda}}(a)\int_a^\infty \varphi_{\underline{\lambda}}(y)\,dy\right\}\right)^{-1}.$$

Proof From Itô's formula, we get

$$\begin{aligned}\psi_{\underline{\lambda}}(x) &= \mathbb{E}_x\big(e^{\underline{\lambda}(t\wedge T\wedge T_M)}\psi_{\underline{\lambda}}(X_{t\wedge T\wedge T_M})\big)\\ &= \psi_{\underline{\lambda}}(M)\mathbb{E}_x\big(e^{\underline{\lambda} T_M}, T_M < t\wedge T\big) + e^{\underline{\lambda} t}\mathbb{E}_x\big(\psi_{\underline{\lambda}}(X_t), t < T\wedge T_M\big).\end{aligned}$$

Now,

$$\psi_{\underline{\lambda}}(M)\mathbb{E}_x\big(e^{\underline{\lambda} T_M}, T_M < t\wedge T\big) \le e^{\underline{\lambda} t}\psi_{\underline{\lambda}}(M)\mathbb{P}_x(T_M < T) = e^{\underline{\lambda} t}\psi_{\underline{\lambda}}(M)\frac{\Lambda(x)}{\Lambda(M)}.$$

This term tends to 0 as $M \to \infty$ (see Lemma 6.18). Then,

$$\psi_{\underline{\lambda}}(x) = e^{\underline{\lambda} t}\mathbb{E}_x\big(\psi_{\underline{\lambda}}(X_t), t < T\big) \ge \psi(a)\mathbb{P}_x(X_t \ge a | t < T)e^{\underline{\lambda} t}\mathbb{P}_x(t < T).$$

The result follows from Theorem 6.26. □

In the case $\eta = 0$, we have some extra information.

Proposition 6.30 *If* $\eta = 0$ *then*

(i) $\Lambda(X_s)1_{s<T}$ *is a* $\mathbb{P}_x$*-martingale*;
(ii) *For any* $x > 0$, *we have* $\mathbb{P}_x(T^Z < \infty \text{ or } \lim_{a\nearrow\infty} T_a^Z < \infty) = 0$, *where* Z *is the* *h-transform defined in Theorem* 6.26 *part* (iv);
(iii) *For all* $s \ge 0$, *all* $x > 0$ *and* $A \in \mathcal{B}(\mathcal{C}([0, s]))$,

$$\lim_{t\to\infty}\mathbb{P}_x(X \in A | t < T) = \lim_{M\to\infty}\mathbb{P}_x(X \in A | T_M < T) = \mathbb{P}_x(Z \in A).$$

Proof (i) We notice that $w(x) = \mathbb{E}_x(w(X_s), s < T)$ (see (6.33)), but $w = \Lambda/\Lambda(1)$, proving that

$$\Lambda(x) = \mathbb{E}_x\big(\Lambda(X_s), s < T\big).$$

The strong Markov property then shows the result.

(ii) Let $T_a^Z = \inf\{t \ge 0 : Z_t = a\}$, $S^Z = \lim_{a\nearrow\infty} T_a^Z$, and let $G(y) = \frac{1}{\Lambda(y)}$ for $y > 0$. A straightforward computation shows that $\frac{1}{2}G'' + (\frac{\Lambda'}{\Lambda} - \alpha)G' = 0$. Therefore, by Itô's formula, if $0 < \epsilon < x < a$,

$$G(x) = \mathbb{E}_x\big(G(Z_{T_a^Z\wedge T_\epsilon^Z})\big) = G(\epsilon)\mathbb{P}_x\big(T_\epsilon^Z < T_a^Z\big) + G(a)\mathbb{P}_x\big(T_a^\epsilon < T_\epsilon^Z\big),$$

from which we deduce that $\mathbb{P}_x(T_a^Z < T_\epsilon^Z) = \frac{G(x)-G(\epsilon)}{G(a)-G(\epsilon)}$. Since $G(0+) = \infty$, we obtain $\mathbb{P}_x(T_a^Z < T_0^Z) = 1, \forall a > x$. Hence

$$\mathbb{P}_x\big(S^Z \le T^Z\big) = 1.$$

An application of Girsanov's Theorem leads to

$$\mathbb{E}_x\left(\frac{\Lambda(X_s)}{\Lambda(x)}, X \in A, s < T_a, s < T_\epsilon\right) = \mathbb{P}_x\big(Z \in A, s < T_a^Z, s < T_\epsilon^Z\big)$$

for $A \in \mathcal{B}(\mathcal{C}([0,s]))$, $0 < \epsilon < x < a$. Now, let $\epsilon \to 0$, $a \to \infty$ to obtain

$$\mathbb{E}_x\left(\frac{\Lambda(X_s)}{\Lambda(x)}, X \in A, s < T\right) = \mathbb{P}_x\big(Z \in A, s < S^Z\big).$$

On the other hand,

$$\Lambda(x) = \Lambda(M)\mathbb{P}_x(T_M \le s, T_M < T) + \mathbb{E}_x\big(\Lambda(X_s), s < T_M \wedge T\big)$$

and $\mathbb{P}_x(T_M < T) = \frac{\Lambda(x)}{\Lambda(M)}$ imply that

$$\mathbb{P}_x(s < T_M | T_M < T) = \mathbb{E}_x\left(\frac{\Lambda(X_s)}{\Lambda(x)},\ s < T_M \wedge T\right).$$

The Monotone Convergence Theorem allows us to conclude

$$\lim_{M\to\infty} \mathbb{P}_x(s < T_M | T_M < T) = \mathbb{E}_x\left(\frac{\Lambda(X_s)}{\Lambda(x)},\ s < T\right) = \mathbb{P}_x\big(s < S^Z\big).$$

The middle term is equal to 1 by (i), and we get $\mathbb{P}_x(s < S^Z \le T_0^Z) = 1$ for any $s > 0$. Thus (ii) is proved.

(iii) Consider

$$\begin{aligned}
\mathbb{P}_x(X \in A, s < T_M | T_M < T) &= \mathbb{E}_x\left(\frac{\mathbb{P}_{X_s}(T_M < T)}{\mathbb{P}_x(T_M < T)}, X \in A, s < T_M, s < T\right)\\
&= \mathbb{E}_x\left(\frac{\Lambda(X_s)}{\Lambda(x)}, X \in A, s < T_M, s < T\right)\\
&= \mathbb{P}_x\big(Z \in A, s < T_M^Z\big).
\end{aligned}$$

Finally, $\lim_{M\to\infty} \mathbb{P}_x(X \in A, s < T_M | T_M < T) = \mathbb{P}_x(Z \in A, s < S^Z) = \mathbb{P}_x(Z \in A)$, and $\lim_{M\to\infty} \mathbb{P}_x(X \in A, T_M \le s | T_M < T) \le \lim_{M\to\infty} \mathbb{P}_x(T_M \le s | T_M < T) = 0$, which allows us to conclude the result. □

Example 6.17. (Continuation) When the drift is a negative constant, that is, $\alpha > 0$, we have $\eta = \underline{\lambda} = \alpha^2/2$ and Theorem 6.26 was shown in Martínez and San Martín (1994). In this case, we have $\lim_{t\to\infty} \mathbb{P}_x(X_t \in E | t < T) = \int_E \alpha^2 y e^{-\alpha y}\, dy$ for any $E \subset \mathbb{R}_+$.

Example 6.31 Consider the case where $\alpha(x) = x$, that is, X is a Ornstein–Uhlenbeck process. Then $\Lambda(\infty) = \infty$, $\underline{\lambda} = 1$, the measure μ is finite and moreover the operator $\mathcal{L}$ has a discrete spectrum on $L^2(d\mu)$ (see, for example, Mandl 1961). Nevertheless, it has a continuum family of QSDs indexed by $\lambda \in (0, 1]$, which have densities with respect to the Lebesgue measure of the form $e^{-x}U_\lambda$, where U_λ is a parabolic cylindric function (see Lladser and San Martín 2000 for the domain of attraction of each QSD).

The fact that $\underline{\lambda} = \eta$, under the hypothesis of Theorem 6.26, allows us to get some information about the dependence of $\underline{\lambda}$ on α. For this purpose, we denote Λ^α, $\underline{\lambda}(\alpha)$ and $\eta(\alpha)$ the quantities Λ, $\underline{\lambda}$ and η associated to α.

Corollary 6.32 *Assume that $\alpha \geq \beta$ are $\mathcal{C}^1$ functions that satisfy* **H**, **H1**. *Then*

$$\underline{\lambda}(\alpha) \geq \underline{\lambda}(\beta).$$

In particular, if $\alpha(x) \geq k \geq 0$, for all $x \geq 0$, where k is a constant, then $\underline{\lambda}(\alpha) \geq \frac{k^2}{2}$.

Proof The main ingredient of the proof is the fact $\underline{\lambda}(\alpha) = \eta(\alpha)$ and $\underline{\lambda}(\beta) = \eta(\beta)$. On the other hand, let

$$dX_t = dB_t - \alpha(X_t)\,dt, \qquad X_0 = x,$$
$$dY_t = dB_t - \beta(Y_t)\,dt, \qquad Y_0 = x.$$

Since $\alpha \geq \beta$, we get $X_t \leq Y_t$ for every t, and so $T(X) \leq T(Y)$. Therefore, for every $s \geq 0$,

$$\mathbb{P}_x\big(T(X) > s\big) \leq \mathbb{P}_x\big(T(Y) > s\big),$$

which implies that $\underline{\lambda}(\alpha) \geq \underline{\lambda}(\beta)$. □

Corollary 6.33 *Let α be a nonnegative $\mathcal{C}^1$ function that satisfies* **H**, **H1**. *Then $\underline{\lambda}(\alpha) \leq (\limsup_{x\to\infty} \alpha(x))^2/2$. In particular, if α is also decreasing $\underline{\lambda}(\alpha) = (\lim_{x\to\infty} \alpha(x))^2/2$.*

Proof The result is obvious if $\limsup_{x\to\infty} \alpha(x) = \infty$. Let $dX_t = dB_t - \alpha(X_t)\,dt$ with $X_0 = x$. Consider $\beta > \limsup_{x\to\infty} \alpha(z)$ and take x_0 large enough such that for all $z \geq x_0$ it holds $\alpha(z) \leq \beta$. Also, we consider the process $dY_t = dB_t - \beta\,dt$ with the initial condition $Y_0 = x$.

Assume that $x > x_0$. Then $Y_t \leq X_t$ for all $t \leq T_{x_0}(X)$. Therefore, for any such x and all $s \geq 0$,

$$\mathbb{P}_x\big(T(X) > s\big) \geq \mathbb{P}_x\big(T_{x_0}(X) > s\big) \geq \mathbb{P}_x\big(T_{x_0}(Y) > s\big) = \mathbb{P}_{x-x_0}\big(T(Y) > s\big).$$

Then $\underline{\lambda}(\alpha) \leq \frac{\beta^2}{2}$ and the first part follows. For the second part, it is enough to notice that $\alpha \geq \lim_{z\to\infty} \alpha(z)$. □

6.3 Characterization of the QSD Family

We shall study the set of QSDs associated to the process X. We recall that a probability measure ν supported on $(0,\infty)$ is a QSD if there exists a parameter $\beta=\beta(\nu)\in[0,\infty)$ such that for any measurable set $B\subset\mathbb{R}_+$, any $t\ge 0$ it holds (see Theorem 2.2)

$$\int \mathbb{E}_x(X_t\in B, t<T)\,d\nu(x)=e^{-\beta t}\nu(B).$$

The following result gives the complete classification of the QSDs.

Theorem 6.34 *The family of QSDs is empty when $\eta=0$. This is the case when $\Lambda(\infty)<\infty$ or $\Lambda(\infty)=\infty$, $\underline{\lambda}=0$. If $\eta>0$, that is, $\Lambda(\infty)=\infty$ and $\underline{\lambda}=\eta$, then for every QSD ν we have $\beta(\nu)\in(0,\underline{\lambda}]$ and*

$$d\nu=\frac{\psi_\beta\,d\mu}{\int\psi_\beta(x)\,d\mu(x)}.$$

Proof Consider a QSD ν. If $\mu(B)=0$ then for all $x>0$ we have $\mathbb{E}_x(X_t\in B, t<T)=\int_B r_t(x,y)\,d\mu(y)=0$, and we deduce that ν has a density with respect to μ. Let us call this density h. Using Fubini's Theorem, we obtain that h satisfies the following equation (dx-a.s.)

$$h(x)=e^{\beta t}\int r_t(z,x)h(z)\,d\mu(z).$$

Let us fix $t>0$ and consider a dense set of $E\subset\mathbb{R}_+$ such that for all $y\in E$ $\int r_{t+1}(x,y)h(x)\,d\mu(x)<\infty$. Using Harnack's Inequality, we conclude that for every $y>0$ there exists a constant $C=C(t,y)$ such that

$$\sup_{\{z:|z-y|\le y/2\}} r_t(x,z)\le Cr_{t+1}\big(x,y'\big),$$

where $|y-y'|\le y/2$ and $y'\in E$. This shows that h is locally bounded in $(0,\infty)$ and it has a continuous version there. So in what follows, we assume that h is continuous and positive on $(0,\infty)$. The symmetry of r shows that for $x>0$

$$h(x)=e^{\beta t}\int r_t(x,z)h(z)\,d\mu(z)=\mathbb{E}_x\big(e^{\beta(t\wedge T)}h(X_{t\wedge T})\big),$$

where we have defined $h(0)=0$ (we still do not know if h is bounded near 0 or continuous at 0). The strong Markov property shows that $h(X_{t\wedge T})$ is a martingale, and then Doob's Sampling Theorem yields for $x\in(\epsilon,M)$ the relation

$$h(x)=\mathbb{E}_x\big(e^{\beta(T_M\wedge T_\epsilon\wedge n)}h(X_{T_M\wedge T_\epsilon\wedge n})\big).$$

Since h is positive on $[\epsilon,M]$, we deduce that

$$\mathbb{E}_x\big(e^{\beta(T_M\wedge T_\epsilon\wedge n)}\big)\le\frac{h(x)}{\min\{h(z):z\in[\epsilon,M]\}},$$

and the Monotone Convergence Theorem implies

$$\mathbb{E}_x\big(e^{\beta(T_M \wedge T_\epsilon)}\big) \le \frac{h(x)}{\min\{h(z) : z \in [\epsilon, M]\}}.$$

Finally, the Dominated Convergence Theorem shows

$$h(x) = h(M)\mathbb{E}_x\big(e^{\beta T_M}, T_M < T_\epsilon\big) + h(\epsilon)\mathbb{E}_x\big(e^{\beta T_\epsilon}, T_\epsilon < T_M\big). \tag{6.34}$$

On the other hand, if we take $B = \mathbb{R}_+$ in the definition of a QSD, we get

$$e^{-\beta t} = \int \mathbb{P}_x(t < T)\, d\nu(x) \ge \nu\big([x_0, \infty)\big)\mathbb{P}_{x_0}(t < T).$$

We consider $x_0 > 0$ such that $\nu([x_0, \infty)) > 0$. Using Theorems 6.24 and 6.26, we deduce $0 \le \beta \le \underline{\lambda}$. Since $\mathbb{P}_x(t < T) < 1$ for all $x \ge 0$, we deduce that $\beta > 0$. This shows the result when $\underline{\lambda} = 0$. So in what follows, we assume implicitly that $\underline{\lambda} > 0$.

Let us define $g(x) = \frac{h(1)}{\psi_\beta(1)}\psi_\beta(x)$. The result is proved as soon as we show $h = g$. Using Itô's formula, we deduce that

$$\begin{aligned} g(x) &= \mathbb{E}_x\big(e^{\beta(T_M \wedge T_\epsilon)} g(X_{T_M \wedge T_\epsilon})\big) \\ &= g(M)\mathbb{E}_x\big(e^{\beta T_M}, T_M < T_\epsilon\big) + g(\epsilon)\mathbb{E}_x\big(e^{\beta T_\epsilon}, T_\epsilon < T_M\big). \end{aligned}$$

We consider $x = 1$ and we take $\epsilon \to 0$ to conclude that

$$h(1) = g(1) = g(M)\mathbb{E}_1\big(e^{\beta T_M}, T_M < T\big).$$

This also shows that the limit exists in (6.34) and is given by

$$h(1) = h(M)\mathbb{E}_1\big(e^{\beta T_M}, T_M < T\big) + h(0+)\mathbb{E}_1\big(e^{\beta T}, T < T_M\big).$$

This proves that $h(0+)$ exists and is finite. Since $h(0+) \ge 0$ and $g(1) = h(1)$, we deduce that $h(M) \le g(M)$ for any $M \ge 1$. In other words, the function $G(x) = g(x) - h(x)$ is nonnegative for $x \ge 1$. Assume that $G(x_0) < 0$ for some $x_0 \in (0, 1)$, then

$$\min\big\{G(x) : x \in (0, 1)\big\}\mathbb{P}_{x_0}(s < T) \le \mathbb{E}_{x_0}\big(G(X_s), s < T\big) = G(x_0) < 0.$$

If we take $s \to \infty$ in this inequality, we get $0 \le G(x_0) < 0$, which is a contradiction, and therefore $G \ge 0$, or equivalently, $h \le g$.

Finally, $0 = G(1) = \mathbb{E}_1(G(X_s), s < T)$ and we deduce that $G(x) = 0$ almost surely for the measure $\mathbb{P}_1(X_s \in dx, s < T)$. This measure is given by

$$r_s(1, x)e^{-\gamma(x)}\, dx$$

which has a positive density in $(0, \infty)$. This fact, together with the continuity of G, shows that $G = 0$ on $(0, \infty)$ (in particular, $h(0+) = g(0+) = 0$). We conclude that $h = g$, proving the result. □

6.4 h-Processes Associated to X

In this section, we show how the different h-processes associated to each $\lambda \in [0, \underline{\lambda}]$ appear in a limit procedure for the conditional process. When $\Lambda(\infty) < \infty$ or $\Lambda(\infty) = \infty$ but $\underline{\lambda} = 0$ (that is, $\eta = 0$) this is essentially studied in Theorem 6.16, part (iv), and Theorem 6.30, part (iii).

For the rest of this section, we assume that α satisfies **H**, **H1** and $\underline{\lambda} > 0$. Hence, we have $\eta = \underline{\lambda}$ and $\psi_\lambda \in L^1(d\mu)$ for $\lambda \in (0, \underline{\lambda}]$. Also, each of these ψ_λ induces a QSD given by the probability measure whose density with respect to μ is proportional to ψ_λ.

According to Theorem 6.26, part (iv), for all $s \geq 0$, all $x > 0$ and $A \in \mathcal{B}(\mathcal{C}([0, s]))$, the limit

$$\lim_{t\to\infty} \mathbb{P}_x(X \in A | t < T) = \mathbb{E}_x\left(e^{\underline{\lambda} s} \frac{\psi_{\underline{\lambda}}(X_s)}{\psi_{\underline{\lambda}}(x)}, X \in A, s < T \right)$$

exists and it is an h-transform of X. The aim of this section is to show how to get a similar result for each $\lambda \in [0, \underline{\lambda})$ with a suitable conditioning that approaches "$T = \infty$".

For that purpose, let us consider an increasing sequence of stopping times (S_M^λ) associated to ψ_λ, defined as follows:

$$S_M^\lambda = \inf\{s > 0 : F_\lambda(X_s, s) \geq M\},$$

where $F_\lambda(x, s) = e^{\lambda s}\psi_\lambda(x)$ for $x \geq 0, s > 0$. The main result of this section is

Theorem 6.35 *Assume α satisfies* **H**, **H1** *and $\underline{\lambda} > 0$. Then, for any $\lambda \in (0, \underline{\lambda})$, $s \geq 0$, $x > 0$ and $A \in \mathcal{B}(\mathcal{C}([0, s]))$,*

$$\lim_{M\to\infty} \mathbb{P}_x\big(X \in A | S_M^\lambda < T\big) = \mathbb{E}_x\left(e^{\lambda s} \frac{\psi_\lambda(X_s)}{\psi_\lambda(x)}, X \in A, s < T \right).$$

To prove this result, we need the following lemmas.

Lemma 6.36 *For every $\lambda < \underline{\lambda}$, we have that ψ_λ is a strictly increasing positive function on $\mathbb{R}_+$ and*

$$\lim_{M\to\infty} \psi_\lambda(M) = \infty.$$

Proof In the case $\lambda = 0$, we have $\psi_0 = \Lambda$ and the result is verified in this case. Hence in the rest of the proof we restrict ourselves to the case $\lambda \neq 0$ and $\lambda < \underline{\lambda}$.

First, let us show $\psi = \psi_\lambda$ is not bounded. On the contrary assume $\forall x \in \mathbb{R}_+$ that $|\psi(x)| \leq K$. Consider $F(x, t) = \psi(x)e^{\lambda t}$. Then, by Itô's formula, we have $\psi(x) = \mathbb{E}_x(F(X_s, s), s < T_M \wedge T) + \psi(M)\mathbb{E}_x(e^{\lambda T_M}, T_M < s \wedge T)$.

Since $e^{\lambda T_M} 1_{T_M < s\wedge T} \leq (e^{\lambda s} \vee 1) 1_{T_M < s}$, we obtain

$$\big|\psi(x)\big| \leq K e^{\lambda s}\mathbb{P}_x(s < T) + K\big(e^{\lambda s} \vee 1\big)\mathbb{P}_x(T_M < s).$$

Letting $M\to\infty$, we get $|\psi(x)|\le Ke^{\lambda s}\mathbb{P}_x(s<T)$. From Theorem 6.24 and since $\lambda<\eta=\underline{\lambda}$, we obtain $e^{\lambda s}\mathbb{P}_x(s<T)\to 0$ as $s\to\infty$. This yields $\psi=0$, which is a contradiction, so we deduce that ψ is not bounded.

Now, let us show that ψ only vanishes at $x=0$. On the contrary let $x_0>0$ be such that $\psi(x_0)=0$ and $\psi(x)>0$ for $x\in(0,x_0)$. By Itô's formula, we obtain

$$\bigl|\psi(x)\bigr|=\bigl|\mathbb{E}_x\bigl(\psi(X_s)e^{\lambda s},s<T_{x_0}\wedge T\bigr)\bigr|\le\max_{y\in[0,x_0]}\bigl|\psi(y)\bigr|e^{\lambda s}\mathbb{P}_x(s<T)\to 0$$

as $s\to\infty$. Therefore, we get that $\psi(x)$ is strictly positive for $x>0$.

Let us now prove that ψ is increasing. Assume there exist $x<y$ for which $\psi(x)>\psi(y)$. In the case $\lambda<0$, take $z<x$ such that $\psi(z)=\psi(y)$. Denote by $\bar{x}\in(z,y)$ a point verifying $\psi(\bar{x})=\max_{r\in[z,y]}\psi(r)$. Then

$$\frac{1}{2}\psi''(\bar{x})=\frac{1}{2}\psi''(\bar{x})-\alpha(\bar{x})\psi'(\bar{x})=-\lambda\psi(\bar{x})>0,$$

which is a contradiction. Assume now $\lambda>0$. Since ψ is not bounded, there exists $z>y$ such that $\psi(x)=\psi(z)$. Consider $\bar{x}\in(x,z)$ such that $\psi(\bar{x})=\min_{r\in[x,z]}\psi(r)>0$, then

$$\frac{1}{2}\psi''(\bar{x})=\frac{1}{2}\psi''(\bar{x})-\alpha(\bar{x})=-\lambda\psi(\bar{x})<0,$$

which again is a contradiction. Hence ψ is increasing.

Finally, if $\psi(x)=\psi(y)$ for $x<y$, we see that ψ is constant on $[x,y]$, and therefore,

$$0\ne-\lambda\psi(z)=\frac{1}{2}\psi''(z)-\alpha(z)\psi'(z)=0$$

for $z\in(x,y)$. The result follows from this. □

Lemma 6.37 *For any $\lambda\in(0,\underline{\lambda}]$, we have*

(i) $\psi_\lambda(x)=\lambda\mathbb{E}_x(\int_0^T\psi_\lambda(X_s)\,ds)$, *for all* $x\ge 0$.
(ii) $\mathbb{E}_x(\psi_\lambda(X_s),s<T)=e^{-\lambda s}\psi_\lambda(x)$ *for all* $x\ge 0, s>0$.

Proof (i) Using Itô's formula, we get

$$\begin{aligned}\psi(x)&=\psi(M)\mathbb{P}_x(T_M<T)+\lambda\mathbb{E}_x\int_0^{T_M\wedge T}\psi(X_s)\,ds\\&=\psi(M)\frac{\Lambda(x)}{\Lambda(M)}+\lambda\mathbb{E}_x\int_0^{T_M\wedge T}\psi(X_s)\,ds.\end{aligned}$$

From Lemma 6.18 and formula (6.24), the proof is completed for part (i).

(ii) Again from Itô's formula, we get

$$\psi(x)=\mathbb{E}_x\bigl(\psi(X_s)e^{\lambda s},s<T_M\wedge T\bigr)+\psi(M)\mathbb{E}\bigl(e^{\lambda T_M},T_M<s\wedge T\bigr).$$

Since

$$\psi(M)\mathbb{E}_x\big(e^{\lambda T_M}, T_M < s < T\big) \le \psi(M)e^{\lambda s}\mathbb{P}_x(T_M < T) = \frac{\psi(M)}{\Lambda(M)}\Lambda(x)e^{\lambda s},$$

the last term converges to 0 as $M \to \infty$, and the result follows. □

Lemma 6.38 *For* $0 < \lambda < \underline{\lambda}$, *we have*:

(i) $\mathbb{P}_x(S_M^\lambda < T) = \frac{\psi_\lambda(x)}{M}$ *if* $\psi_\lambda(x) \in (0, M)$.
(ii) $\lim_{M\to\infty} \frac{\mathbb{P}_x(S_M^\lambda < T \wedge s)}{\mathbb{P}_x(S_M^\lambda < T)} = 0$.

Proof (i) We shall write S_M instead of S_M^λ and ψ instead of ψ_λ. Take x such that $\psi(x) \in (0, M)$. If $t \le S_M \wedge T$ then $X_t \in [0, \psi^{-1}(M)]$. Therefore, from Itô's formula, we get

$$\begin{aligned}\psi(x) &= \mathbb{E}_x\big(F(X_{S_M\wedge T\wedge s}, S_M \wedge T \wedge s)\big)\\ &= M\mathbb{P}_x(S_M < T \wedge s) + \mathbb{E}_x\big(F(X_s, s), s < S_M \wedge T\big).\end{aligned}$$

Now $\mathbb{E}_x(F(X_s, s), s < S_M \wedge T) \le M\mathbb{P}_x(s < T)$, which converges to 0 as $s \to \infty$, proving the first part of the lemma.

(ii) We can assume $\psi(x) < M$. On the set $\{S_M < T \wedge s\}$, we have $\psi(X_{S_M}) = Me^{-\lambda S_M} \ge Me^{-\lambda s}$. Therefore, $T_{\psi^{-1}(Me^{-\lambda s})} \le S_M$.

Hence,

$$\frac{\mathbb{P}_x(S_M < T \wedge s)}{\mathbb{P}_x(S_M < T)} \le \frac{\mathbb{P}_x(T_{\psi^{-1}(Me^{-\lambda s})} < T)}{\mathbb{P}_x(S_M < T)} = \frac{M}{\psi(x)}\frac{\Lambda(x)}{\Lambda(\psi^{-1}(Me^{-\lambda s}))}.$$

Put $N = \psi^{-1}(Me^{-\lambda s})$, then $\psi(N) = Me^{-\lambda s}$ and N converges to ∞ with M.

Thus, $\frac{\mathbb{P}_x(S_M<T\wedge s)}{\mathbb{P}_x(S_M<T)} \le e^{\lambda s}\frac{\Lambda(x)}{\psi(x)}\frac{\psi(N)}{\Lambda(N)}$ converges to 0 as $M \to \infty$. □

Proof (Theorem 6.35) Let θ_s be the shift operator by s units of time. We denote $S_\bullet$ instead of $S_\bullet^\lambda$. It can be checked that $S_{Me^{-\lambda s}} \circ \theta_s = S_M - s$ on the set $\{s \le S_M < \infty\}$. Observe that on this set $\psi(X_s) \le Me^{-\lambda s}$, therefore from Lemma 6.38(i), we have $\mathbb{P}_{X_s}(S_{Me^{-\lambda s}} < T) = \frac{\psi(X_s)}{M}e^{\lambda s}$. Now, by using the Markov property, we get

$$\begin{aligned}\frac{\mathbb{P}_x(X \in A, s \le S_M < T)}{\mathbb{P}_x(S_M < T)} &= \mathbb{E}_x\left(\frac{\mathbb{P}_{X_s}(S_{Me^{-\lambda s}} < T)}{\mathbb{P}_x(S_M < T)}, X \in A, s < T, s \le S_M\right)\\ &= \mathbb{E}_x\left(e^{\lambda s}\frac{\psi_\lambda(X_s)}{\psi_\lambda(x)}, X \in A, s < T, s \le S_M\right).\end{aligned}$$

This last term converges, as $M \to \infty$, to

$$\mathbb{E}_x\left(e^{\lambda s}\frac{\psi_\lambda(X_s)}{\psi_\lambda(x)}, X \in A, T > s\right)$$

because $1_{\{S_M \ge s\}} \ge 1_{\{T_{\psi_\lambda^{-1}(Me^{-\lambda s})} \ge s\}} \nearrow 1_{\{T_\infty \ge s\}} = 1$ $\mathbb{P}_x$-a.e.

To finish the proof, we use Lemma 6.38, part (ii), to get the result. Indeed,

$$\left| \mathbb{P}_x(X \in A | S_M < T) - \frac{\mathbb{P}_x(X \in A, s \leq S_M < T)}{\mathbb{P}_x(S_M < T)} \right| \leq \frac{\mathbb{P}_x(S_M < T \wedge s)}{\mathbb{P}_x(S_M < T)},$$

which converges to 0 as M tends to ∞. □

6.5 R-Positivity for X

We give necessary and sufficient conditions in order that a one-dimensional diffusion X killed at 0 be R-positive. This means that the process Z whose law is the conditional law of X to never hit the origin is positive recurrent. Our conditions are stated in terms of the function $\underline{\lambda}$, where $\underline{\lambda}(z)$ is the bottom of the spectrum of the eigenvalue problem associated to the diffusion killed at z.

Let $T_z^X = T_z = \inf\{t > 0 : X_t = z\}$ be the hitting time of z. We are mainly interested in the case $z = 0$ and we denote $T = T_0$. As usual, X^T corresponds to X killed at 0. Most of the functions and parameters we consider in this section will depend on α. To avoid overburdening notation, we shall explicitly denote such dependence only if it is necessary. We will consider the diffusion X killed at different points. In this sense, it is useful to introduce the notation $\alpha^{(z)}$ which is the restriction of α to the region $[z, \infty)$. Since most of the time we will deal with the process X killed at 0, we shall use α as synonymous to $\alpha^{(0)}$, when there is no possible confusion.

Throughout this section, we shall assume that α satisfies **H** and most of the time we shall assume also **H1**, which is $\Lambda(\infty) = \infty$, that is, absorption is certain.

Fix some $z \in \mathbb{R}_+$. The eigenvalue problem $\frac{1}{2}v''(x) - \alpha(x)v'(x) = -\lambda v(x)$ $v(z) = 0$, $v'(z) = 1$ has a unique solution in $[z, \infty)$ denoted by $\psi_{z,\lambda;\alpha}$. When there is no possible confusion about α, we shall use the simple notation $\psi_{z,\lambda}$. This unique solution satisfies for $x \geq z$

$$\psi'_{z,\lambda}(x) = e^{\gamma(x)-\gamma(z)}\left(1 - 2\lambda \int_0^x \psi_{z,\lambda}(\xi) e^{\gamma(z)-\gamma(\xi)}\, d\xi\right),$$

$$\psi_{z,\lambda}(x) = \int_0^x e^{\gamma(y)-\gamma(z)}\left(1 - 2\lambda \int_0^y \psi_{z,\lambda}(\xi) e^{\gamma(z)-\gamma(\xi)}\, d\xi\right) dy.$$

The functions $\psi_{z,\lambda}(x)$, $\psi'_{z,\lambda}(x)$ are continuous on (z, λ, x).

We denote by $\underline{\lambda}_\alpha(z)$, or simply by $\underline{\lambda}(z)$ if there is no possible confusion, the value given by $\underline{\lambda}(z) = \max\{\lambda : \psi_{z,\lambda}$ is positive in $(z, \infty)\}$. We know that $\underline{\lambda}(z)$ is characterized by $\underline{\lambda}(z) = \max\{\lambda : \psi_{z,\lambda}$ is increasing on $[z, \infty)\}$. We point out that $\psi_{z,\underline{\lambda}(z)}$ is strictly increasing.

Under **H**, **H1**, we have the following useful characterization for any $x > z$:

$$\underline{\lambda}(z) = \lim_{t\to\infty} -\frac{\log(\mathbb{P}_x(t < T_z))}{t}.$$

From this representation, we conclude that the function $\underline{\lambda}(\bullet)$ is increasing on $\mathbb{R}_+$. We point out that a simple coupling argument shows that $\underline{\lambda}_\alpha$ is increasing also in

α, that is, if $\alpha \geq \beta$ on $[z, \infty)$ and both functions satisfy Hypotheses **H** and **H1** then $\underline{\lambda}_\alpha(z) \geq \underline{\lambda}_\beta(z)$ (see Corollary 6.32).

According to Theorem 6.26, the following limit exists for $x > 0$ fixed and defines a diffusion Z

$$\lim_{t\to\infty} \mathbb{P}_x(X_s \in A | t < T) = e^{\underline{\lambda}(0)s}\mathbb{E}_x\big(\psi_{0,\underline{\lambda}(0)}(X_s)\psi_{0,\underline{\lambda}(0)}(x), X_s \in A, s < T\big)$$
$$= \mathbb{P}_x(Z_s \in A).$$

This diffusion satisfies the SDE (see (6.31))

$$dZ_t = dB_t - \phi(Z_t)\,dt, \quad \text{where } \phi(y) = \alpha(y) - \frac{\psi'_{0,\underline{\lambda}(0)}(y)}{\psi_{0,\underline{\lambda}(0)}(y)}$$

and takes values on $(0, \infty)$. In fact, it never reaches 0 because its drift is of order $1/x$ for x near 0. The transition density for Z is (with respect to the Lebesgue measure)

$$p^Z(t, x, y) = \frac{\psi_{0,\underline{\lambda}(0)}(y)}{\psi_{0,\underline{\lambda}(0)}(x)} e^{\underline{\lambda}(0)t} p_t(x, y).$$

From Remark 6.10, we get for $x > 0$, $y > 0$

$$p^Z(t, x, y)\,dy = \frac{\psi_{0,\underline{\lambda}(0)}(y)}{\psi_{0,\underline{\lambda}(0)}(x)} e^{-\int_x^y \alpha(\xi)\,d\xi}\mathbb{E}_x\big(e^{-1/2\int_0^t h_\alpha(B_s)\,ds}, B_t \in dy, t < T\big),$$

where $h_\alpha = \alpha^2 - \alpha' - 2\underline{\lambda}_\alpha(0)$. This function h_α will be used below to compare the qualitative behavior of the diffusion Z for different drifts.

When $\Lambda(\infty) < \infty$, a similar result holds according to Theorem 6.16, by replacing $\underline{\lambda}(0)$ by $\eta = 0$ and $\psi_{0,\underline{\lambda}(0)}$ by Λ.

Our main tool is to kill the process X at different levels. Let us introduce some notation. Consider a fixed $x > 0$, and we kill X at x. The associated limiting process Z^x has a drift given by (see (6.31))

$$-\phi_x(y) = \frac{\psi'_{x,\underline{\lambda}(x)}(y)}{\psi_{x,\underline{\lambda}(x)}(y)} - \alpha(y) \quad \text{for } y > x.$$

A straightforward computation yields the following relation between the eigenfunction for Z^x killed at z, for $z > x$, and the two eigenfunctions for X one killed at z and the other at x. For any $\lambda \in \mathbb{R}$ and any $y \geq z$,

$$\psi_{z,\lambda-\underline{\lambda}(x);\phi_x}(y) = \frac{\psi_{z,\lambda;\alpha}(y)\psi_{x,\underline{\lambda}(x);\alpha}(z)}{\psi_{x,\underline{\lambda}(x);\alpha}(y)}, \tag{6.35}$$

where we have put $\underline{\lambda}(x) = \underline{\lambda}_\alpha(x)$. This relation also holds if we replace $\underline{\lambda}(x)$ by any $\beta \leq \underline{\lambda}(x)$. From this relation, we get $\underline{\lambda}_{\phi_x}(z) = \underline{\lambda}_\alpha(z) - \underline{\lambda}_\alpha(x)$. The following two results give some basic information about the limiting process Z.

Theorem 6.39 *Assume α satisfies* **H** *and* **H1**. *Then*

$$\phi(x) = \alpha(x) - \psi'_{0,\underline{\lambda}(0);\alpha}(x)/\psi_{0,\underline{\lambda}(0);\alpha}(x)$$

satisfies **H** *on $[z, \infty)$ for all $z > 0$. This means*

$$\int_z^\infty e^{-\gamma^Z(x)} \int_z^x e^{\gamma^Z(\xi)}\, d\xi\, dx = \int_z^\infty e^{\gamma^Z(x)} \int_z^x e^{-\gamma^Z(\xi)}\, d\xi\, dx = \infty,$$

where $\gamma^Z(y) = 2\int_a^y (\alpha(\xi) - \psi'_{0,\underline{\lambda}(0)}(\xi)/\psi_{0,\underline{\lambda}(0)}(\xi))\, d\xi$ and $a > z$ is a fixed constant.

The proof of this theorem is based on the following lemma, for which we assume neither **H** nor **H1**.

Lemma 6.40 *Assume α is locally bounded and measurable. Let $\lambda < 0$, then the following two conditions are equivalent*:

(i) $\psi_{0,\lambda}$ *is unbounded*;
(ii) $\int_0^\infty e^{\gamma(x)} \int_0^x e^{-\gamma(y)}\, dy\, dx = \infty$.

Proof We denote by $\psi = \psi_{0,\lambda}$. Since

$$\psi(x) = \Lambda(x) - 2\lambda \int_0^x e^{\gamma(y)} \int_0^y \psi(z) e^{-\gamma(z)}\, dz\, dy$$

and the fact that $\lambda < 0$, we get that ψ is strictly increasing.

Hence, if $\Lambda(\infty) = \infty$, both conditions (i) and (ii) are satisfied. Therefore, for the rest of the proof we can assume $\Lambda(\infty) < \infty$.

Suppose that (ii) holds. For $x > 1$, $\psi - \Lambda$ can be bounded from below by

$$\begin{aligned}
&-2\lambda \int_1^x e^{\gamma(y)} \int_1^y \psi(z) e^{-\gamma(z)}\, dz\, dy \\
&\geq -2\lambda\psi(1) \int_1^x e^{\gamma(y)} \int_1^y e^{-\gamma(z)}\, dz\, dy \\
&\geq -2\lambda\psi(1) \Bigg[\int_0^x e^{\gamma(y)} \int_0^y e^{-\gamma(z)}\, dz\, dy \\
&\quad - \int_0^1 e^{\gamma(y)} \int_0^y e^{-\gamma(z)}\, dz\, dy - \Lambda(x) \int_0^1 e^{-\gamma(z)}\, dz\Bigg].
\end{aligned}$$

Then, ψ is unbounded.

Now, assume $\int_0^\infty e^{\gamma(x)} \int_0^x e^{-\gamma(y)}\, dy\, dx < \infty$. We shall prove that ψ is bounded. Indeed, take a large x_0 such that $-2\lambda \int_{x_0}^\infty e^{\gamma(y)} \int_0^y e^{-\gamma(z)}\, dz\, dy \leq 1/2$. For $x > x_0$, we have

$$\psi(x) \le v(x_0) + \Lambda(\infty) - 2\lambda\psi(x)\int_{x_0}^{x} e^{\gamma(y)}\int_0^y e^{-\gamma(z)}\,dz\,dy$$
$$\le \psi(x_0) + \Lambda(\infty) + \psi(x)/2.$$

Therefore, ψ is bounded by $2(\psi(x_0) + \Lambda(\infty))$. □

Proof (Theorem 6.39) We denote by $\psi = \psi_{0,\underline{\lambda}(0);\alpha}$. We also recall the notation $\gamma^Z(y) = 2\int_a^y \phi(\xi)\,d\xi = \gamma(y) - \gamma(a) - 2\log(\psi(y)/\psi(a))$, for some $a > 0$ fixed. Then,

$$\int_z^\infty e^{-\gamma^Z(y)}\int_a^y e^{\gamma^Z(\xi)}\,d\xi\,dy = \int_z^\infty \frac{\psi^2(y)}{\psi^2(a)}e^{-\gamma(y)}\int_a^y \frac{\psi^2(a)}{\psi^2(\xi)}e^{\gamma(\xi)}\,d\xi\,dy$$
$$\ge \int_z^\infty e^{-\gamma(y)}\int_a^y e^{\gamma(\xi)}\,d\xi\,dy = \infty,$$

where we have used the monotonicity of ψ and Hypotheses **H**, **H1** for α.

For the other integral involved in condition **H**, we consider two different situations. In the first one, we assume $\underline{\lambda}(0) = 0$. In this case, $\psi = \Lambda$ and $\phi = \alpha - \Lambda'/\Lambda$. Since $d\Lambda(y) = e^{\gamma(y)}\,dy$, an integration by parts yields

$$\int_z^x e^{\gamma^Z(y)}\int_a^y e^{-\gamma^Z(\xi)}\,d\xi\,dy$$
$$= \int_z^x \frac{e^{\gamma(y)}}{\Lambda^2(y)}\int_a^y \Lambda^2(\xi)e^{-\gamma(\xi)}\,d\xi\,dy$$
$$= \int_a^z \Lambda^2(y)e^{-\gamma(y)}\left(\frac{1}{\Lambda(z)} - \frac{1}{\Lambda(x)}\right)dy + \int_z^x \Lambda(y)e^{-\gamma(y)}\left(1 - \frac{\Lambda(y)}{\Lambda(x)}\right)dy.$$

Since Λ increases to ∞, we can take $x_n \uparrow \infty$ such that $\Lambda(x_n) = \Lambda(n)/2$. Then,

$$\int_z^\infty e^{\gamma^Z(y)}\int_a^y e^{-\gamma^Z(\xi)}\,d\xi\,dy$$
$$\ge \int_a^z \Lambda^2(y)e^{-\gamma(y)}\left(\frac{1}{\Lambda(z)} - \frac{1}{\Lambda(x)}\right)dy + \int_z^n \Lambda(y)e^{-\gamma(y)}\left(1 - \frac{\Lambda(y)}{\Lambda(x_n)}\right)dy$$
$$\ge \int_a^z \Lambda^2(y)e^{-\gamma(y)}\left(\frac{1}{\Lambda(z)} - \frac{1}{\Lambda(x)}\right)dy + \frac{1}{2}\int_z^n \Lambda(y)e^{-\gamma(y)}\,dy,$$

which converges to infinity because α satisfies **H**.

We are left with the case $\underline{\lambda}(0) > 0$. In this situation, we deduce that (see Corollary 6.21)

$$\int_0^\infty \psi_{0,\underline{\lambda}(0);\alpha}(y)e^{-\gamma(y)}\,dy < \infty \quad \text{and} \quad \int_0^\infty e^{-\gamma(y)}\,dy < \infty.$$

The last conclusion follows from the fact that $\psi_{0,\underline{\lambda}(0);\alpha}(\bullet)$ is nonnegative and strictly increasing on $[0,\infty)$.

Now, consider $w(y)=\psi_{z,0;\alpha}(y)=e^{-\gamma(z)}(\Lambda(y)-\Lambda(z))$. Using (6.35), we have

$$\psi_{z,-\underline{\lambda}(z);\phi}(y)=\frac{w(y)\psi(z)}{\psi(y)}.$$

From Lemma 6.40, the proof will be finished as soon as we prove w/ψ is unbounded. So, let us assume $w/\psi\le D$ on $[z,\infty)$. Then,

$$\int_z^\infty w(y)e^{-\gamma(y)}\,dy\le D\int_z^\infty \psi(y)e^{-\gamma(y)}\,dy$$

which is finite. On the other hand,

$$\int_z^\infty w(y)e^{-\gamma(y)}\,dy=e^{-\gamma(z)}\int_z^\infty \Lambda(y)e^{-\gamma(y)}\,dy-\Lambda(z)\int_z^\infty e^{-\gamma(y)}\,dy.$$

This quantity is infinite because α satisfies **H**. We arrive at a contradiction, and the result is proved. □

The second result supplies the recurrence classification of Z in terms of integrability properties of $\psi_{0,\eta}$. Recall that if $\Lambda(\infty)<\infty$ we have $\eta=0$, and if $\Lambda(\infty)=\infty$ then $\eta=\underline{\lambda}(0)$.

Theorem 6.41 *Assume* α *satisfies* **H**. *The process* Z *is*

(i) *Positive recurrent if and only if* $\int_0^\infty \psi_{0,\eta}^2(x)e^{-\gamma(x)}\,dx<\infty$;
(ii) *Null recurrent if and only if* $\int_0^\infty \psi_{0,\eta}^2(x)e^{-\gamma(x)}\,dx=\infty$, $\int_a^\infty \psi_{0,\eta}^{-2}(x)e^{\gamma(x)}\,dx=\infty$ *for some* $a>0$ *(or all* $a>0$*)*;
(iii) *Transient if and only if* $\int_a^\infty \psi_{0,\eta}^{-2}(x)e^{\gamma(x)}\,dx<\infty$ *for some* $a>0$ *(or all* $a>0$*)*.

Proof Let $v=\psi_{0,\eta;\alpha}$ and consider, for $a>0$, the function

$$\Lambda^Z(y)=\int_a^y e^{\gamma^Z(z)}\,dz=v^2(a)\int_a^y v^{-2}(z)e^{\gamma(z)-\gamma(a)}\,dz.$$

We first notice that $\Lambda^Z(0+)=-\infty$ because $v(x)=x+O(x^2)$ for x near 0. On the other hand, if $\Lambda^Z(\infty)=\infty$ then Z is recurrent (see 5.5.22 in Karatzas and Shreve 1988). In the case $\Lambda^Z(\infty)<\infty$, for any $x>0$, it holds

$$\mathbb{P}_x\Big(\lim_{t\uparrow S}Z_t=\infty\Big)=\mathbb{P}_x\Big(\inf_{0\le t<S}Z_t>0\Big)=1,$$

where S is the explosion time of Z. In this case, the process Z is transient. Hence, Z is transient if and only if $\Lambda^Z(\infty)<\infty$, which is equivalent to

$$\int_a^\infty v^{-2}(z)e^{\gamma(z)}\,dz<\infty.$$

Let T_b^Z be the hitting time of $b > 0$ for the process Z. The process Z is positive recurrent when $\mathbb{E}_x(T_b^Z) < \infty$, for any $x, b \in (0, \infty)$. Using the formulas on p. 353 in Karatzas and Shreve (1988) and the fact that the speed measure for Z is given by

$$dm(x) = 2\frac{e^{-\gamma(a)}}{\upsilon^2(a)}\upsilon^2(x)e^{-\gamma(x)}\,dx,$$

we deduce that Z is positive recurrent if and only if $\int_0^\infty \upsilon^2(x)e^{-\gamma(x)}\,dx < \infty$. □

6.5.1 R-Classification

The classification of Z induces the R-classification of the killed diffusion X^T. We shall state necessary and sufficient conditions for α to be R-positive in terms of the function $\underline{\lambda}$. We introduce a concept of a *gap* and we will prove that there exists a gap when the diffusion is R-positive. This concept of a gap is analogous to the condition already used in Ferrari et al. (1996) to show R-positivity of Markov chains on countable spaces.

We also point out that the notion of R-positivity for diffusions extends the standard definition of R-positivity introduced by Vere-Jones (see Vere-Jones 1967) for nonnegative matrices, which in terms of the Perron–Frobenius theory reduces to the fact that the inner product of the left and right positive eigenvectors is finite (see Seneta 1981, Theorem 6.4). Hence, this notion turns out to be nontrivial only for processes taking values on infinite spaces. In the context of one-dimensional statistical mechanics with an infinite number of states, R-positivity of the transfer matrix associated to the Hamiltonian was shown to be a necessary and sufficient condition for the existence of a unique Gibbs state Kesten (1976). The material of this section is taken from Martínez and San Martín (2004).

We start with the R-classification of X.

Definition 6.42 The process X^T, or equivalently, α, is said to be R-positive (respectively, R-recurrent, R-null, or R-transient) if the process Z is positive recurrent (respectively, recurrent, null recurrent, or transient).

Recall that Corollary 6.21 states that under both **H**, **H1** we have

$$\underline{\lambda}(0) > 0 \quad \Leftrightarrow \quad \int_0^\infty \psi_{0,\underline{\lambda}(0)}(x)e^{-\gamma(x)}\,dx < \infty.$$

Using that $\psi_{0,\underline{\lambda}(0)}$ is an increasing function, we deduce that $\underline{\lambda}(0) > 0$ is a necessary condition for R-positivity. Moreover, whenever X^T is R-positive, it holds

$$\int_0^\infty e^{-\gamma(x)}\,dx < \infty.$$

The probabilistic meaning of this fact is that the process W whose drift in $\mathbb{R}$ is $-\alpha(|x|)$ is positive recurrent. In fact, the invariant probability measure of W has a density proportional to $e^{-\gamma(|x|)}$.

We introduce a new concept that will be important in the R-classification.

Definition 6.43 α has a *gap* at x with respect to $y < x$ if $\underline{\lambda}(x) > \underline{\lambda}(y)$.

We are mainly interested in gaps with respect to $y = 0$, in which case we just say that α has a gap at x. We notice that if α has a gap at x so does at any $z > x$.

Definition 6.44 Consider $\underline{\lambda}(\infty) = \lim_{x\to\infty} \underline{\lambda}(x)$. We define $\bar{x} = \inf\{x \geq 0 : \underline{\lambda}(x) = \underline{\lambda}(\infty)\} \leq \infty$.

Theorem 6.45 *Assume* **H**, **H1** *hold.*

(i) *If α has a gap at some $z > 0$ then α is R-positive and α has a gap at any $x > 0$.*
(ii) *If for some $z > 0$ the function $\alpha^{(z)}$ is R-positive then $\alpha^{(y)}$ is R-positive for $0 \leq y \leq z$ and $\underline{\lambda}$ is strictly increasing on $[0, z]$. In particular, α has a gap at z.*
(iii) *If α does not have a gap then $\alpha^{(z)}$ is R-transient for any $z > 0$.*

Remark 6.46 We notice that if $\underline{\lambda}(\infty) = \infty$ then necessarily α has a gap, which implies that α is R-positive. We also point out that if α is R-transient then $\bar{x} = 0$ and $\underline{\lambda}(0) = \underline{\lambda}(\infty)$.

Before we prove this theorem, we need some lemmas. We first investigate some extra properties about the eigenfunctions $\psi_{z,\lambda}$. A useful tool will be supplied by the Wronskian $W[f, g]$, between two $\mathcal{C}^1$ functions f and g, which we recall is given by $W[f, g](x) = f'(x)g(x) - f(x)g'(x)$. Once f and g are fixed, we shall simply write $W(x)$ instead of $W[f, g](x)$.

Lemma 6.47 *For any $a > 0$ there exists $\tilde{\lambda} > \underline{\lambda}(0)$ such that $\psi_{0,\lambda}$ is strictly increasing on $[0, a]$ for any $\lambda \in [\underline{\lambda}(0), \tilde{\lambda}]$.*

Proof The result follows from the facts that $\psi'_{0,\lambda}(x)$ is jointly continuous and $\psi_{0,\underline{\lambda}(0)}$ is strictly increasing on $[0, \infty)$. □

Lemma 6.48 *Assume $\psi_{0,\lambda}$ is increasing on $[0, a]$. Then, for all $\tau < \lambda$ the function $\psi_{0,\tau}$ is also increasing on $[0, a]$. Moreover, for $x \in (0, a]$ it holds: $\psi_{0,\tau}(x) > \psi_{0,\lambda}(x)$; $\psi'_{0,\tau}(x) > \psi'_{0,\lambda}(x)$ and the ratio $\psi'_{0,\tau}(x)/\psi_{0,\tau}(x)$ is a strictly decreasing continuous function of τ on the region $(-\infty, \lambda]$. In particular, the above properties hold for $\lambda = \underline{\lambda}(0)$ on $(0, \infty)$.*

Proof We first notice that if $\psi_{0,\lambda}$ is increasing on $[0, a]$ then it is strictly increasing in the same interval. In fact, from the equality $\psi'_{\lambda}(x) = e^{\gamma(x)}(1 -$

$2\lambda\int_0^x \psi_\lambda(z)e^{-\gamma(z)}\,dz)$, we conclude that $\psi'_{0,\lambda} > 0$ on $[0,a)$. Consider the Wronskian $W(x) = W[\psi_{0,\lambda},\psi_{0,\tau}](x) = \psi'_{0,\lambda}(x)\psi_{0,\tau}(x) - \psi_{0,\lambda}(x)\psi'_{0,\tau}(x)$. A straightforward computation shows that $W(0) = 0$ and

$$W' = 2\alpha W - 2(\lambda-\tau)\psi_{0,\lambda}\psi_{0,\tau},$$

or equivalently,

$$W(x) = -2(\lambda-\tau)e^{\gamma(x)}\int_0^x e^{-\gamma(\xi)}\psi_{0,\lambda}(\xi)\psi_{0,\tau}(\xi)\,d\xi.$$

If $\psi_{0,\lambda}, \psi_{0,\tau}$ are increasing on $[0,b]$ then $W(x) < 0$ on this interval, and therefore

$$\frac{\psi'_{0,\lambda}(x)}{\psi_{0,\lambda}(x)} < \frac{\psi'_{0,\tau}(x)}{\psi_{0,\tau}(x)} \quad \text{for } x \in (0,b].$$

This implies that $\psi_{0,\tau}$ is increasing in $[0,a]$ (otherwise take the first $x^* < a$ where $\psi'_{0,\tau}(x^*) = 0$ to arrive at a contradiction). We deduce

$$\frac{\psi'_{0,\lambda}(x)}{\psi_{0,\lambda}(x)} < \frac{\psi'_{0,\tau}(x)}{\psi_{0,\tau}(x)} \quad \text{for } x \in (0,a]. \tag{6.36}$$

Moreover, by integrating (6.36), we get for any $\varepsilon > 0$

$$\psi_{0,\lambda}(x) < \frac{\psi_{0,\lambda}(\varepsilon)}{\psi_{0,\tau}(\varepsilon)}\psi_{0,\tau}(x).$$

Since $\lim_{\varepsilon\downarrow 0}\psi_{0,\lambda}(\varepsilon)/\psi_{0,\tau}(\varepsilon) = 1$, we obtain

$$\psi_{0,\lambda}(x) \le \psi_{0,\tau}(x) \quad \forall x \in (0,a],$$

which, together with (6.36), implies $\psi'_{0,\lambda}(x) < \psi'_{0,\tau}(x)$. Finally, the ratio $\psi'_{0,\tau}(x)/\psi_{0,\tau}(x)$ is clearly continuous on τ for any $x \in (0,a]$. □

Let $z \ge x \ge 0$ be fixed. Consider the Wronskian $W = W[\psi_{x,\lambda},\psi_{z,\tau}]$ in the region $[z,\infty)$, which is given by $W(y) = \psi'_{x,\lambda}(y)\psi_{z,\tau}(y) - \psi_{x,\lambda}(y)\psi'_{z,\tau}(y)$. One has $W(z) = -\psi_{x,\lambda}(z)$ and $W' = 2\alpha W + 2(\tau-\lambda)\psi_{x,\lambda}\psi_{z,\tau}$. Therefore, for $y \ge z$,

$$\begin{aligned} W(y) &= e^{2\int_z^y \alpha(\xi)\,d\xi}\left(W(z) + 2(\tau-\lambda)\int_z^y \psi_{x,\lambda}(\eta)\psi_{z,\tau}(\eta)e^{-2\int_z^\eta \alpha(\xi)\,d\xi}\,d\eta\right) \\ &= e^{2\int_z^y \alpha(\xi)\,d\xi}\left(-\psi_{x,\lambda}(z) + 2(\tau-\lambda)\int_z^y \psi_{x,\lambda}(\eta)\psi_{z,\tau}(\eta)e^{-2\int_z^\eta \alpha(\xi)\,d\xi}\,d\eta\right). \end{aligned} \tag{6.37}$$

Lemma 6.49 *Assume that for $x < z$ fixed, $\underline{\lambda}(x) < \underline{\lambda}(z)$. Then, for $\tau \in (\underline{\lambda}(x),\underline{\lambda}(z)]$ and $y \in [z,\infty)$, we have*

$$W[\psi_{x,\underline{\lambda}(x)},\psi_{z,\tau}](y) < 0. \tag{6.38}$$

In particular, for $y \in [z,\infty)$,

$$\frac{\psi'_{x,\underline{\lambda}(x)}(y)}{\psi_{x,\underline{\lambda}(x)}(y)} \le \frac{\psi'_{z,\underline{\lambda}(z)}(y)}{\psi_{z,\underline{\lambda}(z)}(y)}. \tag{6.39}$$

Furthermore,

$$2\big(\underline{\lambda}(z)-\underline{\lambda}(x)\big)\int_z^{\infty}\psi_{x,\underline{\lambda}(x)}(\eta)\psi_{z,\underline{\lambda}(z)}(\eta)e^{-2\int_z^{\eta}\alpha(\xi)\,d\xi}\,d\eta=\psi_{x,\underline{\lambda}(x)}(z). \tag{6.40}$$

Proof Let $\underline{\lambda}(x)<\tau\le\underline{\lambda}(z)$. Assume that (6.38) does not hold, that is, for some finite y_0, the following strict inequality holds:

$$2\big(\tau-\underline{\lambda}(x)\big)\int_z^{y_0}\psi_{x,\underline{\lambda}(x)}(\eta)\psi_{z,\tau}(\eta)e^{-2\int_z^{\eta}\alpha(\xi)\,d\xi}\,d\eta>\psi_{x,\underline{\lambda}(x)}(z).$$

By Lemma 6.47 and an argument based on continuity, we get the existence of $\tilde{\lambda}\in(\underline{\lambda}(x),\tau)$ such that

(a) $\psi_{x,\tilde{\lambda}}$ is increasing on $[x,y_0]$;
(b) $2(\tau-\tilde{\lambda})\int_z^{y_0}\psi_{x,\tilde{\lambda}}(\eta)\psi_{z,\tau}(\eta)e^{-2\int_z^{\eta}\alpha(\xi)\,d\xi}\,d\eta>\psi_{x,\tilde{\lambda}}(z)$.

From (6.37), we have

$$W[\psi_{x,\tilde{\lambda}},\psi_{z,\tau}](y_0)=\psi'_{x,\tilde{\lambda}}(y_0)\psi_{z,\tau}(y_0)-\psi_{x,\tilde{\lambda}}(y_0)\psi'_{z,\tau}(y_0)>0.$$

Since $\psi_{z,\tau}$ is increasing (see Lemma 6.48), we get $\psi'_{x,\tilde{\lambda}}(y_0)>0$, and therefore $\psi_{x,\tilde{\lambda}}$ is strictly increasing on a small interval $[y_0,y_0+\delta]$. If there exists a point $y^*>y_0$ such that $\psi'_{x,\tilde{\lambda}}(y^*)=0$, we arrive at a contradiction. In fact, consider the smallest possible y^*. From (6.37) and relation (b), we get $W[\psi_{x,\tilde{\lambda}},\psi_{z,\tau}](y^*)>0$, and therefore $\psi'_{x,\tilde{\lambda}}(y^*)>0$. The conclusion is that $\psi_{x,\tilde{\lambda}}$ is strictly increasing on $[x,\infty)$, but this is again a contradiction because $\tilde{\lambda}>\underline{\lambda}(x)$. Therefore,

$$2\big(\tau-\underline{\lambda}(x)\big)\int_z^{\infty}\psi_{x,\underline{\lambda}(x)}(\eta)\psi_{z,\tau}(\eta)e^{-2\int_z^{\eta}\alpha(\xi)\,d\xi}\,d\eta\le\psi_{x,\underline{\lambda}(x)}(z)$$

holds, and (6.38) and (6.39) follow.

Now, let us prove (6.40). Take a large t_0 and find a $\tilde{\tau}>\underline{\lambda}(z)$, close enough to $\underline{\lambda}(z)$, such that $\psi_{z,\tilde{\tau}}$ is increasing on $[z,t_0]$. Since $\tilde{\tau}>\underline{\lambda}(z)$, there exists $t_1>t_0$, the closest value to t_0, where $\psi'_{z,\tilde{\tau}}(t_1)=0$, then $W[\psi_{x,\underline{\lambda}(x)},\psi_{z,\tilde{\tau}}](t_1)>0$. From (6.37), we get

$$2\big(\tilde{\tau}-\underline{\lambda}(x)\big)\int_z^{t_1}\psi_{x,\underline{\lambda}(x)}(\eta)\psi_{z,\tilde{\tau}}(\eta)e^{-2\int_z^{\eta}\alpha(\xi)\,d\xi}\,d\eta>\psi_{x,\underline{\lambda}(x)}(z).$$

Using Lemma 6.48, the inequality $\psi_{z,\tilde{\tau}} \le \psi_{z,\underline{\lambda}(z)}$ holds on $[z, t_1]$. Therefore, we obtain

$$2(\tilde{\tau} - \underline{\lambda}(x))\left(\int_z^{\infty} \psi_{x,\underline{\lambda}(x)}(\eta)\psi_{z,\underline{\lambda}(z)}(\eta)e^{-2\int_z^{\eta}\alpha(\xi)\,d\xi}\,d\eta\right) > \psi_{x,\underline{\lambda}(x)}(z).$$

Thus, (6.40) is proved by passing to the limit $\tilde{\tau} \to \underline{\lambda}(z)$. □

Proof (Theorem 6.45) (i) Let us prove that the existence of a gap at some $z > 0$ is sufficient for α to be R-positive. From Lemma 6.49, by integrating inequality (6.39) (where $x = 0$), we get

$$\psi_{0,\underline{\lambda}(0)}(y) \le \frac{\psi_{0,\underline{\lambda}(0)}(y_0)}{\psi_{z,\underline{\lambda}(z)}(y_0)}\psi_{z,\underline{\lambda}(z)}(y), \qquad \text{for } 0 < z < y_0 < y.$$

From this inequality and (6.40), we get

$$\begin{aligned}\int_{y_0}^{\infty} \psi_{0,\underline{\lambda}(0)}^2(y)e^{-\gamma(y)}\,dy &\le \frac{\psi_{0,\underline{\lambda}(0)}(y_0)}{\psi_{z,\underline{\lambda}(z)}(y_0)}\int_{y_0}^{\infty} \psi_{0,\underline{\lambda}(0)}(y)\psi_{z,\underline{\lambda}(z)}(y)e^{-\gamma(y)}\,dy \\ &\le \frac{\psi_{0,\underline{\lambda}(0)}(y_0)}{\psi_{z,\underline{\lambda}(z)}(y_0)}\frac{\psi_{0,\underline{\lambda}(0)}(z)}{2(\underline{\lambda}(z) - \underline{\lambda}(0))}e^{-\gamma(z)} < \infty.\end{aligned}$$

This shows that α is R-positive.

Now we prove that if α has a gap at $z > 0$ then it has a gap at any $x > 0$. Without loss of generality, we can assume that $x < z$. If there is no gap at x, we have $\underline{\lambda}(0) = \underline{\lambda}(x)$. For the sake of simplicity, we denote $\lambda = \underline{\lambda}(0)$. Using (6.37), the Wronskian $W = W[\psi_{0,\lambda}, \psi_{x,\lambda}]$ is

$$W(y) = \psi_{0,\lambda}'(y)\psi_{x,\lambda}(y) - \psi_{0,\lambda}(y)\psi_{x,\lambda}'(y) = -\psi_{0,\lambda}(x)e^{2\int_x^y \alpha(\xi)\,d\xi} \quad \text{for } x \le y.$$

Therefore, we get

$$\left(\frac{\psi_{0,\lambda}}{\psi_{x,\lambda}}\right)'(y) = \frac{W(y)}{\psi_{x,\lambda}^2(y)} = -\psi_{0,\lambda}(x)\frac{e^{2\int_x^y \alpha(\xi)\,d\xi}}{\psi_{x,\lambda}^2(y)}.$$

Consider $x < y_0$ and integrate the above equality on $[y_0, y]$ to obtain

$$\psi_{0,\lambda}(y) = \psi_{x,\lambda}(y)\left(\frac{\psi_{0,\lambda}(y_0)}{\psi_{x,\lambda}(y_0)} - \psi_{0,\lambda}(x)\int_{y_0}^{y}\frac{e^{2\int_x^{\eta}\alpha(\xi)\,d\xi}}{\psi_{x,\lambda}^2(\eta)}\,d\eta\right). \tag{6.41}$$

The assumption of having a gap at $z > x$ and the assumption $\underline{\lambda}(0) = \underline{\lambda}(x)$ ensure that $\underline{\lambda}(z) > \underline{\lambda}(x)$ and that $\alpha^{(x)}$ has a gap at z with respect to x. Therefore, using the part of the theorem already proved, $\alpha^{(x)}$ is R-positive. So far we have the statement

$$\alpha^{(x)} \text{ is } R\text{-positive} \quad \text{and} \quad \underline{\lambda}(x) = \underline{\lambda}(0). \tag{6.42}$$

We shall prove that this leads to a contradiction. We first remark that the following integral is finite

$$\int_x^\infty \psi^2_{x,\underline{\lambda}(x)}(\eta) e^{-2\int_x^\eta \alpha(\xi)\,d\xi}\,d\eta < \infty,$$

which implies for any $y_0 \geq x$ (we use that $\int 1\,dx = \int f/f\,dx$)

$$\int_{y_0}^\infty \frac{e^{2\int_x^\eta \alpha(\xi)\,d\xi}}{\psi^2_{x,\underline{\lambda}(x)}(\eta)}\,d\eta = \infty.$$

This is a contradiction to (6.41) because at some large y we obtain (recall that $\lambda = \underline{\lambda}(0)$)

$$\psi_{0,\lambda}(x) \int_{y_0}^y \frac{e^{2\int_x^\eta \alpha(\xi)\,d\xi}}{\psi^2_{x,\lambda}(\eta)}\,d\eta > \frac{\psi_{0,\lambda}(y_0)}{\psi_{x,\lambda}(y_0)},$$

and therefore $\psi_{0,\underline{\lambda}(0)}(y) < 0$. Thus, we have proved that α has a gap at x.

(ii) Take $y < z$. If $\alpha^{(z)}$ is R-positive and $\underline{\lambda}(z) = \underline{\lambda}(y)$, we get a contradiction as we have done for (6.42). Thus, $\underline{\lambda}(z) > \underline{\lambda}(y)$, so $\alpha^{(y)}$ has a gap at z with respect to y, which implies that $\alpha^{(y)}$ is R-positive, and $\underline{\lambda}$ is strictly increasing on $[0, z]$.

(iii) We notice that α does not have a gap at any $x > 0$, and therefore $\underline{\lambda}(0) = \underline{\lambda}(x)$. From (6.41), we find

$$\int_{y_0}^\infty \frac{e^{2\int_x^\eta \alpha(\xi)\,d\xi}}{\psi^2_{x,\underline{\lambda}(0)}(\eta)}\,d\eta < \infty.$$

Therefore, $\alpha^{(x)}$ is R-transient. □

In the next result, we investigate further the R-classification of $\alpha^{(x)}$, relating this problem to the behavior of $\underline{\lambda}$.

Theorem 6.50 *Assume* **H, H1** *hold. Then, the function $\underline{\lambda}$ is strictly increasing on $[0, \bar{x})$, and $\alpha^{(x)}$ is R-positive for $x \in [0, \bar{x})$. $\underline{\lambda}$ is constant on $[\bar{x}, \infty)$ and $\alpha^{(x)}$ is R-transient on $(\bar{x}, \infty)$. $\underline{\lambda}$ is continuous in $[0, \infty)$; it is C^1 on $[0, \infty)$ except perhaps at $\bar{x}$. Moreover, $\underline{\lambda}'$ satisfies, for $x \in [0, \bar{x})$,*

$$\int_x^\infty \psi^2_{x,\underline{\lambda}(x)}(y) e^{-2\int_x^y \alpha(\xi)\,d\xi}\,dy = \frac{1}{2\underline{\lambda}'(x)}. \tag{6.43}$$

In particular, $\underline{\lambda}'(x) > 0$ on $[0, \bar{x})$.

Finally, when $0 < \bar{x} < \infty$, we have that $\alpha^{(\bar{x})}$ is R-recurrent. It is R-null if and only if $\underline{\lambda}'(\bar{x}-) = 0$, and it is R-positive if and only if $\underline{\lambda}'(\bar{x}-) > 0$ (that is, if $\underline{\lambda}'$ is discontinuous at $\bar{x}$).

It is worth noticing that a formula similar to (6.43) holds for $\underline{\lambda}(x)$

$$\int_x^\infty \psi_{x,\underline{\lambda}(x)}(y)e^{-2\int_x^y \alpha(\xi)\,d\xi}\,dy = \frac{1}{2\underline{\lambda}(x)}.$$

This is a particular case of the relation (6.25) established for any $\lambda \in (0, \underline{\lambda}(x)]$.

Lemma 6.51 *Assume that α is R-transient. Then, there exists $\epsilon > 0$ such that any solution of the problem $v'' - 2\alpha v' = -2\underline{\lambda}(0)v$ whose initial conditions satisfy $0 \le v(0) \le \epsilon$, $|v'(0) - 1| \le \epsilon$ is positive on $(0, \infty)$.*

Proof We begin by fixing some constants used in the proof. Let $a_1 > 1$ be the smallest solution of $\log(a_1)/a_1 = (4e)^{-1}$ and $a^* > a_1$ the smallest solution of $\log(a^*)/a^* = (2e)^{-1}$. We notice that $a^* < e$, and for any $a^* < a < e$ we have $(2e)^{-1} < \log(a)/a < e^{-1}$.

We denote by $w = \psi_{0,\underline{\lambda}(0)}$ and we choose $\epsilon > 0$ small enough such that the following conditions are satisfied: v is positive on $(0, 1]$,

$$\max\{w(1)/v(1), v(1)/w(1)\} \le a_1 \quad \text{and} \quad \epsilon \int_1^\infty w^{-2}(x)e^{\gamma(x)}\,dx \le (4e)^{-1}.$$

For $a \in (a^*, e)$, we shall prove that $v(x) > w(x)/a$ on $[1, \infty)$. Suppose the contrary. Since $v(1)/w(1) \ge 1/a_1 > 1/a$, we obtain that

$$1 < x(a) := \inf\{x > 1 : v(x) \le w(x)/a\} < \infty.$$

Consider the Wronskian $W = W[w, v]$. It is straightforward to prove that $W(x) = v(0)e^{\gamma(x)}$. Since v is positive on the interval $[1, x(a)]$, we obtain

$$w(x) = \frac{w(1)}{v(1)}v(x)\exp\left(\int_1^x \frac{W(y)}{w(y)v(y)}\,dy\right) \quad \text{for } x \in [1, x(a)].$$

Using the relations $w(x(a)) = v(x(a))a > 0$ and $v(x) \ge w(x)/a$ on $[1, x(a)]$, we obtain

$$av(x(a)) \le \frac{w(1)}{v(1)}v(x(a))\exp\left(v(0)a\int_1^{x(a)} \frac{e^{\gamma(y)}}{w^2(y)}\,dy\right).$$

Therefore,

$$\frac{\log(a)}{a} \le \frac{\log(w(1)/v(1))}{a} + \epsilon\int_1^\infty \frac{e^{\gamma(y)}}{w^2(y)}\,dy \le (2e)^{-1},$$

which is a contradiction. Thus, we have proved that $v \ge w/a^*$ on $[1, \infty)$; in particular, v is positive. □

Corollary 6.52 *Assume that $\alpha^{(0)}$ is R-transient and $\tilde{\alpha}$ is an extension of $\alpha^{(0)}$. Then, there is $\delta > 0$ such that $\underline{\lambda}_{\tilde{\alpha}}(x) = \underline{\lambda}_\alpha(0)$ for $x \in [-\delta, 0]$.*

Proof Consider $\epsilon > 0$ given by Lemma 6.51. If $\delta > 0$ is sufficiently small, we have, for fixed $x \in [-\delta, 0)$, that $v = \psi_{x,\underline{\lambda}_\alpha(0);\tilde{\alpha}}$ satisfies $0 \le v(0) \le \epsilon$, $|v'(0) - 1| \le \epsilon$ and v is positive on $(x, 0]$. Therefore, from the previous lemma, v is positive on (x, ∞), which implies that $\underline{\lambda}_{\tilde{\alpha}}(x) \ge \underline{\lambda}_\alpha(0)$. The opposite inequality follows from the fact that $\underline{\lambda}_{\tilde{\alpha}}$ is an increasing function. □

Proof (Theorem 6.50) From Theorem 6.45, it follows that $\underline{\lambda}$ is strictly increasing on $[0, \bar{x})$, and in the same interval $\alpha^{(x)}$ is R-positive. Also $\alpha^{(x)}$ is R-transient in the region $(\bar{x}, \infty)$.

Let us now prove that $\underline{\lambda}$ is continuous on $[0, \infty)$. We use the continuity of $\psi_{x,\lambda}(y)$ on x, λ, y. Consider $x \in [0, \bar{x})$. As z decreases to x, the right-hand side of (6.40) converges to 0 and the integral on the left-hand side stays bounded away from zero. Therefore, we deduce the right continuity of $\underline{\lambda}$ at x. For $x \in (0, \bar{x}]$, we obtain the left continuity of $\underline{\lambda}$ in the same way. The only thing left to prove is the right continuity at $\bar{x}$. If it were the case that $\underline{\lambda}(\bar{x}) < \underline{\lambda}(\infty)$, we would get a contradiction with (6.40) by letting z decrease to $\bar{x}$ because for all $z > \bar{x}$ we have $\underline{\lambda}(\infty) = \underline{\lambda}(z)$.

An application of the Dominated Convergence Theorem leads us to conclude from (6.40) that

$$\int_x^\infty \psi^2_{x,\underline{\lambda}(x)}(y) e^{-2\int_x^y \alpha(\xi)\,d\xi}\,dy = \frac{1}{2\underline{\lambda}'(x)}, \tag{6.44}$$

and we deduce that $\underline{\lambda}$ is C^1 on $[0, \bar{x})$.

Let $0 < \bar{x} < \infty$. From the definition of $\bar{x}$, we have $\underline{\lambda}(y) < \underline{\lambda}(\bar{x})$ for any $y < \bar{x}$, and according to Corollary 6.52, we obtain that $\alpha^{(\bar{x})}$ is R-recurrent.

From (6.39), if $x < z < \bar{x} < y_0 \le y$, we have

$$\frac{\psi^2_{x,\underline{\lambda}(x)}(y)}{\psi^2_{x,\underline{\lambda}(x)}(y_0)} \le \frac{\psi^2_{z,\underline{\lambda}(z)}(y)}{\psi^2_{z,\underline{\lambda}(z)}(y_0)}.$$

Using the Monotone Convergence Theorem in (6.44), we can pass to the limit to $\bar{x}$ and conclude that

$$\int_{\bar{x}}^\infty \psi^2_{\bar{x},\underline{\lambda}(\bar{x})}(y) e^{-2\int_{\bar{x}}^y \alpha(\xi)\,d\xi}\,dy = \lim_{x\uparrow\bar{x}} \frac{1}{2\underline{\lambda}'(x)}.$$

Therefore, $\alpha^{(\bar{x})}$ is R-positive if and only if $\lim_{x\uparrow\bar{x}} \underline{\lambda}'(x) > 0$. □

The R-classification already obtained for $\alpha^{(x)}$, $x > 0$, can be put in terms of points of increase from the left for the function $\underline{\lambda}$. In fact, $\alpha^{(x)}$ is R-recurrent (respectively, R-transient) if and only if x is a point of increase (respectively, constancy) from the left for $\underline{\lambda}$. The distinction between R-null and R-positive is done by the left derivative. Thus, in order to obtain the R-classification of $\alpha^{(0)}$, we rely on extensions of $\alpha^{(0)}$ to the left of 0. Clearly, the classification of $\alpha^{(0)}$ does not depend on the chosen extension. To fix notations, $\tilde{\alpha}$ is said to be an extension of $\alpha^{(0)}$ if $\tilde{\alpha}$ is defined on $[-\epsilon, \infty)$ for some $\epsilon > 0$ and $\tilde{\alpha}^{(0)} = \alpha^{(0)}$. From Theorems 6.45 and 6.50, we directly obtain the following characterization.

Theorem 6.53 *Assume* **H**, **H1** *hold.*

(i) $\alpha^{(0)}$ *is R-transient if and only if for some (any) extension $\tilde{\alpha}$, 0 is a point of constancy for $\underline{\lambda}_{\tilde{\alpha}}$.*
(ii) $\alpha^{(0)}$ *is R-positive if and only if for some (any) extension $\tilde{\alpha}$ one has $\underline{\lambda}'_{\tilde{\alpha}}(0-) > 0$.*
As a matter of completeness,
(iii) $\alpha^{(0)}$ *is R-null if and only if for some (any) extension $\tilde{\alpha}$, 0 is a point of increase for $\underline{\lambda}_{\tilde{\alpha}}$ and $\underline{\lambda}'_{\tilde{\alpha}}(0-) = 0$.*

Remark 6.54 As a straightforward consequence, we get that α is R-transient whenever it is periodic and satisfies **H**, **H1**. A slight generalization is the following one. Consider a subperiodic function α in $[0,\infty)$, that is, $\alpha(x+a) \le \alpha(x)$ for some $a > 0$ and for all $x \ge 0$, which also satisfies **H**, **H1**. A simple comparison argument gives $\underline{\lambda}(a) \le \underline{\lambda}(0)$, thus $\underline{\lambda}$ is a constant function. Take $\tilde{\alpha}$ any subperiodic extension of α. Again a comparison argument shows that 0 is a point of constancy for $\underline{\lambda}_{\tilde{\alpha}}$, implying that α is R-transient.

Furthermore, the following result is verified.

Proposition 6.55 *Let $x \ge 0$ and assume $\alpha^{(x)}$ is R-positive. Then, for any $z > x$, the function ϕ_x satisfies hypotheses* **H** *and* **H1** *on $[z,\infty)$. Moreover, α has a gap at y with respect to z if and only if ϕ_x has a gap at y with respect to z, where $y > z > x$. This last condition ensures that Z^x killed at z is R-positive; in particular, $\underline{\lambda}_{\phi_x}(z) > 0$.*

Proof From Theorem 6.39, the function ϕ_x satisfies Hypothesis **H** in the region $[z,\infty)$, for $z > x$. Hypothesis **H1** for ϕ_x in $[z,\infty)$ follows from

$$\int_z^\infty e^{2\int_z^y \phi_x(\xi)\,d\xi}\,dy = \psi^2_{x,\underline{\lambda}(x)}(z) \int_z^\infty \frac{e^{2\int_z^y \alpha(\xi)\,d\xi}}{\psi^2_{x,\underline{\lambda}(x)}(y)}\,dy.$$

The last quantity is infinite because $\alpha^{(x)}$ is R-positive. The rest of the proof follows immediately from relation (6.35). □

We now establish a comparison criterion to study R-positivity.

Theorem 6.56 *Assume the functions α, β satisfy* **H**, **H1** *and any one of the following three conditions:*

(C1) *α,β are C^1 and $h_\alpha = \alpha^2 - \alpha' - 2\underline{\lambda}_\alpha(0) \ge h_\beta = \beta^2 - \beta' - 2\underline{\lambda}_\beta(0)$ on $[0,\infty)$;*
(C2) *$\alpha \ge \beta$ and $\underline{\lambda}_\alpha(\infty) = \underline{\lambda}_\beta(\infty)$;*
(C3) *$\alpha \le \beta$ and $\underline{\lambda}_\alpha(0) = \underline{\lambda}_\beta(0)$.*

Then, the following properties hold:

(i) *If β is R-transient then α is R-transient;*
(ii) *If α is R-positive then β is R-positive.*

Proof We first assume α and β verify condition (C1). We denote by $\lambda = \underline{\lambda}_\alpha(0)$, $\tau = \underline{\lambda}_\beta(0)$, $v = \psi_{0,\lambda;\alpha}$ and $w = \psi_{0,\tau;\beta}$. Now, consider the function $H = v'w - vw' - (\alpha - \beta)vw$. A simple computation yields

$$H' = (\alpha + \beta)H + vw(h_\alpha - h_\beta).$$

By hypothesis, the function $h_\alpha - h_\beta$ is nonnegative, which implies

$$H(x) = e^{\int_0^x (\alpha(\xi)+\beta(\xi))\,d\xi} \int_0^x v(y)w(y)\big(h_\alpha(y) - h_\beta(y)\big)e^{-\int_0^y (\alpha(z)+\beta(z))\,dz}\,dy \ge 0.$$

Therefore, we get $v'/v - \alpha \ge w'/w - \beta$ on $(0,\infty)$. Integrating this inequality and using the relation $\lim_{\varepsilon\downarrow 0} v(\varepsilon)/w(\varepsilon) = 1$, we obtain

$$w^2(x)e^{-2\int_0^x \beta(z)\,dz} \le v^2(x)e^{-2\int_0^x \alpha(z)\,dz}.$$

Then, properties (i) and (ii) follow from the criteria given in Theorem 6.41.

We now assume (C2) holds. Let $\tilde\beta$ and $\tilde\alpha$ be any pair of extensions of β and α, respectively, defined on $[-\varepsilon,\infty)$ for some $\varepsilon > 0$ and satisfying $\tilde\beta \le \tilde\alpha$. By comparison, we have the inequality $\underline{\lambda}_{\tilde\beta}(x) \le \underline{\lambda}_{\tilde\alpha}(x)$, valid for all $x \ge -\varepsilon$.

Let us prove relation (i). Since β is R-transient, we have $\underline{\lambda}_{\tilde\beta}(x) = \underline{\lambda}_\beta(0) = \underline{\lambda}_\beta(\infty)$, for all $x < 0$ close enough to 0. By hypothesis and comparison, we get

$$\underline{\lambda}_\alpha(\infty) = \underline{\lambda}_\beta(\infty) = \underline{\lambda}_{\tilde\beta}(x) \le \underline{\lambda}_{\tilde\alpha}(x) \le \underline{\lambda}_\alpha(\infty),$$

which implies that 0 is a point of constancy for $\underline{\lambda}_{\tilde\alpha}$, proving that α is R-transient.

Now let us prove (ii). If β has a gap then it is R-positive. So for the rest of the proof, we can assume that $\underline{\lambda}_\beta(0) = \underline{\lambda}_\beta(\infty)$. By hypothesis and comparison, we have $\underline{\lambda}_\alpha(\infty) = \underline{\lambda}_\beta(\infty) = \underline{\lambda}_\beta(0) \le \underline{\lambda}_\alpha(0) \le \underline{\lambda}_\alpha(\infty)$, so $\underline{\lambda}_\beta(0) = \underline{\lambda}_\alpha(0)$. Since $\underline{\lambda}_\alpha(0) = \underline{\lambda}_\alpha(\infty)$ and α is assumed to be R-positive, Theorem 6.53(ii) implies that $\underline{\lambda}'_{\tilde\alpha}(0-) > 0$. From $\underline{\lambda}_{\tilde\beta}(x) \le \underline{\lambda}_{\tilde\alpha}(x)$, we get $\underline{\lambda}'_{\tilde\beta}(0-) \ge \underline{\lambda}'_{\tilde\alpha}(0-) > 0$. By using again Theorem 6.53(ii), we conclude that β is R-positive.

The proof that (C3) implies (i) and (ii) is similar to the previous one. □

We remark that among the conditions of Theorem 6.56, (C2) is the easiest one to verify. The other two conditions depend on $\underline{\lambda}_\alpha(0), \underline{\lambda}_\beta(0)$ which, in general, are not simple to compute.

Two special cases are studied in the following result.

Corollary 6.57 *Assume α satisfies* **H, H1**.

(i) *If*

$$\underline{\alpha}(\infty) := \liminf_{x\to\infty} \alpha(x) > \sqrt{2\underline{\lambda}(0)}, \tag{6.45}$$

then the process X^T is R-positive. A sufficient condition for (6.45) *to hold is for some $b > 0$*

$$\underline{\alpha}(\infty) \ge \Big(\big(\sup\{\alpha(x) : x \in [0,b]\}\big)^2 + (\pi/b)^2\Big)^{\frac{1}{2}}. \tag{6.46}$$

In particular, the condition $\lim_{x\to\infty} \alpha(x) = \infty$ implies that X^T is R-positive.

(ii) *Assume the following limit exists* $\alpha(\infty) := \lim_{x\to\infty} \alpha(x) \geq 0$. *If* $\alpha(\infty) \leq \alpha(x)$ *for all* $x \geq 0$ *then* $\underline{\lambda}(0) = \alpha(\infty)^2/2$, *and the process* X^T *is* R*-transient. In particular, this holds whenever* α *is a nonnegative decreasing function.*

Lemma 6.58 *Let* $b > 0$ *and consider*

$$\hat{\lambda}(b) = \sup\{\lambda : \psi_{0,\lambda} \text{ is increasing on } [0, b]\}.$$

Then,

$$\underline{\lambda}(0) < \hat{\lambda}(b) < \left(D^2 + (\pi/b)^2\right)/2, \quad \text{where } D = \sup\{\alpha(x) : x \in [0, b]\}. \tag{6.47}$$

Proof The first inequality in (6.47) follows from Lemma 6.47. For proving the second inequality, consider the function $g(x) = e^{Dx} \sin(\pi x/b)$. g is positive on $(0, b)$, it verifies $g(0) = g(b) = 0$ and the equation $g'' - 2Dg' = -2\lambda g$, where $\lambda = (D^2 + (\pi/b)^2)/2$. Assume that $v = \psi_{0,\lambda}$ is increasing on $[0, b]$. Using the Wronskian $W = W[v, g]$, we deduce that $W' = 2DW + 2v'g(\alpha - D)$, and therefore,

$$0 < W(b) = -g'(b)v(b) = 2e^{2Db} \int_0^b e^{-2Dx} v'(x)g(x)\left(\alpha(x) - D\right) dx \leq 0,$$

which is a contradiction. Therefore, $\psi_{0,\lambda}$ cannot be increasing on $[0, b]$, proving that $\hat{\lambda}(b) < (D^2 + (\pi/b)^2)/2$. □

Proof (Corollary 6.57) The proof is based on a comparison with the constant drift case. To prove (i), we notice that (6.45) implies $\underline{\lambda}(x) > \underline{\lambda}(0)$ for any large enough x. Therefore, α has a gap, which ensures that α is R-positive. The fact that condition (6.46) is sufficient for (6.45) follows from property (6.47) in Lemma 6.58.

We now prove (ii). For any $\epsilon > 0$, there exists x_0 large enough and such that $\underline{\lambda}(x) \leq (\alpha(\infty) + \epsilon)^2/2$ for $x \geq x_0$, proving that $\underline{\lambda}(\infty) \leq \alpha(\infty)^2/2$. On the other hand, the condition $0 \leq \alpha(\infty) \leq \alpha(x)$ for all $x \geq 0$ ensures that $\underline{\lambda}(0) \geq \alpha(\infty)^2/2$, proving that $\underline{\lambda}(x) = \alpha(\infty)^2/2$ for all $x \geq 0$. The rest of the prove is based on Theorem 6.56. Indeed, take β to be the constant function $\alpha(\infty)$. Condition (C2) in Theorem 6.56 is satisfied, and since β is R-transient we get α is also R-transient. □

One is tempted to believe that α is R-positive whenever it is increasing, nonconstant, and eventually positive. This is the case when α is unbounded, but in the bounded case α is not R-positive, in general. In this direction, the following result gives a sufficient integral condition in order for the X^T to be R-transient.

Proposition 6.59 *Assume that* α *is bounded on* $[0, \infty)$, *it satisfies* **H**, **H1** *and* $\alpha(\infty) = \lim_{y\to\infty} \alpha(y) \geq \alpha(x)$ *for all* $x \geq 0$. *Also we assume that* $\alpha(\infty) > 0$. *If*

$$\int_0^\infty \left(\alpha(\infty) - \alpha(x)\right)\left(\alpha(\infty)x + 1\right) dx < \frac{1}{2e},$$

then X^T *is* R*-transient, that is,* $\underline{\lambda}(0) = \underline{\lambda}(\infty) = \frac{\alpha(\infty)^2}{2}$.

Example 6.60 Let $\alpha(x)=1-K/(1+x)^3$. From condition (6.46) in Corollary 6.57, it follows that for large values of K, α is R-positive. In an opposite way, from Proposition 6.59, we find that for small values of K, α is R-transient.

Proof (Proposition 6.59) Consider the nonnegative function $f(x)=\alpha(\infty)-\alpha(x)$. Let β be the constant function $\beta=\alpha(\infty)$. Denote by $\tau=\underline{\lambda}_\beta(0)$ the bottom of its spectrum, which is $\tau=\alpha(\infty)^2/2$. We shall prove $\underline{\lambda}_\alpha(0)=\tau$. Put $v=\psi_{0,\tau;\alpha}$ and $w=\psi_{0,\tau;\beta}$. We notice that $w(x)=xe^{\beta x}$. At this point, we do not know if v is nonnegative.

From $w''-2\beta w'=-2\tau w$ and $v''-2(\beta-f(x))v'=-2\tau v$, we deduce that the Wronskian $W=W[w,v]$ is given by

$$W(x)=2e^{2\beta x-2\int_0^x f(y)\,dy}\int_0^x f(z)w'(z)v(z)e^{-2\beta z+2\int_0^z f(y)\,dy}\,dz.$$

Since w' and f are nonnegative, if v is positive on some interval $(0,x_0]$ then W is nonnegative in that interval. This implies the inequality $v(x)\le w(x)$ for all $x\in[0,x_0]$. Hence, using the explicit form for w, we obtain the following upper bound for W

$$W(x)\le 2e^{2\beta x}\int_0^x f(z)w'(z)w(z)e^{-2\beta z}\,dz=2e^{2\beta x}\int_0^x f(z)z(\beta z+1)\,dz. \tag{6.48}$$

On the other hand, for $x\in(0,x_0]$, we have the equality

$$w(x)=v(x)\exp\left(\int_0^x \frac{W(y)}{w(y)v(y)}\,dy\right).$$

Now, consider the function $g(a)=\log(a)/(2a)$, which is nonnegative for $a\ge 1$ and attains its maximum at $a=e$, with $g(e)=1/(2e)$. Moreover, g is strictly increasing on $[1,e)$ and strictly decreasing on $(e,\infty]$. From the hypothesis $\int_0^\infty f(z)(\beta z+1)\,dz<1/(2e)$, there exists a unique $\bar a\in[1,e)$ such that

$$\int_0^\infty f(z)(\beta z+1)\,dz=\frac{\log(\bar a)}{2\bar a}.$$

We shall prove that $v\ge w/\bar a$. For this purpose, take any $a>\bar a$, sufficiently close to $\bar a$, in order to have $g(a)>g(\bar a)$. Assume that $x(a):=\inf\{x>0: v(x)<w(x)/a\}$ is finite. Notice that $x(a)>0$. Since v is positive on $(0,x(a)]$, we have

$$av\big(x(a)\big)=w\big(x(a)\big)=v\big(x(a)\big)\exp\left(\int_0^{x(a)}\frac{W(y)}{w(y)v(y)}\,dy\right).$$

Therefore, since $v(x)\ge w(x)/a$ on $[0,x(a)]$, we get from (6.48)

$$\log(a) = \int_0^{x(a)} \frac{W(y)}{w(y)v(y)}\,dy \le a \int_0^{x(a)} \frac{W(y)}{w^2(y)}\,dy$$
$$\le 2a \int_0^{x(a)} \frac{e^{2\beta y}}{w^2(y)} \int_0^y f(z)z(\beta z+1)\,dz$$
$$\le 2a \int_0^{\infty} \frac{1}{y^2} \int_0^y f(z)z(\beta z+1)\,dz = 2a \int_0^{\infty} f(z)(\beta z+1)\,dz.$$

This implies that

$$g(a) = \frac{\log(a)}{2a} \le \int_0^{\infty} f(z)(\beta z+1)\,dz = g(\bar{a}),$$

obtaining a contradiction. Thus, $x(a) = \infty$.

We have proved that

$$\psi_{0,\alpha(\infty)^2/2;\alpha} \ge w/\bar{a},$$

implying that $\psi_{0,\alpha(\infty)^2/2;\alpha}$ is nonnegative. Hence, $\underline{\lambda}_\alpha(0) \ge \alpha(\infty)^2/2$. The opposite inequality follows from a comparison with the constant case $\alpha(\infty)$. Thus, $v = \psi_{0,\underline{\lambda}(0);\alpha} \ge w/\bar{a}$.

Finally, since $\alpha \le \alpha(\infty)$ we get

$$\psi_{0,\underline{\lambda}(0);\alpha}(x)^{-2} e^{2\int_0^x \alpha(\xi)\,d\xi} \le \bar{a}^2 w(x)^{-2} e^{2\alpha(\infty)x} = (\bar{a}/x)^2,$$

and α is R-transient from Theorem 6.41, part (iii). □

6.6 Domain of Attraction for a QSD

In this section, we study the following problem. Given a QSD ν, what can we say about the probability measures ζ with support on $(0,\infty)$ that are attracted to ν under the conditional evolution. More precisely, we introduce the following concept.

Definition 6.61 Given a QSD ν, we say that the probability measure ζ supported on $(0,\infty)$ is in *the domain of attraction* of ν if

$$\lim_{t\to\infty} \mathbb{E}_\zeta\big(f(X_t)|T > t\big) = \int f(x)\,d\nu(x),$$

for all continuous and bounded functions $f : (0,\infty) \to \mathbb{R}$.

It is a difficult problem to give general conditions under which a probability measure ζ is in the domain of attraction of a QSD. We notice that each QSD ν is in its domain of attraction since ν is a fixed point for the conditional iteration. The existence of the Yaglom limit $\underline{\nu}$ can be rephrased as: For every $x > 0$, the Dirac measure at x is in the domain of attraction of $\underline{\nu}$.

In this section, we shall assume that X solves the SDE

$$dX_t = dB_t - \alpha\, dt, \qquad X_0 = x_0 > 0, \tag{6.49}$$

where $\alpha > 0$ is constant. That is, X is a Brownian motion with negative drift. We shall present here mainly the results of Martínez et al. (1998).

The relevant quantities are $\gamma(x) = 2\alpha x$, $\Lambda(x) = \frac{1}{2\alpha}(e^{2\alpha x} - 1)$ and the speed measure $d\mu(y) = e^{-2\alpha y}\, dy$. The drift satisfies Hypothesis **H** and $\Lambda(\infty) = \infty$, and it is straightforward to show that $\underline{\lambda} = \alpha^2/2$. For every $\lambda \in (0, \alpha^2/2)$, the right eigenfunction ψ_λ is given by (recall that we use the normalization $\psi_\lambda(0) = 0$, $\psi_\lambda'(0) = 1$)

$$\psi_\lambda(x) = e^{\alpha x} \sinh\bigl(x\sqrt{\alpha^2 - 2\lambda}\bigr), \quad x \ge 0,$$

and for the extremal value $\underline{\lambda} = \alpha^2/2$ we have

$$\psi_{\underline{\lambda}}(x) = x e^{\alpha x}, \quad x \ge 0.$$

Thus, for every $\lambda \in (0, \alpha^2/2]$ there is a QSD ν_λ, which is moreover absolutely continuous with respect to the Lebesgue measure and its density is proportional to $\varphi_\lambda = e^{-\gamma(y)}\psi_\lambda(y)$, that is,

$$d\nu_\lambda(y) = \begin{cases} M_\lambda e^{-\alpha y} \sinh(y\sqrt{\alpha^2 - 2\lambda})\, dy, & y \ge 0 \quad \text{if } \lambda \in (0, \alpha^2/2), \\ M_\lambda y e^{-\alpha y}\, dy, \quad y \ge 0 & \text{if } \lambda = \alpha^2/2, \end{cases}$$

where in each case M_λ is a normalizing constant. In what follows, we denote by $\underline{\nu} = \nu_{\alpha^2/2}$, the QSD associated to $\underline{\lambda}$.

We shall denote by $q_t(x, y)$ the transition kernel of X with respect to the Lebesgue measure, which is well known and given by

$$\mathbb{P}_x(X_t \in dy, t < T) = q_t(x, y)\, dy = e^{-\frac{\alpha^2}{2}t} e^{-\alpha(y-x)} p_t^-(x, y)\, dy, \quad x > 0, y > 0,$$

where $p_t^-(x, y) = \frac{1}{\sqrt{2\pi t}}(e^{-\frac{(x-y)^2}{2t}} - e^{-\frac{(x+y)^2}{2t}})$ is the transition kernel of a killed Brownian motion (this corresponds to $\alpha = 0$). We shall denote by $p_t(x, y) = \frac{1}{\sqrt{2\pi t}} e^{-\frac{(x-y)}{2t}}$ the transition density for the standard Brownian motion.

Integrating p^- over $y \in \mathbb{R}_+$, we obtain one of the main formulas we require in what follows

$$\mathbb{P}_x(t < T) = e^{\alpha x - \frac{\alpha^2}{2}t} \int_0^\infty e^{-\alpha y} p_t^-(x, y)\, dy = \int_t^\infty \frac{x}{\sqrt{2\pi u^3}} e^{-\frac{(x-\alpha u)^2}{2u}}\, du. \tag{6.50}$$

Using the explicit formula of q, we can refine the results obtained in Theorem 6.26.

Proposition 6.62 *The QSD $\underline{\nu}$ whose density is proportional to $\varphi_{\underline{\lambda}}(y) = ye^{-\alpha y}$ is the Yaglom limit. On the other hand, we have the limits*

$$\begin{aligned}\lim_{t\to\infty} t^{\frac{3}{2}} e^{\frac{\alpha^2}{2}t} q_t(x,y) &= \frac{2xy}{\sqrt{2\pi}} e^{-\alpha(y-x)},\\ \lim_{t\to\infty} t^{\frac{3}{2}} e^{\frac{\alpha^2}{2}t} \mathbb{P}_x(t<T) &= \frac{2x}{\alpha^2\sqrt{2\pi}} e^{\alpha x}.\end{aligned} \tag{6.51}$$

Remark 6.63 It is straightforward to show that the limits in (6.51) are uniform in x on compact sets of $[0,\infty)$ and in α on compact sets of $(0,\infty)$.

One important tool for studying the domain of attraction of a QSD is the following characterization of a QSD.

Lemma 6.64 *Assume that ν is a probability measure supported on $[0,\infty)$ such that for all $t \ge 0$, $\mathbb{P}_\nu(t<T) = e^{-ct}$ holds for some finite c. Then $c \in (0, \alpha^2/2]$ and $\nu = \nu_c$, that is, ν is a QSD.*

Proof Since ν is supported on $[0,\infty)$, we conclude that $\mathbb{P}_\nu(t<T) = \int \mathbb{P}_x(t < T)\, d\nu(x) < 1$, and therefore $c > 0$. Clearly, $\nu\{0\} < 1$, otherwise $c = \infty$. Now, take any $x_0 > 0$ such that $\nu([x_0,\infty)) > 0$ to conclude that $\mathbb{P}_\nu(t<T) \ge \mathbb{P}_{x_0}(t < T)\nu([x_0,\infty))$. Here we have used the obvious monotonicity of $\mathbb{P}_x(t<T)$ on the initial condition x. From the asymptotic given by (6.51), we get $c \le \alpha^2/2$.

We have to show that $\nu = \nu_c$. For that purpose, we notice that

$$\mathbb{P}_\nu(t<T) = e^{-ct} = \mathbb{P}_{\nu_c}(t<T),$$

which is equivalent to having for all $t \ge 0$

$$\begin{aligned} e^{-ct} &= \int_t^\infty \int_0^\infty \frac{1}{\sqrt{2\pi u^3}} e^{-\frac{x^2}{2u}} x e^{\alpha x}\, d\nu_c(x)\, du\\ &= \int_t^\infty \int_0^\infty \frac{1}{\sqrt{2\pi u^3}} e^{-\frac{x^2}{2u}} x e^{\alpha x}\, d\nu(x)\, du. \end{aligned}$$

In particular, this implies that for all $u > 0$ the measures $e^{-\frac{x^2}{2u}} x e^{\alpha x}\, d\nu_c(x)$ and $e^{-\frac{x^2}{2u}} x e^{\alpha x}\, d\nu(x)$ are finite (here we have used the monotonicity of both measures with respect to u). Differentiating with respect to t, we obtain for all $t > 0$

$$\int_0^\infty e^{-\frac{x^2}{4t}} \left[e^{-\frac{x^2}{4t}} x e^{\alpha x} \left(d\nu_c(x) - d\nu(x)\right)\right] = 0.$$

Integrating by parts, we get for all $t > 0$

$$\int_0^\infty G(x) e^{-\frac{x^2}{4t}}\, dx = 0,$$

with $G(x)=x\int_0^x e^{-\frac{v^2}{4t}} ve^{\alpha v}\,(d\nu_c(v)-d\nu(v))$. Using that $(e^{-\frac{x^2}{4t}}:t>0)$ is a complete family (see Brown 1968, pp. 42–43), we deduce that $G\equiv 0$, and the result is shown. □

In the next result, we study the domain of attraction for the QSDs that are not extremal.

Theorem 6.65 *Assume that ζ is a probability measure with support contained in $(0,\infty)$ and with a density ρ that satisfies*

$$\lim_{x\to\infty} -\frac{1}{x}\log\rho(x)=\beta\in(0,\alpha).$$

Then ζ is in the domain of attraction of ν_λ where $\lambda=\lambda(\beta)=\alpha\beta-\beta^2/2$.

Proof We shall prove that the family of probability measures $\zeta_t(\bullet):=\mathbb{P}_\zeta(X_t\in\bullet|t<T)$ for $t\geq 0$ is tight in $(0,\infty)$. Then, we shall prove that any possible weak limit ν satisfies

$$\mathbb{P}_\nu(t<T)=e^{-\lambda(\beta)t},$$

and the result will follows from the previous lemma.

Tightness. The proof is quite technical and we refer the reader to Martínez et al. (1998) for details. The first step is to obtain a crude lower estimate of $\mathbb{P}_\zeta(t<T)=\int_0^\infty \mathbb{P}_x(t<T)\rho(x)\,dx$. For that purpose and for each fixed $\epsilon>0$ such that $\epsilon<\epsilon_0=\min\{\frac{\alpha-\beta}{2\beta},\frac{1}{2}\}$, consider $\beta^+=\beta(\epsilon)=\beta(1+\epsilon)<\alpha$. Let $\lambda^+=\lambda(\beta^+)=\alpha\beta^+-(\beta^+)^2/2$, in particular $\alpha-\sqrt{\alpha^2-2\lambda^+}=\beta^+$. Then, the density of ν_{λ^+} is equivalent to $M_{\lambda^+}e^{-\beta^+x}$ for large values of x. Hence, there exists $x_0=x_0(\epsilon)$ such that for all $x\geq x_0$

$$\rho(x)\geq M_{\lambda^+}e^{-\alpha x}\sinh\bigl(x\sqrt{\alpha^2-2\lambda^+}\bigr)=M_{\lambda^+}\varphi_{\lambda^+}(x).$$

Thus, we obtain

$$\begin{aligned}\mathbb{P}_\zeta(t<T)&\geq\int_0^\infty \mathbb{P}_x(t<T)M_{\lambda^+}\varphi_{\lambda^+}(x)\,dx-\int_0^{x_0}\mathbb{P}_x(t<T)M_{\lambda^+}\varphi_{\lambda^+}(x)\,dx\\&=e^{-\lambda^+t}-\int_0^{x_0}\mathbb{P}_x(t<T)M_{\lambda^+}\varphi_{\lambda^+}(x)\,dx\\&\geq e^{-\lambda^+t}-\mathbb{P}_{x_0}(t<T).\end{aligned}$$

Now using the asymptotic behavior of $\mathbb{P}_{x_0}(t<T)$ given by (6.51), we find a $t_0=t_0(\epsilon)$ such that for all $t\geq t_0$

$$\mathbb{P}_\zeta(t<T)\geq\frac{1}{2}e^{-\lambda^+t}. \tag{6.52}$$

To prove tightness for the family $(\zeta_t : t > 0)$ as $t \to \infty$, it is enough to find $\theta > 0$ and a constant C such that for all large t we have $\mathbb{E}_\zeta(e^{\theta X_t}, t < T) \le C\mathbb{P}_\zeta(t < T)$. We start by noticing that

$$\mathbb{E}_\zeta\big(e^{\theta X_t}, t < T\big) = \int_0^\infty \mathbb{E}_x\big(e^{\theta X_t}, t < T\big)\rho(x)\,dx$$
$$= \int_0^\infty \int_0^\infty e^{\theta y + \alpha(x-y) - \frac{\alpha^2}{2}t} p_t^-(x,y)\,dy\rho(x)\,dx.$$

The integral in the x variable is divided into four regions: $[0, M]$, $[M, at]$, $[at, bt]$, $[bt, \infty)$, and we call them successively I, II, III, IV. We first choose $M = M(\epsilon)$ such that $K = K(\epsilon) = \sup\{\rho(x)e^{-\beta^- x} : x \ge M\} < \infty$, where $\beta^- = \beta^-(\epsilon) = \beta(1-\epsilon)$. In what follows, we also denote by $\lambda^- = \lambda(\beta^-) = \alpha\beta^- - (\beta^-)^2/2$.

Of the four integrals, the simplest to bound is I, and we consider $0 < \theta \le \theta_I = \alpha/2$ for which

$$I = \int_0^M \mathbb{E}_x\big(e^{\theta X_t}, t < T\big)\rho(x)\,dx \le \mathbb{E}_M\big(e^{\theta X_t}, t < T\big);$$

here we have used the monotonicity of X on its initial position. This last integral is

$$I \le \int_0^\infty e^{\theta y} e^{\alpha(M-y) - \frac{\alpha^2}{2}t} p_t^-(M,y)\,dy$$
$$= e^{\theta M - (\alpha\theta - \theta^2/2)t} \int_0^\infty e^{(\alpha-\theta)(M-y) - (\alpha-\theta)^2 t/2} p_t^-(M,y)\,dy$$
$$= e^{\theta M - (\alpha\theta - \theta^2/2)t}\mathbb{P}_M(t < \tilde{T}),$$

where $\tilde{T}$ is the time to hit zero for a Brownian motion with constant drift $-(\alpha - \theta)$. The probability $\mathbb{P}_M(t < \tilde{T})$ decays like $t^{-3/2}e^{-(\alpha-\theta)^2 t/2}$ (uniformly on θ, see Remark 6.63). Hence, there exists $t_I = t_I(\epsilon) \ge t_0(\epsilon)$ and $A_I = A_I(\epsilon) < \infty$ such that for all $t > t_I$ we have

$$I \le e^{\theta M} A_I e^{-\frac{\alpha^2}{2}t} \le 2e^{\theta M} A_I \frac{e^{-\lambda^+ t}}{2} \le 2e^{\theta M} A_I \mathbb{P}_\zeta(t < T).$$

The last inequality follows from (6.52). Defining $C_I = C_I(\epsilon, \theta) = 2e^{\theta M} A_I$, we obtain the desired bound for I. We also define $\epsilon_I = \epsilon_0$.

Let us now consider II. We bound in this case p^- by p. We assume that $0 < a < \alpha - \beta < \alpha - \beta^-$ and $0 < \theta \le \theta_{II} = \frac{\beta}{3} < \beta^-$ to get (we also assume that $at > M$)

$$II \le \int_M^{at} \int_0^\infty e^{\theta y + \alpha(x-y) - \frac{\alpha^2}{2}t} p_t(x,y)\,dy\rho(x)\,dx$$
$$= \int_M^{at} \int_0^\infty e^{\theta y} p_t(x, y + \alpha t)\,dy\rho(x)\,dx$$

$$
\begin{aligned}
&= e^{-\lambda^- t} \int_0^\infty e^{(\theta-\beta^-)y} \int_M^{at} \rho(x) e^{\beta^- x} p_t\big(x, y + (\alpha - \beta^-)t\big)\, dx\, dy \\
&\le K e^{-\lambda^- t} \int_0^\infty e^{(\theta-\beta^-)y} \int_{-\infty}^{at} p_t\big(x, y + (\alpha - \beta^-)t\big)\, dx\, dy \\
&\le \frac{K}{\sqrt{2\pi}} e^{-\lambda^- t} \int_0^\infty e^{(\theta-\beta^-)y} \int \mathbf{1}\left(u \le \frac{at - y - (\alpha - \beta^-)t}{\sqrt{t}}\right) e^{-\frac{u^2}{2}}\, du\, dy \\
&\le \frac{K}{\sqrt{2\pi}} e^{-\lambda^- t} \int_0^\infty e^{(\theta-\beta^-)y}\, dy \int \mathbf{1}\big(u \ge (\alpha - \beta^- - a)\sqrt{t}\big) e^{-\frac{u^2}{2}}\, du.
\end{aligned}
$$

For $z > 0$, we use the standard Gaussian bound $\int_z^\infty e^{-\frac{u^2}{2}}\, du \le \int_z^\infty \frac{u}{z} e^{-\frac{u^2}{2}}\, du = \frac{1}{z} e^{-\frac{z^2}{2}}$. Thus we get the upper bound

$$
II \le \frac{K}{(\beta^- - \theta)(\alpha - \beta^- - a)\sqrt{2\pi t}} e^{-(\lambda^- + (\alpha - \beta^- - a)^2/2)t}.
$$

We impose that $\lambda^- + (\alpha - \beta^- - a)^2/2 \ge \lambda^+$, which is equivalent to

$$
\alpha\beta(1-\epsilon) - \beta^2(1-\epsilon)^2/2 + \big(\alpha - \beta(1-\epsilon) - a\big)^2/2 \ge \alpha\beta(1+\epsilon) - \beta^2(1+\epsilon)^2/2,
$$

which turns out to be equivalent to

$$
\beta\epsilon + 2\sqrt{(\alpha - \beta)\beta\epsilon} \le \alpha - \beta - a.
$$

Hence, we take $\epsilon_{II} = \max\{\epsilon : \beta\epsilon + 2\sqrt{(\alpha - \beta)\beta\epsilon} \le \alpha - \beta - a\} \wedge \epsilon_0$ and also consider $t_{II} = t_{II}(\epsilon, a) = 1 \vee \frac{M(\epsilon)}{a} \vee t_0(\epsilon)$ to get for all $t \ge t_{II}$

$$
II \le C_{II} \mathbb{P}_\zeta(t < T),
$$

where $C_{II} = C_{II}(\epsilon, a, \theta) = \frac{2K(\epsilon)}{(\beta^- - \theta)(\alpha - \beta^- - a)\sqrt{2\pi}}$.

Let us now bound III. Once again we replace p^- by p and obtain

$$
\begin{aligned}
III &\le \int_{at}^{bt} \int_0^\infty e^{\theta y} p_t(x, y + \alpha t)\, dy \rho(x)\, dx \\
&= \sqrt{2\pi t} \int_{at}^{bt} p_t(x, \alpha t) \int_0^\infty e^{\theta y} p_t(y, 0) e^{y(x - \alpha t)/t}\, dy \rho(x)\, dx \\
&\le \int_{at}^{bt} \frac{x}{at} p_t(x, \alpha t) \int_0^\infty e^{y(\theta + b - \alpha)}\, dy \rho(x)\, dx \\
&= \frac{1}{a(\alpha - (b + \theta))} \int_{at}^{bt} \frac{x}{t} p_t(x, \alpha t) \rho(x)\, dx.
\end{aligned}
$$

Obviously, we have imposed that $a < b$ and

$$
b + \theta < \alpha. \tag{6.53}
$$

We take $\theta_{III}(b) = \frac{1}{2}(\alpha - b)$.

On the other hand, using (6.50), we get

$$\mathbb{P}_\zeta(t < T) \geq \int_{at}^{bt} \frac{x}{t+1} \int_t^{t+1} p_u(x, \alpha u)\, du \rho(x)\, dx.$$

The idea is to replace $p_u(x, \alpha u)$ in the last integral by $p_t(x, \alpha t)$. For that purpose, we study the function $g(u) = \sqrt{u} p_u(x, \alpha u)$ in the region $at < x < bt < \alpha t < \alpha u$. The derivative of this function is

$$g'(u) = g(u) \frac{(x^2 - \alpha^2 u^2)}{2u^2} \geq -\frac{\alpha^2}{2} g(u).$$

This inequality implies that

$$\sqrt{t+1} \int_t^{t+1} p_u(x, \alpha u)\, du \geq \int_t^{t+1} \sqrt{u} p_u(x, \alpha u)\, du \geq \frac{\sqrt{t}}{1 + \alpha^2/2} p_t(x, \alpha t).$$

Hence, if we take $t_{III} = 1$, we obtain the lower bound for all $t \geq t_{III}$

$$\mathbb{P}_\zeta(t < T) \geq \left(\frac{t}{t+1}\right)^{3/2} \frac{1}{1 + \alpha^2/2} \int_{at}^{bt} \frac{x}{t} p_t(x, \alpha t) \rho(x)\, dx,$$

which implies that

$$III \leq C_{III} \mathbb{P}_\zeta(t < T),$$

with $C_{III} = C_{III}(\epsilon, a, b, \theta) = 2^{3/2} \frac{1+\alpha^2/2}{a(\alpha - (b+\theta))}$. We also take $\epsilon_{III} = \epsilon_0$.

We are left we the term IV, on which we also replace p^- by p. We take $t_{IV} = t_{IV}(\epsilon, b) \geq t_0(\epsilon) \vee 1$, large enough such that for all $t \geq t_{IV}, x \geq bt$ one has $\rho(x) \leq e^{-\beta^- x}$. Thus, we get

$$\begin{aligned} IV &\leq \int_{bt}^{\infty} \int_0^{\infty} e^{\theta y + \alpha(x-y) - \beta^- x} p_t(x, y + \alpha t)\, dy\, dx \\ &= e^{-\lambda^- t} \int_0^{\infty} e^{(\theta - \beta^-) y} \int_{bt}^{\infty} p_t\big(x, y + \big(\alpha - \beta^-\big)t\big)\, dx\, dy. \end{aligned}$$

We also consider $0 < \theta < \theta_{IV} = \beta/3$. We split the integral over $[0, \infty)$ in two pieces $\mathcal{S}_1$ and $\mathcal{S}_2$. Take $\delta > 0$ and assume that $b > \delta + \alpha - \beta^-$, which is compatible with the restriction given in (6.53) if $\delta < \beta^- - \theta$. Notice that $\beta^- - \theta \geq \beta/2 - \beta/3 = \beta/6 > 0$. Hence, if $y \leq \delta t$ then $y + (\alpha - \beta^-)t \leq (\delta + \alpha - \beta^-)t \leq bt$, and therefore we get

$$\begin{aligned} \mathcal{S}_1 &= e^{-\lambda^- t} \int_0^{\delta t} e^{(\theta - \beta^-) y} \int_{bt}^{\infty} p_t\big(x, y + \big(\alpha - \beta^-\big)t\big)\, dx\, dy \\ &\leq e^{-\lambda^- t} \int_0^{\delta t} e^{(\theta - \beta^-) y}\, dy \int_{bt}^{\infty} p_t\big(x, \big(\delta + \alpha - \beta^-\big)t\big)\, dx \\ &\leq \frac{e^{-\lambda^- t}}{(\beta^- - \theta)\sqrt{2\pi t}(b - (\delta + \alpha - \beta^-))} e^{-(b - (\delta + \alpha - \beta^-))^2 t/2}. \end{aligned}$$

Once again we use that $\lambda^+ - \lambda^- = 2(\alpha-\beta)\beta\epsilon$, so if we assume that $b \geq \delta + \alpha - \beta^- + \sqrt{4(\alpha-\beta)\beta\epsilon}$, and since $t \geq 1$, we get

$$\mathcal{S}_1 \leq \frac{1}{(\beta^- - \theta)\sqrt{2\pi}(b - (\delta + \alpha - \beta^-))} e^{-\lambda^+ t},$$

which is dominated by $\mathbb{P}_\zeta(t < T)$ for large t according to (6.52). Hence we need to impose $0 < \delta \leq \beta^- - \theta - \sqrt{4(\alpha-\beta)\beta\epsilon}$. For this restriction to be feasible, we require that $\beta^- - \theta - \sqrt{4(\alpha-\beta)\beta\epsilon} > 0$, which will be satisfied because below we impose the strongest condition on ϵ.

The final term to study is

$$\begin{aligned}\mathcal{S}_2 &= e^{-\lambda^- t} \int_{\delta t}^{\infty} e^{(\theta-\beta^-)y} \int_{bt}^{\infty} p_t\big(x, y + (\alpha-\beta^-)t\big)\, dx\, dy \\ &\leq e^{-\lambda^- t} \int_{\delta t}^{\infty} e^{(\theta-\beta^-)y}\, dy = \frac{1}{\beta^- - \theta} e^{-(\lambda^- + (\beta^- - \theta)\delta)t}.\end{aligned}$$

We impose that $\lambda^- + (\beta^- - \theta)\delta \geq \lambda^+$, or equivalently, $(\beta^- - \theta)\delta \geq 2(\alpha-\beta)\beta\epsilon$. Therefore, we need ϵ to verify

$$\frac{2(\alpha-\beta)\beta\epsilon}{\beta^- - \theta} \leq \beta^- - \theta - \sqrt{4(\alpha-\beta)\beta\epsilon}.$$

A stronger requirement is obtained by imposing that $\theta \leq \theta_{IV} = \beta/3$ and

$$\frac{2(\alpha-\beta)\beta\epsilon}{\beta/6} \leq \beta/6 - \sqrt{4(\alpha-\beta)\beta\epsilon}.$$

So, we take $\epsilon_{IV} = \max\{\epsilon : 12(\alpha-\beta)\epsilon + \sqrt{4(\alpha-\beta)\beta\epsilon} \leq \beta/6\}$ and for any $0 < \epsilon \leq \epsilon_{IV}$ consider $\delta = \delta(\epsilon) = \frac{2(\alpha-\beta)\beta\epsilon}{\beta^- - \theta}$, $\theta \leq \theta_{IV} = \beta/3$. Thus, if $t \geq t_{IV}$, we get

$$IV \leq C_{IV} \mathbb{P}_\zeta(t < T),$$

where $C_{IV} = C_{IV}(\epsilon, b, \theta) = 2\big(\frac{1}{\beta^- - \theta} + \frac{1}{(\beta^- - \theta)\sqrt{2\pi}(b - (\delta + \alpha - \beta^-))}\big)$.

Finally, we decide to take a fixed $0 < \epsilon < \min\{\epsilon_I, \epsilon_{II}, \epsilon_{III}, \epsilon_{IV}\}$, $a := \frac{1}{2}(\alpha - \beta) \leq \alpha - \beta^- < b := \delta(\epsilon) + \alpha - \beta^- + \sqrt{4(\alpha-\beta)\beta\epsilon}$, $0 < \theta \leq \min\{\theta_I, \theta_{II}, \theta_{III}(b), \theta_{IV}\}$ and $C = C_I(\epsilon,\theta) + C_{II}(\epsilon, a, \theta) + C_{III}(\epsilon, a, b, \theta) + C_{IV}(\epsilon, b, \theta)$. Then, for all $t \geq \bar{t} = \max\{t_I(\epsilon), t_{II}(\epsilon, a), t_{III}, t_{IV}(\epsilon, b)\}$, we have

$$\mathbb{E}_\zeta\big(e^{\theta X_t} | t < T\big) \leq C,$$

and then $(\zeta_t(\bullet) = \mathbb{P}_\zeta(X_t \in \bullet | t < T) : t \geq \bar{t})$ is tight.

Uniqueness of the limit measure. Let us prove now that for all $s > 0$ one has

$$\lim_{t\to\infty} \left| \frac{\partial}{\partial s} \log \mathbb{P}_\zeta(t + s < T) + \lambda(\beta) \right| = 0.$$

Consider $\ell_u = \int_0^\infty \frac{x}{u} p_u(x, \alpha u)\rho(x)\,dx$, then $\mathbb{P}_\zeta(t+s<T) = \int_{t+s}^\infty \ell_u\,du$. Hence, the quantity to be studied is

$$\frac{\partial}{\partial s}\log \mathbb{P}_\zeta(t+s<T) = \frac{-\ell_{t+s}}{\mathbb{P}_\zeta(t+s<T)}.$$

Using l'Hôpital Rule, to justify the claim it is enough to show that

$$\lim_{u\to\infty} \frac{\frac{\partial}{\partial u}\ell_u}{\ell_u} = -\lambda(\beta).$$

In order to prove this, we split the integral defining ℓ in two pieces:

$$\mathcal{J} = \int_0^M \frac{x}{u} p_u(x, \alpha u)\rho(x)\,dx \quad \text{and} \quad \mathcal{L} = \int_M^\infty \frac{x}{u} p_u(x, \alpha u)\rho(x)\,dx.$$

It is straightforward to prove that $\ell_u \ge Ce^{-\lambda^+(u)}$ and $|\frac{\partial}{\partial s}\mathcal{J}| \le Ce^{-\alpha^2 u/2}$. Also, it is quite simple to show that $\mathcal{J}/\ell_u$ converges to 0. Differentiating $\mathcal{L}$, we obtain

$$\frac{\partial}{\partial s}\mathcal{L} = \left(-\frac{3}{2u} - \lambda(\beta)\right)\mathcal{L} + \mathcal{R}.$$

It is possible, and tedious, to show that $\mathcal{R}/\ell_u$ converges to 0 (see Martínez et al. 1998). Using the Intermediate Value Theorem for a ratio, it follows that $\frac{\partial}{\partial s}\log \mathbb{P}_\zeta(t+s<T)$ is bounded for all large values of t, uniformly on compact sets of s.

Now, we conclude the proof of the theorem. Assume that $(\mathbb{P}_\zeta(X_{t_k} \in \bullet | t_k < T) : k \ge 1)$ converges weakly to ν, which is, of course, a probability measure supported on $[0,\infty)$. For $s \ge 0$ fixed, consider the function $h(x) = \mathbb{P}_x(s<T)$. h is a continuous and bounded function, then

$$\begin{aligned} e^{-\lambda(\beta)s} &= \lim_{k\to\infty} \frac{\mathbb{P}_\zeta(t_k+s<T)}{\mathbb{P}_\zeta(t_k<T)} = \lim_{k\to\infty} \mathbb{E}_\zeta\big(h(X_{t_k}) \,|\, t_k<T\big) \\ &= \int h(x)\,d\nu(x) = \mathbb{P}_\nu(s<T). \end{aligned}$$

We conclude that $\nu = \nu_{\lambda(\beta)}$. Thus, there is a unique weak limit for $\mathbb{P}_\zeta(X_t \in \bullet \,|\, t<T)$ as t tends to infinity, and this limit is the QSD $\nu_{\lambda(\beta)}$, proving the result. □

In the next result, we study the domain of attraction for the extremal QSD $\underline{\nu}$.

Theorem 6.66 *Assume that ζ is a probability measure with support contained in $(0,\infty)$ that satisfies*

$$\liminf_{x\to\infty} -\frac{1}{x}\log\zeta\big([x,\infty)\big) \ge \alpha.$$

Then ζ is in the domain of attraction of $\underline{\nu}$.

Proof The first part of the proof consists in proving that we can assume that ζ has a continuous density that satisfies

$$\liminf_{t\to\infty} -\frac{\log \rho(x)}{x} \geq \alpha. \tag{6.54}$$

For that purpose, consider the density $h(x)$ given by

$$h(x)\,dx = \mathbb{P}_\zeta(X_1 \in dx | 1 < T) = c\int_0^\infty e^{\alpha(w-y-\alpha^2/2)} p^-(1, w, y)\,d\zeta(w)\,dx,$$

where c is a normalizing constant. Replacing p^- by p, we obtain the upper bound

$$\begin{aligned} h(x) &\leq \frac{c}{\sqrt{2\pi}} \int_0^\infty e^{-\frac{(w-x-\alpha)^2}{2}}\,d\zeta(w) \\ &= \frac{c}{\sqrt{2\pi}}\left(e^{-\frac{(x+\alpha)^2}{2}} + \int_0^\infty \zeta([w,\infty))\big(-(w-x-\alpha)\big)e^{-\frac{(w-x-\alpha)^2}{2}}\,dw\right). \end{aligned}$$

For the last equality, we have used integration by parts. Consider $\theta \in \mathbb{R}$ for which there exists w_0 large enough, such that for all $w \geq w_0$ we have $\zeta([w,\infty)) \leq e^{\theta w}$. Replacing this upper bound, we get for $x \geq w_0$

$$\begin{aligned} h(x) &\leq \frac{c}{\sqrt{2\pi}}\left(e^{-\frac{(w_0-x-\alpha)^2}{2}} + \int_{w_0}^\infty e^{\theta w}|w-x-\alpha|e^{-\frac{(w-x-\alpha)^2}{2}}\,dw\right) \\ &\leq \frac{c}{\sqrt{2\pi}}\left(e^{-\frac{(w_0-x-\alpha)^2}{2}} + e^{\frac{\theta^2}{2}+\theta\alpha}e^{\theta x}\int_{-\infty}^\infty |w-x-\alpha|e^{-\frac{(w-x-\alpha-\theta)^2}{2}}\,dw\right) \\ &\leq \frac{c}{\sqrt{2\pi}}\big(e^{-\frac{(w_0-x-\alpha)^2}{2}} + e^{\frac{\theta^2}{2}+\theta\alpha}e^{\theta x}\big(|\theta|\sqrt{2\pi}+2\big)\big). \end{aligned}$$

We deduce that h satisfies $\liminf_{x\to\infty} -\frac{\log h(x)}{x} \geq -\theta$, implying that

$$\liminf_{t\to\infty} -\frac{\log h(x)}{x} \geq \liminf_{t\to\infty} -\frac{\log \zeta([x,\infty))}{x} \geq \alpha.$$

On the other hand, if η is the probability measure whose density is h, we get for all $t \geq 0$

$$\mathbb{P}_\zeta(X_{t+1} \in \bullet | t+1 < T) = \mathbb{P}_\eta(X_t \in \bullet | t < T),$$

showing we can assume that ζ has a continuous density ρ verifying (6.54).

The case where $\liminf_{x\to\infty} -\frac{\log \rho(x)}{x} > \alpha$ is straightforward to study due to the inequality $\frac{\mathbb{P}_x(t<T)}{\mathbb{P}_1(t<T)} \leq Cxe^{\alpha x}$ (see Martínez and San Martín 1994) and the fact that $\rho(x) \leq e^{-(\alpha+a_0)x}$, for some $a_0 > 0$ and large x. Indeed, from the Dominated Convergence Theorem and (6.51), we get

$$\lim_{t\to\infty} \frac{\mathbb{P}_\zeta(t<T)}{\mathbb{P}_1(t<T)} = \int_0^\infty xe^{\alpha(x-1)}\,d\zeta(x),$$

and again from (6.51)

$$\lim_{t\to\infty} t^{3/2} e^{\frac{\alpha^2}{2}t} \mathbb{P}_\zeta(t < T) = \frac{2}{\alpha^2\sqrt{2\pi}} \int_0^\infty x e^{\alpha x}\, d\zeta(x).$$

Let us denote by ρ_t the density at time t of the conditional evolution

$$\rho_t(y)\, dy = \mathbb{P}_\zeta(X_t \in dy | t < T) = \frac{e^{-\frac{\alpha^2}{2}t}}{\mathbb{P}_\zeta(t < T)} \int_0^\infty e^{\alpha(x-y)} p_t^-(x, y)\, d\zeta(x).$$

For a fixed y, we have $p_t^-(x, y) \le C e^{xy/t}$, and the integrand is dominated by $e^{(\alpha+a_0)x}$ for large x and $t \ge y/a_0$. Again the Dominated Convergence Theorem yields

$$\lim_{t\to\infty} t^{3/2} \int_0^\infty e^{\alpha(x-y)} p_t^-(x, y)\, d\zeta(x) = \frac{2y e^{-\alpha y}}{\sqrt{2\pi}} \int_0^\infty x e^{\alpha x}\, d\zeta(x).$$

Hence, we get the limit

$$\lim_{t\to\infty} t^{3/2} \rho_t(y) = \frac{y e^{-\alpha y}}{\alpha^2} = \rho_\infty(y),$$

which is a density in $(0, \infty)$. Scheffe's Lemma allows us to show that ρ_t converges to ρ_∞ in $L^1(dx)$ and a fortiori the measures $\mathbb{P}_\zeta(X_t \in \bullet | t < T)$ converge weakly to $\underline{\nu}$, showing the result in this case.

For the rest of the proof, we assume that ρ is the density of ζ which is a continuous function and there exist $x_0 > 0$ and a continuous positive and strictly decreasing function $\epsilon : [x_0, \infty) \to \mathbb{R}_+$ such that $\lim_{x\to\infty} \epsilon(x) = 0$ and

$$\rho(x) \le e^{-\alpha x + \epsilon(x) x},$$

for all $x \ge x_0$.

For large t, we define $\overline{\Psi}(t) = \sup\{s : \frac{s}{t} = 4\epsilon(s)\}$ and $\Psi(t) = \overline{\Psi}(t) \vee t^\gamma$, with $\gamma \in (1/2, 1)$ being a fixed number. This function satisfies the following properties:

$$\lim_{u\to\infty} \Psi(u) = \infty, \ \lim_{u\to\infty} \frac{\Psi(u)}{u} = 0, \quad \text{and for large } t, \quad \frac{\Psi(t)}{t} \ge 4\epsilon(t).$$

We define $h(x) = \rho(x) e^{\alpha x}$ and consider the following quantities

$$K(t, y) = \int_0^{\Psi(t)} h(x) e^{-\frac{x^2+y^2}{2t}} \sinh \frac{xy}{t}\, dx,$$

$$L(t, y) = \int_{\Psi(t)}^\infty h(x) e^{-\frac{x^2+y^2}{2t}} \sinh \frac{xy}{t}\, dx,$$

$$M(t) = \int_0^{\Psi(t)} h(x) e^{-\frac{x^2}{2t}} \sinh \frac{xy}{t}\, dx.$$

With these definitions, we get

$$\rho_t(y) = \frac{K(t,y) + L(t,y)}{\int_0^\infty e^{-\alpha z}(K(t,z) + L(t,z))\,dz}.$$

The result will be proved as soon as we show the following relations:

$$\lim_{t\to\infty} \frac{K(t,y)}{M(t)} = y, \qquad \lim_{t\to\infty} \frac{L(t,y)}{M(t)} = 0, \tag{6.55}$$

$$\lim_{t\to\infty} \int_0^\infty e^{-\alpha z}\frac{K(t,z)}{M(t)}\,dz = \int_0^\infty z e^{-\alpha z}\,dz, \qquad \lim_{t\to\infty} \int_0^\infty e^{-\alpha z}\frac{L(t,z)}{M(t)}\,dz = 0. \tag{6.56}$$

Using that $z \le \sinh z \le z e^z$, we obtain

$$e^{-\frac{z^2}{2t}} z M(t) \le K(t,z) \le e^{\frac{\Psi(t) z}{t}} z M(t),$$

and then (6.55) and (6.56) follow. For the other two relations, we shall study $L(t,y)$ more closely. Consider a large t such that $\Psi(t) \ge x_0$. In what follows, we denote by $\epsilon_t = \epsilon(\Psi(t))$, then

$$L(t,y) \le \int_{\Psi(t)}^\infty e^{\epsilon_t} e^{-\frac{(x-y)^2}{2t}}\,dx = e^{\epsilon^2 t/2 + \epsilon_t y} \int_{\Psi(t)}^\infty e^{-\frac{(x-y-\epsilon_t t)^2}{2t}}\,dx.$$

If t is large enough such that $\Psi(t)/t \le \alpha/2$ and $\epsilon_t \le \alpha/2$, we obtain that for some constant A (see details in Martínez et al. 1998)

$$\int_0^\infty e^{-\alpha z} L(t,z)\,dz \le A\left(\frac{t}{\Psi(t)} e^{-\frac{\Psi(t)^2}{4t}} + \sqrt{t} e^{-\frac{\alpha \Psi(t)}{12}}\right).$$

If we use that $\Psi(t) \ge t^\gamma$, we get the upper bound

$$\int_0^\infty e^{-\alpha z} L(t,z)\,dz \le A\left(t^{1-\gamma} e^{-t^{2\gamma-1}} + \sqrt{t} e^{-\frac{\alpha t^\gamma}{12}}\right).$$

Finally, for every K and every large t, we have $M(t) \ge \frac{e^{-\frac{K^2}{2t}}}{t} \int_0^K h(x) x\,dx$. If we take K such that $\int_0^K h(x) x\,dx > 0$, we obtain the desired two relations, and the proof is concluded. □

We finish this section with two examples that show how difficult it is to characterize the domain of attraction of each QSD. The first example considers as the initial measure the mixture $\zeta_\epsilon = \epsilon \underline{\nu} + (1-\epsilon)\nu_\lambda$, where $0 < \lambda < \underline{\lambda}$. For every $0 < \epsilon < 1$, the probability measure ζ_ϵ is in the domain of attraction of $\underline{\nu}$; nevertheless, ζ_ϵ can be arbitrarily close to ν_λ, in bounded variation, by taking small ϵ.

The other example is more sophisticated and the details can be found in Martínez et al. (1998). It is possible to construct a probability measure ζ and a sequence of times $(t_n : n \in \mathbb{N})$ converging to infinity, with the following properties

$$0 < \beta_1 = \liminf_{x\to\infty} -\frac{\log \zeta([x,\infty))}{x} < \beta_2 = \limsup_{x\to\infty} -\frac{\log \zeta([x,\infty))}{x} < \alpha,$$

$$\lim_{n\to\infty} \mathbb{P}_\zeta(X_{t_{2n}} \in \bullet | t_{2n} < T) = \nu_{\lambda(\beta_1)}(\bullet),$$

$$\lim_{n\to\infty} \mathbb{P}_\zeta(X_{t_{2n+1}} \in \bullet | t_{2n+1} < T) = \nu_{\lambda(\beta_2)}(\bullet).$$

We refer to reader to the article Lladser and San Martín (2000) where the authors study the domain of attraction of a QSD in the context of an Ornstein–Uhlenbeck process.

6.7 Examples

In this section, we examine some interesting special cases.

6.7.1 Drift Eventually Constant

We now study the case when the drift is eventually constant, where further explicit computations can be made. The setting is $\alpha(x) = \theta$ for all $x \ge \ell$, for some $\ell \ge 0$. When $\theta > 0$, conditions **H**, **H1** and $\mu(\mathbb{R}_+) < \infty$ hold.

Proposition 6.67 *Assume α is eventually constant with $\theta > 0$. Then, there exists $\underline{\theta} = \underline{\theta}(\ell)$ such that X^T is R-positive if and only if $\theta > \underline{\theta}$, X^T is R-null if and only if $\theta = \underline{\theta}$, and X^T is R-transient if and only if $\theta < \underline{\theta}$. The value $\underline{\theta}$ is the unique solution of*

$$\frac{\psi'_{0,\underline{\theta}^2/2}(\ell)}{\psi_{0,\underline{\theta}^2/2}(\ell)} = \underline{\theta}.$$

The condition $\theta > \underline{\theta}$ is equivalent to $\underline{\lambda}(0) < \theta^2/2$. Moreover, $\underline{\lambda}(0)$ admits the following representation

$$\underline{\lambda}(0) = \sup\left\{\lambda \le \min\bigl(\hat{\lambda}(\ell), \theta^2/2\bigr) : \frac{\psi'_{0,\lambda}(\ell)}{\psi_{0,\lambda}(\ell)} + \sqrt{\theta^2 - 2\lambda} \ge \theta\right\},$$

where $\hat{\lambda}(\ell) = \sup\{\lambda : \psi_{0,\lambda}$ is increasing on $[0,\ell]\}$.

In the special case $\alpha(x)=\theta_0 1_{\{x<\ell\}}+\theta 1_{\{x\ge\ell\}}$ *with* $\theta>\theta_0$, *the critical value* $\underline{\theta}$ *is given by the implicit formula* $\underline{\theta}=\theta_0+\chi\cot(\chi\ell)$, *where* χ *is uniquely determined by* $\theta_0=-\chi\cot(2\chi\ell)$ *and* $\chi\in[\pi/4\ell,\pi/2\ell)$.

In the latter special case, the dependence of $\underline{\theta}$ on ℓ,θ_0 verifies the homogeneity condition

$$\underline{\theta}(\ell,\theta_0)=\frac{\underline{\theta}(1,\ell\theta_0)}{\ell}.$$

It can be proved easily that $\underline{\theta}$ is increasing in θ_0 and decreasing in ℓ, with asymptotic values

$$\lim_{\ell\to 0}\underline{\theta}(\ell,\theta_0)=\infty,\qquad \lim_{\ell\to\infty}\underline{\theta}(\ell,\theta_0)=\theta_0,$$

$$\lim_{\theta_0\to 0}\underline{\theta}(\ell,\theta_0)=\pi/4\ell,\qquad \lim_{\theta_0\to\infty}\underline{\theta}(\ell,\theta_0)=\infty.$$

Moreover, using the inequality $\pi\le 4x\cot(x)(1-2x\cot(2x))\le\pi^2$ for $x\in[\pi/4,\pi/2)$, we obtain

$$\theta_0+\frac{\pi}{4\ell(1+2\ell\theta_0)}\le\underline{\theta}\le\theta_0+\frac{\pi^2}{4\ell(1+2\ell\theta_0)}.$$

If $\ell=0$, that is, the drift is constant on $[0,\infty)$, the process X^T is R-transient. We observe that the above criterion gives $\underline{\theta}=\infty$.

Proof Since for a constant drift $-\theta$ the bottom of the spectrum is $\theta^2/2$, we get $\underline{\lambda}(\ell)=\theta^2/2$, and a simple computation yields $\psi_{\ell,\underline{\lambda}(\ell)}(x)=(x-\ell)e^{\theta(x-\ell)}$. In particular, $u_{\ell,\underline{\lambda}(\ell)}^{-2}(x)e^{2\theta(x-\ell)}=(x-\ell)^{-2}$, which is integrable near ∞. Therefore, $\alpha^{(\ell)}$ is R-transient, and the result follows when $\ell=0$. In the sequel, we shall assume that $\ell>0$. We observe that $\underline{\lambda}(0)\le\theta^2/2$.

We denote by $\hat{\lambda}=\hat{\lambda}(\ell)=\sup\{\lambda:\psi_{0,\lambda}$ is increasing on $[0,\ell]\}$. From Lemma 6.58, we have

$$\underline{\lambda}(0)<\hat{\lambda}<\big(D^2+(\pi/\ell)^2\big)/2,\quad\text{where } D=\sup\big\{\alpha(x):x\in[0,\ell]\big\}.$$

We notice that $\psi'_{0,\hat{\lambda}}(\ell)=0$, otherwise for some $\lambda>\hat{\lambda}$ we would have that $\psi_{0,\lambda}$ is increasing on $[0,\ell]$, contradicting the maximality of $\hat{\lambda}$.

The mapping $\psi'_{0,\tau^2/2}(\ell)/\psi_{0,\tau^2/2}(\ell)-\tau$, as a function of τ, is continuous and strictly decreasing on $[0,\sqrt{2\hat{\lambda}}]$, positive at 0 and negative at $\sqrt{2\hat{\lambda}}$. Therefore, there exists a unique root of this function in $(0,\sqrt{2\hat{\lambda}})$ which we denote by $\underline{\theta}$. This root verifies

$$\frac{\psi'_{0,\underline{\theta}^2/2}(\ell)}{\psi_{0,\underline{\theta}^2/2}(\ell)}=\underline{\theta}\quad\text{and}\quad\left[\theta\le\underline{\theta}\Leftrightarrow\left(\frac{\psi'_{0,\theta^2/2}(\ell)}{\psi_{0,\theta^2/2}(\ell)}\ge\theta\text{ and }\theta\le\sqrt{2\hat{\lambda}}\right)\right].$$

Let us take

$$\lambda^* = \sup\left\{\lambda \le \min(\hat{\lambda}, \theta^2/2) : \frac{\psi'_{0,\lambda}(\ell)}{\psi_{0,\lambda}(\ell)} + \sqrt{\theta^2 - 2\lambda} \ge \theta\right\}. \tag{6.57}$$

As before, one can easily prove that λ^* satisfies $0 < \lambda^* < \hat{\lambda}$.

The equivalence $\theta > \underline{\theta} \Leftrightarrow \lambda^* < \theta^2/2$ plays an important role in the sequel, and it follows from

$$\lambda^* = \theta^2/2 \quad \Leftrightarrow \quad \left(\frac{\psi'_{0,\theta^2/2}(\ell)}{\psi_{0,\theta^2/2}(\ell)} \ge \theta \text{ and } \theta \le \sqrt{2\hat{\lambda}}\right) \quad \Leftrightarrow \quad \theta \le \underline{\theta}.$$

We shall now prove that $\lambda^* = \underline{\lambda}(0)$. Take any $\lambda \le \min(\hat{\lambda}, \theta^2/2)$. The function $\psi_{0,\lambda}$ is increasing on $[0, \ell]$. The question is to determine the values of λ for which $\psi_{0,\lambda}$ is increasing in (ℓ, ∞). For this purpose, consider the solution of

$$\frac{1}{2} f''(x) - \theta f'(x) = -\lambda f(x), \qquad x \in [\ell, \infty),$$

with boundary conditions $f(\ell) = \psi_{0,\lambda}(\ell)$, $f'(\ell) = \psi'_{0,\lambda}(\ell)$. Obviously, $f = \psi_{0,\lambda}$ on $[\ell, \infty)$. For the analysis of this solution, we consider two possible cases. When $\rho = \sqrt{\theta^2 - 2\lambda} > 0$, the solution is given by

$$f(x) = e^{\theta(x-\ell)}\big(A \sinh\big(\rho(x-\ell)\big) + B \cosh\big(\rho(x-\ell)\big)\big).$$

The condition for having an increasing (nonnegative solution) is equivalent to $A \ge -B$. From the boundary conditions, we obtain

$$0 < f(\ell) = \psi_{0,\lambda}(\ell) = B, \qquad f'(\ell) = \psi'_{0,\lambda}(\ell) = \theta B + \rho A.$$

We need $A \ge -B$, which is equivalent to $\psi'_{0,\lambda}(\ell) \ge (\theta - \sqrt{\theta^2 - 2\lambda})\psi_{0,\lambda}(\ell)$. In other words,

$$\frac{\psi'_{0,\lambda}(\ell)}{\psi_{0,\lambda}(\ell)} + \sqrt{\theta^2 - 2\lambda} \ge \theta.$$

On the other hand, in the case $\lambda = \theta^2/2$ (then necessarily $\theta^2/2 \le \hat{\lambda}$), the solution is

$$f(x) = \big(C(x-\ell) + B\big)e^{\theta(x-\ell)},$$

where $B = \psi_{0,\theta^2/2}(\ell) > 0$ and $C = \psi'_{0,\theta^2/2}(\ell) - \theta\psi_{0,\theta^2/2}(\ell)$. The condition for having a nonnegative solution is $C \ge 0$, which is equivalent to

$$\frac{\psi'_{0,\theta^2/2}(\ell)}{\psi_{0,\theta^2/2}(\ell)} \ge \theta.$$

In summary, we have shown that f is positive if and only if $\lambda \le \lambda^*$. In particular, ψ_{0,λ^*} is positive on $[0, \infty)$, proving that $\lambda^* \le \underline{\lambda}(0)$. On the other hand, since $\psi_{0,\underline{\lambda}(0)}$

is positive and $\underline{\lambda}(0) \le \min(\hat{\lambda}, \theta^2/2)$, the argument given above allows us to conclude the equality $\lambda^* = \underline{\lambda}(0)$.

Thus, in the case $\underline{\lambda}(0) < \theta^2/2$, from (6.57) one gets

$$\frac{\psi'_{0,\underline{\lambda}(0)}(\ell)}{\psi_{0,\underline{\lambda}(0)}(\ell)} + \sqrt{\theta^2 - 2\underline{\lambda}(0)} = \theta,$$

which in the previous notation amounts to $A = -B$. Therefore, the solution $\psi_{0,\underline{\lambda}(0)}$ is, for $x > \ell$,

$$\psi_{0,\underline{\lambda}(0)}(x) = \psi_{0,\underline{\lambda}(0)}(\ell) e^{(\theta - \sqrt{\theta^2 - 2\underline{\lambda}(0)})(x-\ell)}.$$

In particular, $\psi^2_{0,\underline{\lambda}(0)}(x) e^{-\gamma(x)} = \psi^2_{0,\underline{\lambda}(0)}(\ell) e^{-\gamma(\ell)} e^{-2\sqrt{\theta^2 - 2\underline{\lambda}(0)}(x-\ell)}$ for $x > \ell$, which is integrable, and therefore X^T is R-positive.

On the other hand, if $\underline{\lambda}(0) = \theta^2/2$, one has $\psi_{0,\underline{\lambda}(0)} = e^{\theta(x-\ell)}(C(x-\ell) + B)$ for $x \ge \ell$, with $B > 0$ and $C \ge 0$. Then, the function

$$\psi^2_{0,\underline{\lambda}(0)}(x) e^{-\gamma(x)} = \big(C(x-\ell) + B\big)^2 e^{-\gamma(\ell)},$$

is not integrable near ∞.

In summary, X^T is R-positive if and only if $\underline{\lambda}(0) < \theta^2/2$, which we have proved to be equivalent to $\theta > \underline{\theta}$.

Now, we prove that α is R-transient if and only if $\theta < \underline{\theta}$. Remark that $\underline{\lambda}(0) = \theta^2/2$ holds under both conditions of the claimed equivalence. Since $\psi_{0,\theta^2/2}(x) = e^{\theta(x-\ell)}(C(x-\ell) + B)$ for $x \ge \ell$, we get that α is R-transient if and only if $C > 0$, or equivalently,

$$\theta \psi_{0,\theta^2/2}(\ell) < \psi'_{0,\theta^2/2}(\ell),$$

which holds if and only if $\theta < \underline{\theta}$.

Finally, we give an explicit formula for $\underline{\theta}$ when $\alpha(x) = \theta_0 1_{\{x<\ell\}} + \theta 1_{\{x \ge \ell\}}$ and $\theta > \theta_0$. In this case, the solution $\psi_{0,\lambda}$ for $\lambda > \theta_0^2/2$, is

$$\psi_{0,\lambda}(x) = \frac{e^{\theta_0 x}}{\chi} \sin(\chi x),$$

where $\chi = \sqrt{2\lambda - \theta_0^2}$, and therefore $\underline{\theta} = \sqrt{2\lambda}$ is the unique solution of

$$\sqrt{2\lambda} = \frac{\psi'_{0,\lambda}(\ell)}{\psi_{0,\lambda}(\ell)} = \theta_0 + \chi \cot(\chi \ell) = \sqrt{\chi^2 + \theta_0^2}.$$

We obtain the relation $\theta_0 = -\chi \cot(2\chi\ell)$, from which we get the desired value of $\underline{\theta}$. □

6.7.2 $\bar{x}$ Can Be R-Positive

In the ultimately constant case, if $\bar{x} > 0$, $\alpha^{(\bar{x})}$ is always R-null (Proposition 6.67), then the transition from R-positive to R-transient occurs through an R-null point. We show that this is not always the case, that is, we exhibit an example were $0 < \bar{x} < \infty$ and $\alpha^{(\bar{x})}$ is R-positive. Let us construct it. Take a function g verifying the following conditions:

(i) $g > 0$ on $(0, \infty)$, $g(0) = 0$ and $g'(0) = 1$;
(ii) $\int_0^\infty g^2(x)\,dx < \infty$;
(iii) $g + g'$ is positive, $\lim_{x\to\infty} g''(x)/(g(x) + g'(x)) = 0$ and the integrability condition $\int_0^\infty |g''(x)/(g(x) + g'(x))|\,dx < \infty$.

For instance, $g(x) = x/(1 + x)^2$ does the job.

Fix some $a > 0$. Let α be such that $\alpha(x) = 1 + g''(x - a)/(2(g(x - a) + g' \times (x - a)))$ for $x \geq a$. Obviously, we have $\underline{\lambda}(\infty) = \alpha(\infty)^2/2 = 1/2$. Since the function $v(x) = g(x - a)e^{(x-a)}$ solves the problem $v'' - 2\alpha v' = -v$ on (a, ∞) with the boundary conditions $v'(a) = 1, v(a) = 0$ and it is positive on (a, ∞), we get $\underline{\lambda}(a) = \underline{\lambda}(\infty) = 1/2$. On the other hand, from (ii) and (iii), it can be checked that $\alpha^{(a)}$ is R-positive. From Theorem 6.45, part (ii), we conclude $\bar{x} = a$.

6.7.3 $\bar{x} = \infty$ for Some Bounded Drifts

Let us now show that for some bounded drifts we can have $\bar{x} = \infty$. Take a sequence $0 < b_n < 1$ converging towards 1. Consider $x_0 = 0$, $x_{n+1} = x_n + \pi/\sqrt{1 - b_n^2}$ and define $\alpha(x) = b_n$ for $x \in [x_n, x_{n+1})$. We have $\underline{\lambda}(\infty) = 1/2$. Let us prove that $\underline{\lambda}(x_n) < 1/2$. The solution of $v'' - 2\alpha v' = -v$ with $v(x_n) = 0$, $v'(x_n) = 1$ is given by

$$v(x) = \frac{e^{(x-x_n)}}{\sqrt{1 - b_n^2}} \sin\Bigl((x - x_n)\sqrt{1 - b_n^2}\Bigr) \quad \text{for } x \in [x_n, x_{n+1}).$$

Since $v(x_{n+1}) = 0$, we obtain that $\underline{\lambda}(x_n) < 1/2$, and therefore $\bar{x} = \infty$.

Chapter 7
Infinity as Entrance Boundary

7.1 Introduction

In this chapter,[1] we study QSDs for one dimensional diffusions where we allow the drift to have a singularity at the origin and an entrance boundary at $+\infty$. The main motivation for studying this situation comes from models in ecology and economics (see Examples 7.8.1 and 7.8.2 at the end of this chapter).

Consider X to be a drifted Brownian motion

$$dX_t = dB_t - \alpha(X_t)\,dt. \tag{7.1}$$

Of particular interest is the case

$$\alpha(x) = \frac{1}{2x} - \frac{rx}{2} + \frac{c\sigma x^3}{8}$$

obtained after a suitable change of variables in the so-called logistic case (see Example 7.8.1).

It is also worth noticing that the behavior of α at infinity may violate Mandl's conditions. For example, in the logistic case

$$\int_1^\infty e^{-\gamma(z)} \int_1^z e^{\gamma(y)}\,dy\,dz < \infty,$$

where $\gamma(y) := 2\int_1^y \alpha(x)\,dx$.

This unusual situation prevents us from using the results of the previous chapter. A main tool for studying these type of diffusions is the spectral theory of its semi-group, which we discuss in Sect. 7.3, specifically in Theorem 7.3. In Sect. 7.5, we study the Yaglom limit; see Theorem 7.13. Later, in Theorem 7.24 of Sect. 7.7, we

[1]This chapter is based on the work done by us in collaboration with Patrick Cattiaux, Amaury Lambert and Sylvie Méléard in the article *Quasi-stationary distributions and diffusions models in population dynamics*. Annals of Probability **37**, 1926–1969 (2009).

P. Collet et al., *Quasi-Stationary Distributions*, Probability and Its Applications, DOI 10.1007/978-3-642-33131-2_7, © Springer-Verlag Berlin Heidelberg 2013

prove that this limit is the unique QSD when the process *comes down from infinity* (see Definition 7.22).

An important role will be played by the measure μ which is proportional to the speed measure of X and was defined as

$$d\mu(y) := e^{-\gamma(y)}\, dy.$$

We shall obtain $L^2(d\mu)$ estimates for the heat kernel, and the main point is that starting from any $x > 0$, the law of the process at time t is absolutely continuous with respect to μ with a density belonging to $L^2(d\mu)$.

We end the chapter with applications to some biological models and a study of Shiryaev's martingale.

7.2 Basic Properties of X

Since the drift maybe singular at 0, we consider

$$\Lambda(x) = \int_1^x e^{\gamma(y)}\, dy,$$

where $\gamma(y) = \int_1^y 2\alpha(u)\, du$. Also we need to consider

$$\kappa(x) = \int_1^x e^{\gamma(y)} \left(\int_1^y e^{-\gamma(z)}\, dz \right) dy.$$

For most of the results in this chapter, we shall assume sure absorption at zero, that is, for all $x > 0$

$$\mathbf{H1}: \quad \mathbb{P}_x(\tau = T < T_\infty) = 1.$$

It is well known (see, e.g., Ikeda and Watanabe 1988, Chap. VI, Theorem 3.2) that **H1** holds if and only if

$$\Lambda(\infty) = \infty \quad \text{and} \quad \kappa\big(0^+\big) < \infty.$$

We notice that **H1** can be written as $\mathbb{P}_x(\lim_{t\to\infty} X_{t\wedge\tau} = 0) = 1$.

We shall now discuss some properties of the law of X up to T. Using Girsanov's Theorem (see Cattiaux et al. 2009, Proposition 2.2 for details), we obtain that for any bounded Borel function $F : C([0,t]) \to \mathbb{R}$, the expectation $\mathbb{E}_x[F(X)\mathbb{1}_{t<T}]$ is given by

$$\mathbb{E}^{\mathbb{W}_x}\left[F(\omega)\mathbb{1}_{t<T(\omega)} \exp\left(\frac{1}{2}\gamma(x) - \frac{1}{2}\gamma(\omega_t) - \frac{1}{2}\int_0^t \left(\alpha^2 - \alpha'\right)(\omega_s)\, ds \right) \right]$$

where $\mathbb{E}^{\mathbb{W}_x}$ denotes the expectation with respect to the Wiener measure starting from x, and $\mathbb{E}_x$ denotes the expectation with respect to the law of X starting also from x. From this representation, we obtain some properties about the density (or kernel) of X at time $t > 0$.

Theorem 7.1 *Assume* **H1**. *For all $x > 0$ and all $t > 0$, there exists some density $r(t, x, \cdot)$ that verifies*

$$\mathbb{E}_x\big[f(X_t)\mathbb{1}_{t<T}\big] = \int_0^\infty f(y)r(t,x,y)\,d\mu(y)$$

for all bounded Borel functions f.

Moreover, if there exists some $C > 0$ such that $\alpha^2(y) - \alpha'(y) \geq -C$ for all $y > 0$, then for all $t > 0$ and all $x > 0$,

$$\int_0^\infty r^2(t,x,y)\,d\mu(y) \leq \frac{1}{2\pi t} e^{Ct} e^{\gamma(x)}.$$

Proof Consider the density given by the Girsanov's Theorem

$$G(\omega) = \mathbb{1}_{t<T(\omega)} \exp\biggl(\frac{1}{2}\gamma(\omega_0) - \frac{1}{2}\gamma(\omega_t) - \frac{1}{2}\int_0^t \bigl(\alpha^2 - \alpha'\bigr)(\omega_s)\,ds\biggr).$$

Then we get

$$\begin{aligned} \mathbb{E}_x\big[f(X_t)\mathbb{1}_{t<T}\big] &= \mathbb{E}^{\mathbb{W}_x}\big[f(\omega_t)\mathbb{E}^{\mathbb{W}_x}[G|\omega_t]\big] \\ &= \int f(y)\mathbb{E}^{\mathbb{W}_x}[G|\omega_t = y]\frac{1}{\sqrt{2\pi t}}e^{-\frac{(x-y)^2}{2t}}\,dy \\ &= \int_0^\infty f(y)\mathbb{E}^{\mathbb{W}_x}[G|\omega_t = y]\frac{1}{\sqrt{2\pi t}}e^{-\frac{(x-y)^2}{2t}+\gamma(y)}\,d\mu(y), \end{aligned}$$

because $\mathbb{E}^{\mathbb{W}_x}[G|\omega_t = y] = 0$ if $y \leq 0$. Hence, the law of X_t restricted to non-extinction has a density with respect to μ given by

$$r(t,x,y) = \mathbb{E}^{\mathbb{W}_x}[G|\omega_t = y]\frac{1}{\sqrt{2\pi t}}e^{-\frac{(x-y)^2}{2t}+\gamma(y)}.$$

In particular, we obtain

$$\begin{aligned} \int_0^\infty r^2(t,x,y)\,d\mu(y) &= \int_0^\infty \frac{1}{2\pi t}e^{-\frac{(x-y)^2}{t}+2\gamma(y)}\bigl(\mathbb{E}^{\mathbb{W}_x}[G|\omega_t = y]\bigr)^2 e^{-\gamma(y)}\,dy \\ &\leq \int_0^\infty \frac{1}{2\pi t}e^{-\frac{(x-y)^2}{t}+\gamma(y)}\mathbb{E}^{\mathbb{W}_x}\big[G^2|\omega_t = y\big]\,dy, \end{aligned}$$

from where the result follows. □

Notice that the measure μ is a natural one since the kernel of the killed process is symmetric in $L^2(d\mu)$. This property allows us to use the spectral theory for symmetric operators, which is the main objective of the next section.

7.3 Spectral Theory

As usual, $\mathcal{C}_0^\infty((0,\infty))$ denotes the space of infinitely differentiable functions on $(0,\infty)$ with compact support. We introduce the inner product

$$\langle f, g\rangle_\mu = \int_0^\infty f(u)g(u)\,d\mu(u),$$

and consider the symmetric form

$$\mathcal{E}(f,g) = \langle f', g'\rangle_\mu$$

with domain $D(\mathcal{E}) = \mathcal{C}_0^\infty((0,\infty))$. We shall use the theory of Dirichlet forms, for which we recommend Fukushima (1980) or Fukushima et al. (1994), and for details in our particular case Cattiaux et al. (2009). The basic fact is the existence of a nonpositive self adjoint operator $\mathcal{L}$ on $L^2(d\mu)$ with domain $D(\mathcal{L}) \supseteq \mathcal{C}_0^\infty((0,\infty))$ such that for all f and g in $D(\mathcal{E})$ it holds $\mathcal{E}(f,g) = -2\int_0^\infty f(u)\mathcal{L}g(u)\,d\mu(u) = -2\langle f, \mathcal{L}g\rangle_\mu$. For $g \in \mathcal{C}_0^\infty((0,\infty))$, we have

$$\mathcal{L}g = \frac{1}{2}g'' - \alpha g'.$$

$\mathcal{L}$ is the generator of a strongly continuous symmetric semigroup of contractions on $L^2(d\mu)$ which we denote by $(P_t : t \geq 0)$. Under **H1**, this semigroup coincides with the semigroup of X killed at 0, that is, $P_t f(x) = \mathbb{E}[f(X_t^x)\mathbb{1}_{t<T}]$.

Let $(E_\lambda : \lambda \geq 0)$ be the family of spectral projections associated to $-\mathcal{L}$. Since $-\mathcal{L}$ is nonnegative, this family is concentrated on $\mathbb{R}_+$. In particular, we have the representation $\int P_t f g\,d\mu = \int_0^\infty e^{-\lambda t}\,d\langle E_\lambda f, g\rangle_\mu$ for all $t \geq 0$, $f, g \in L^2(d\mu)$. One important fact is the pointwise limit, for any $f \in L^2(d\mu)$,

$$\lim_{t\to\infty} \int (P_t f)^2\,d\mu = 0 \quad \text{for any } f \in L^2(d\mu). \tag{7.2}$$

For that purpose, consider a compact set $K \subset (0,\infty)$ and split the integral in two terms as

$$\begin{aligned}\int (P_t f)^2\,d\mu &= \int \left(P_t(f\mathbb{1}_K + f\mathbb{1}_{K^c})\right)^2 d\mu \\ &\leq 2\int \left(P_t(f\mathbb{1}_K)\right)^2 d\mu + 2\int \left(P_t(f\mathbb{1}_{K^c})\right)^2 d\mu \\ &\leq 2\int \left(P_t(f\mathbb{1}_K)\right)^2 d\mu + 2\int (f\mathbb{1}_{K^c})^2\,d\mu.\end{aligned}$$

For every fixed $\varepsilon > 0$, we may choose K large enough such that the second term, in the latter sum, is bounded by ε. On the other hand, we approximate $f\mathbb{1}_K$ in $L^2(d\mu)$ by $\tilde{f}\mathbb{1}_K$ for some continuous and bounded function $\tilde{f}$, up to ε. Now, thanks to **H1**, we know that $P_t(\tilde{f}\mathbb{1}_K)(x)$ goes to 0 as t goes to infinity for any x. Since

$\int (P_t(\tilde{f}\mathbb{1}_K))^2\,d\mu = \int_K \tilde{f} P_{2t}(\tilde{f}\mathbb{1}_K)\,d\mu$, we prove the claim using the Dominated Convergence Theorem.

Now we shall introduce the main assumption on α for the spectral study of (P_t):

$$\textbf{H2:}\quad C = -\inf_{y\in(0,\infty)} \alpha^2(y) - \alpha'(y) < \infty \quad \text{and} \quad \lim_{y\to\infty} \alpha^2(y) - \alpha'(y) = +\infty.$$

Remark 7.2 Hypothesis **H2** implies that α is unbounded near ∞ (actually, $|\alpha(x)|$ tends to infinity as $x \to \infty$) and $\alpha^-(x)$ or $\alpha^+(x)$ tend to 0 as $x \downarrow 0$ (see Cattiaux et al. 2009). If, in addition, **H1** holds then $\alpha(x) \to \infty$, as $x \to \infty$.

We may now state the main result from the spectral point of view.

Theorem 7.3 *If* **H2** *is satisfied, $-\mathcal{L}$ has a purely discrete spectrum $0 \le \lambda_1 < \lambda_2 < \cdots$. Furthermore, each λ_i ($i \in \mathbb{N}$) is associated to a unique (up to a multiplicative constant) eigenfunction ψ_i of class $\mathcal{C}^2((0,\infty))$ which satisfies the ODE*

$$\frac{1}{2}\psi_i'' - \alpha\psi_i' = -\lambda_i\psi_i. \tag{7.3}$$

The sequence $(\psi_i : i \ge 1)$ is an orthonormal basis of $L^2(d\mu)$, ψ_1 can be chosen to be strictly positive on $(0,\infty)$.

For $g \in L^2(d\mu)$,

$$P_t g = \sum_{i\in\mathbb{N}} e^{-\lambda_i t}\langle \psi_i, g\rangle_\mu \psi_i \tag{7.4}$$

holds in the $L^2(d\mu)$ sense, and then for $f, g \in L^2(d\mu)$,

$$\lim_{t\to\infty} e^{\lambda_1 t}\langle g, P_t f\rangle_\mu = \langle \psi_1, f\rangle_\mu \langle \psi_1, g\rangle_\mu.$$

If, in addition, **H1** *holds, then $\lambda_1 > 0$.*

Proof For $f \in L^2(dx)$, consider the auxiliary semigroup

$$\tilde{P}_t(f) = e^{-\frac{\gamma}{2}} P_t\big(f e^{\frac{\gamma}{2}}\big), \tag{7.5}$$

which is well defined in $L^2(dx)$ since $f e^{\frac{\gamma}{2}} \in L^2(d\mu)$. $(\tilde{P}_t : t \ge 0)$ is then a strongly continuous semigroup in $L^2(dx)$, whose generator $\tilde{\mathcal{L}}$ coincides on $\mathcal{C}_0^\infty((0,\infty))$ with $\frac{1}{2}\frac{d^2}{dx^2} - \frac{1}{2}(\alpha^2 - \alpha')$ since $\mathcal{C}_0^2(0,\infty) \subset D(\mathcal{L})$, and $e^{\frac{\gamma}{2}} \in \mathcal{C}^2(0,\infty)$. The spectral theory of such a Schrödinger operator on the line (or the half-line) is well known. We shall use the results in Chap. 2 of Berezin and Shubin (1991) with the appropriate modifications. In our case, there is a small difficulty because the potential $v = (\alpha^2 - \alpha')/2$ does not necessarily belong to L^∞_{loc} near 0. Nevertheless, since we have assumed that v is bounded from below by $-C/2$, we may consider

$H = \tilde{\mathcal{L}} - (C/2+1)$. This means that we replace v by $w = v + C/2 + 1 \geq 1$, which only produces a translation of the spectrum. For $f \in \mathcal{C}_0^\infty(0,\infty)$, we have

$$-(Hf, f) = -\int_0^\infty Hf(u) f(u)\, du = \int_0^\infty \left(\left|f'(u)\right|^2/2 + w(u) f^2(u)\right) du$$
$$\geq \int_0^\infty f^2(u)\, du, \tag{7.6}$$

and H has a bounded inverse operator. The spectrum of H (hence of $\tilde{\mathcal{L}}$) will be discrete if, for example, H^{-1} is a compact operator. For that it is enough to show that $M = \{f \in D(H); -(Hf, f) \leq 1\}$ is relatively compact. This is proved in Theorem 3.1 of Berezin and Shubin (1991) when w is locally bounded and it has the appropriate asymptotic behavior. In our situation, w may explode to infinity near 0, but this is not a problem because the set M is included into the corresponding one when we truncate w near the origin. Then the spectrum of $\mathcal{L}$ is discrete.

On the other hand, all the eigenvalues of $\tilde{\mathcal{L}}$ are simple (Proposition 3.3 in Berezin and Shubin 1991), and, of course, the corresponding set of normalized eigenfunctions $(\eta_k : k \geq 1)$ is an orthonormal basis of $L^2(dx)$. Hence $(e^{\frac{\gamma}{2}} \eta_k)$ is an orthonormal basis of $L^2(d\mu)$ and each $\psi_k = e^{\frac{\gamma}{2}} \eta_k$ is an eigenfunction of $\mathcal{L}$. We can choose them to be $\mathcal{C}^2((0,\infty))$ and they satisfy (7.3). For every $t > 0$, and for every $g, f \in L^2(d\mu)$, we have $\sum_{k=1}^\infty e^{-\lambda_k t} \langle \psi_k, g\rangle_\mu \langle \psi_k, f\rangle_\mu = \langle g, P_t f\rangle_\mu$.

In addition, if g and f are nonnegative, we get

$$0 \leq \lim_{t\to\infty} e^{\lambda_1 t} \langle g, P_t f\rangle_\mu = \langle \psi_1, f\rangle_\mu \langle \psi_1, g\rangle_\mu,$$

because $\lambda_1 < \lambda_2 \leq \cdots$ and the sum $\sum_{k=1}^\infty |\langle \psi_k, g\rangle_\mu \langle \psi_k, f\rangle_\mu|$ is finite. It follows that $\langle \psi_1, f\rangle_\mu$ and $\langle \psi_1, g\rangle_\mu$ have the same sign. Changing ψ_1 into $-\psi_1$ if necessary, we may assume that $\langle \psi_1, f\rangle_\mu \geq 0$ for any nonnegative f, hence $\psi_1 \geq 0$. Since $P_t \psi_1(x) = e^{-\lambda_1 t} \psi_1(x)$ and ψ_1 is continuous and not trivial, we deduce that $\psi_1(x) > 0$ for all $x > 0$. Using (7.2), we get for $g \in L^2(d\mu)$

$$0 = \lim_{t\to\infty} \langle P_t g, P_t g\rangle_\mu = \lim_{t\to\infty} e^{-2\lambda_1 t} \langle g, \psi_1\rangle_\mu^2,$$

showing that $\lambda_1 > 0$. □

Remark 7.4 Note that λ_1 corresponds to the bottom of the spectrum of $-\mathcal{L}$ (in $L^2(d\mu)$) and this value of the spectrum was denoted as $\underline{\lambda}$ in the previous chapters.

The following result, similar to Lemma 6.9 in the previous chapter, gives some extra information about the kernel r.

Proposition 7.5 *Under* **H1** *and* **H2**, *we have the following expansion*

$$r(t,x,y) = \sum_{k=1}^\infty e^{-\lambda_k t} \psi_k(x) \psi_k(y), \tag{7.7}$$

uniformly on compact sets of $(0, \infty) \times (0, \infty) \times (0, \infty)$.

Therefore, on compact sets of $(0, \infty) \times (0, \infty)$ *we get*

$$\lim_{t \to \infty} e^{\lambda_1 t} r(t, x, y) = \psi_1(x)\psi_1(y). \tag{7.8}$$

Proof Using Theorems 7.1 and 7.3, for every smooth function g compactly supported on $(0, \infty)$, we have

$$\begin{aligned}\sum_{k=1}^{n} e^{-\lambda_k t} \langle \psi_k, g\rangle_\mu^2 &\le \sum_{k=1}^{\infty} e^{-\lambda_k t} \langle \psi_k, g\rangle_\mu^2 \\ &= \int\int g(x)g(y)r(t, x, y)e^{-\gamma(x)-\gamma(y)}\, dx\, dy.\end{aligned}$$

Letting $g(y)\,dy$ tend to the Dirac measure at x and using the regularity of ψ_k and r, we deduce that

$$\sum_{k=1}^{n} e^{-\lambda_k t} \psi_k(x)^2 \le r(t, x, x).$$

The series $\sum_{k=1}^{\infty} e^{-\lambda_k t} \psi_k(x)^2$ converges pointwise, which by the Cauchy–Schwarz Inequality implies the pointwise absolute convergence of

$$\zeta(t, x, y) := \sum_{k=1}^{\infty} e^{-\lambda_k t} \psi_k(x)\psi_k(y)$$

and the bound for all n

$$\sum_{k=1}^{n} e^{-\lambda_k t} \bigl|\psi_k(x)\psi_k(y)\bigr| \le \sqrt{r(t, x, x)}\sqrt{r(t, y, y)}.$$

Using Harnack's Inequality (see, for example, Krylov and Safonov 1981), we get

$$\sqrt{r(t, x, x)}\sqrt{r(t, y, y)} \le C_K r(t, x, y)$$

for any x and y in the compact subset K of $(0, \infty)$. Using the Dominated Convergence Theorem, we obtain that for all Borel functions g, f with compact support in $(0, \infty)$

$$\begin{aligned}&\iint g(x)f(y)\zeta(t, x, y)e^{-\gamma(x)-\gamma(y)}\, dx\, dy \\ &\quad = \iint g(x)f(y)r(t, x, y)e^{-\gamma(x)-\gamma(y)}\, dx\, dy.\end{aligned}$$

Therefore, $\zeta(t, x, y) = r(t, x, y)\, dx\, dy$-a.s., which proves the almost sure version of (7.7).

Since ψ_k are smooth eigenfunctions, we get the pointwise equality

$$e^{-\lambda_k t}\psi_k(x)^2 = \iint r(t/3,x,y)r(t/3,x,z)e^{-\lambda_k t/3}\psi_k(y)\psi_k(z)e^{-\gamma(z)-\gamma(y)}\,dy\,dz$$
$$= e^{-\lambda_k t/3}\langle r(t/3,x,\bullet),\psi_k\rangle_\mu \langle r(t/3,x,\bullet),\psi_k\rangle_\mu$$

which, together with the fact $r(t/3,x,\bullet) \in L^2(d\mu)$ and Theorem 7.1, allows us to deduce that $\sum_{k=1}^{\infty} e^{-\lambda_k t}\psi_k(x)^2$ is given by

$$\iint r(t/3,x,y)r(t/3,x,z)\sum_{k=1}^{\infty} e^{-\lambda_k t/3}\psi_k(y)\psi_k(z)e^{-\gamma(z)-\gamma(y)}\,dy\,dz$$
$$= \iint r(t/3,x,y)r(t/3,x,z)r(t/3,y,z)e^{-\gamma(z)-\gamma(y)}\,dy\,dz = r(t,x,x).$$

Dini's Theorem then proves the uniform convergence in compacts of $(0,\infty)$ for the series

$$\sum_{k=1}^{\infty} e^{-\lambda_k t}\psi_k(x)^2 = r(t,x,x).$$

By the Cauchy–Schwarz Inequality, we have for any n

$$\left|\sum_{k=n}^{\infty} e^{-\lambda_k t}\psi_k(x)\psi_k(y)\right| \le \left(\sum_{k=n}^{\infty} e^{-\lambda_k t}\psi_k(x)^2\right)^{1/2}\left(\sum_{k=n}^{\infty} e^{-\lambda_k t}\psi_k(y)^2\right)^{1/2}.$$

This, together with the Dominated Convergence Theorem, yields (7.8). □

7.4 Sufficient Conditions for Integrability of ψ_1

In this section, we study some properties of the eigenfunctions ψ_i, including their integrability with respect to μ. In particular, we are interested in the integrability of ψ_1 because this function will generate a QSD. First, we relate the integrability of this function with the flux of particles at 0.

Proposition 7.6 *Assume that* **H1** *and* **H2** *are satisfied. Then*

$$\int_1^{\infty} \psi_1 e^{-\gamma}\,dx < \infty,$$

$F(x) = \psi_1'(x)e^{-\gamma(x)}$ *is a nonnegative decreasing function and the following limits exist*

$$F(0^+) = \lim_{x\downarrow 0}\psi_1'(x)e^{-\gamma(x)} \in (0,\infty], \qquad F(\infty) = \lim_{x\to\infty}\psi_1'(x)e^{-\gamma(x)} \in [0,\infty).$$

The function ψ_1 *is increasing and* $\int_1^\infty e^{-\gamma(y)}\,dy < \infty$. *Moreover,* $\int_0^\infty \psi_1(x) \times e^{-\gamma(x)}\,dx = \frac{F(0^+)-F(\infty)}{2\lambda_1}$. *In particular,*

$$\psi_1 \in L^1(d\mu) \quad \text{if and only if} \quad F\big(0^+\big) < \infty.$$

Remark 7.7 Note that $g = \psi_1 e^{-\gamma}$ satisfies the adjoint equation $\frac{1}{2}g'' + (\alpha g)' = -\lambda_1 g$, and then $F(x) = g'(x) + 2\alpha(x)g(x)$ represents the flux at x. Then $\psi_1 \in L^1(d\mu)$, or equivalently, $g \in L^1(dx)$ if and only if the flux at 0 is finite.

Proof Since ψ_1 satisfies $\psi_1''(x) - 2\alpha\psi_1'(x) = -2\lambda_1\psi_1(x)$, we obtain for x_0 and x in $(0,\infty)$

$$\psi_1'(x)e^{-\gamma(x)} = \psi_1'(x_0)e^{-\gamma(x_0)} - 2\lambda_1 \int_{x_0}^x \psi_1(y)e^{-\gamma(y)}\,dy, \tag{7.9}$$

and $F = \psi_1' e^{-\gamma}$ is decreasing. Integrating further gives

$$\psi_1(x) = \psi_1(x_0) + \int_{x_0}^x \left(\psi_1'(x_0)e^{-\gamma(x_0)} - 2\lambda_1 \int_{x_0}^z \psi_1(y)e^{-\gamma(y)}\,dy\right)e^{\gamma(z)}\,dz.$$

If for some $z_0 > x_0$ it holds that $\psi_1'(x_0)e^{-\gamma(x_0)} - 2\lambda_1 \int_{x_0}^{z_0} \psi_1(y)e^{-\gamma(y)}\,dy < 0$, then this inequality holds for all $z > z_0$ since the quantity

$$\psi_1'(x_0)e^{-\gamma(x_0)} - 2\lambda_1 \int_{x_0}^z \psi_1(y)e^{-\gamma(y)}\,dy$$

is decreasing in z. This implies that for large x the function ψ_1 is negative. Indeed, consider $z_1 > z_0$ such that for all $z \ge z_1$ we have $\psi_1'(x_0)e^{-\gamma(x_0)} - 2\lambda_1 \int_{x_0}^z \psi_1(y)e^{-\gamma(y)}\,dy \le -a$, for some $a > 0$. Then, we get the following inequality for all $x \ge z_1$

$$\psi_1(x) \le \psi_1(z_1) - a\big(\Lambda(x) - \Lambda(z_1)\big).$$

Since $\Lambda(\infty) = \infty$, we deduce that ψ_1 is eventually negative. This is a contradiction, and we deduce that for all $x > 0$

$$2\lambda_1 \int_x^\infty \psi_1(y)e^{-\gamma(y)}\,dy \le \psi_1'(x)e^{-\gamma(x)}.$$

Hence, ψ_1 is increasing and, being nonnegative, it is bounded near 0. In particular, $\psi_1(0^+)$ exists. Also we deduce that $F \ge 0$ and that $\int_1^\infty e^{-\gamma(y)}\,dy < \infty$. We can take the limit as $x \to \infty$ in (7.9) to get

$$F(\infty) = \lim_{x\to\infty} \psi_1'(x)e^{-\gamma(x)} \in [0,\infty),$$

and $\psi_1'(x_0)e^{-\gamma(x_0)} = F(\infty) + 2\lambda_1 \int_{x_0}^\infty \psi_1(y)e^{-\gamma(y)}\,dy$. From this equality the result follows. □

In the next results, we give a couple of sufficient conditions for the integrability of the eigenfunctions. The first one is

$$\textbf{H3:} \quad \int_0^1 \frac{1}{\alpha^2(y) - \alpha'(y) + C + 2} e^{-\gamma(y)}\, dy < \infty,$$

where as before $C = -\inf_{x>0}(\alpha^2(x) - \alpha'(x))$.

Proposition 7.8 *Assume that* **H1**, **H2** *and* **H3** *are satisfied. Then* ψ_i *belongs to* $L^1(d\mu)$ *for all* i.

Proof Recall that $\eta_i = e^{-\frac{\gamma}{2}}\psi_i$ is an eigenfunction of the Schrödinger operator H introduced in the proof of Theorem 7.3. Replacing f by η_i in (7.6) thus yields

$$(C/2 + 1 + \lambda_i) \int_0^\infty \eta_i^2(y)\, dy = \int_0^\infty \big(|\eta_i'|^2(y)/2 + w(y)\eta_i^2(y)\big)\, dy.$$

Since the left-hand side is finite, the right-hand side is finite; in particular,

$$\int_0^\infty w(y)\psi_i^2(y)\, d\mu(y) = \int_0^\infty w(y)\eta_i^2(y)\, dy < \infty.$$

As a consequence, using the Cauchy–Schwarz Inequality, we get, on the one hand,

$$\int_0^1 \big|\psi_i(y)\big|\, d\mu(y) \le \left(\int_0^1 w(y)\psi_i^2(y)\, d\mu(y)\right)^{\frac{1}{2}} \left(\int_0^1 \frac{1}{w(y)}\, d\mu(y)\right)^{\frac{1}{2}} < \infty,$$

thanks to **H3**. On the other hand,

$$\int_1^\infty \big|\psi_i(y)\big|\, d\mu(y) \le \left(\int_1^\infty \psi_i^2(y)\, d\mu(y)\right)^{\frac{1}{2}} \left(\int_1^\infty d\mu(y)\right)^{\frac{1}{2}} < \infty,$$

according to Proposition 7.6. We have thus proved that $\psi_i \in L^1(d\mu)$. □

We now obtain sharper estimates using properties of the Dirichlet heat kernel. For this reason, we introduce the hypothesis

$$\textbf{H4:} \quad \int_1^\infty e^{-\gamma(x)}\, dx < \infty \quad \text{and} \quad \int_0^1 x e^{-\frac{\gamma(x)}{2}}\, dx < \infty.$$

Proposition 7.9 *Assume* **H2** *and* **H4** *hold. Then all the eigenfunctions* ψ_k *belong to* $L^1(d\mu)$*, and there is a constant* $K_1 > 0$ *such that for any* $x \in (0, \infty)$ *and any* k

$$\big|\psi_k(x)\big| \le K_1 e^{\lambda_k} e^{\frac{\gamma(x)}{2}}.$$

Moreover, ψ_1 *is strictly positive on* $\mathbb{R}^+$*, and there is a constant* $K_2 > 0$ *such that for any* $x \in (0, 1]$ *and any* k

$$\big|\psi_k(x)\big| \le K_2 x e^{2\lambda_k} e^{\frac{\gamma(x)}{2}}.$$

Recall the semigroup $\tilde{P}_t$ defined in (7.5) is given for $f \in L^2(\mathbb{R}^+, dx)$ by

$$\tilde{P}_t f(x) = \mathbb{E}^{\mathbb{W}_x}\left[f\big(\omega(t)\big)\mathbb{1}_{t<T} \exp\left(-\frac{1}{2}\int_0^t \left(\alpha^2 - \alpha'\right)(\omega_s)\, ds\right)\right],$$

where $\mathbb{E}^{\mathbb{W}_x}$ denotes the expectation w.r.t. the Wiener measure starting from x. The proof of Proposition 7.9 needs some basic estimate on the kernel $\tilde{p}_t(x, y)$, which we summarize in the following lemma (see Cattiaux et al. 2009, Lemma 4.5).

Lemma 7.10 *Assume condition* **H2** *holds. There exists a constant* $K_3 > 0$ *and a continuous increasing function* B *defined on* $[0, \infty)$ *satisfying* $\lim_{z\to\infty} B(z) = \infty$ *and such that for any* $x > 0$, $y > 0$ *we have*

$$0 < \tilde{p}_1(x, y) \le \min\left\{e^{-\frac{(x-y)^2}{4}} e^{-B(\max\{x,y\})}, K_3 p_1^D(x, y)\right\}.$$

Here p_t^D *is the Dirichlet heat kernel in* $\mathbb{R}^+$ *given for* $x, y \in \mathbb{R}^+$ *by* $p_t^D(x, y) = \frac{1}{\sqrt{2\pi t}}(e^{\frac{-(x-y)^2}{2t}} - e^{-\frac{(x+y)^2}{2t}})$.

Proof (Proposition 7.9) As a consequence of the previous lemma, the kernel $\tilde{p}_1(x, y)$ defines a bounded operator $\tilde{P}_1$ from $L^2(\mathbb{R}^+, dx)$ to $L^\infty(\mathbb{R}^+, dx)$. As a byproduct, we get that all the eigenfunctions η_k of $\tilde{P}_1$ are bounded, and more precisely, $|\eta_k| \le K_1 e^{\lambda_k}$.

One deduces also from the previous lemma that the kernel defined for $M > 0$ by $\tilde{p}_1^M(x, y) = \mathbb{1}_{x<M}\mathbb{1}_{y<M}\tilde{p}_1(x, y)$ is a Hilbert–Schmidt operator in $L^2(\mathbb{R}^+, dx)$, in particular, it is a compact operator (see, for example, Conway 1990, pp. 177, 267). Consider also $\tilde{P}_1^M$, the operator with kernel $\tilde{p}_1^M$. From Lemma 7.10, we obtain the following estimate, in the norm of operators acting on $L^2(\mathbb{R}^+, dx)$,

$$\left\| \tilde{P}_1^M - \tilde{P}_1 \right\|_{L^2(\mathbb{R}^+, dx)} \le C' e^{-B(M)}$$

where C' is a positive constant independent of M. Since $\lim_{M\to\infty} B(M) = \infty$, the operator $\tilde{P}_1$ is a limit in norm of compact operators in $L^2(\mathbb{R}^+, dx)$ and hence compact. Since $\tilde{p}_1(x, y) > 0$, the operator $\tilde{P}_1$ is positivity improving (if $f \ge 0$ and f is not the zero function then $\tilde{P}_1 f > 0$), implying that the eigenvector η_1 is positive.

We now claim that $|\eta_k(x)| \le K_2 x e^{2\lambda_k}$ for $0 < x \le 1$. We have from Lemma 7.10 and the explicit expression for $p_1^D(x, y)$ the existence of a constant K_3 such that

$$\begin{aligned} \left|e^{-\lambda_k}\eta_k(x)\right| &\le K_3 \int_0^\infty p_1^D(x, y)\left|\eta_k(y)\right| dy \\ &\le K_3 \|\eta_k\|_\infty \sqrt{\frac{2}{\pi}} e^{-\frac{x^2}{2}} \int_0^\infty e^{-\frac{y^2}{2}} \sinh(xy)\, dy. \end{aligned}$$

We now estimate the integral in the right-hand side. Using the convexity of sinh, we get $\sinh(xy) \le x \sinh(y) \le \frac{x}{2} e^y$, for $x \in [0,1]$, $y \ge 0$ which yields

$$\int_0^\infty e^{-\frac{y^2}{2}} \sinh(xy)\, dy \le \frac{x}{2} \int_0^\infty e^{-\frac{y^2}{2}} e^y\, dy,$$

proving the claim. Together with hypothesis **H4**, this estimate implies that ψ_k belongs to $L^1((0,1), d\mu)$.

Since $\psi_k(x) = \eta_k(x) e^{\frac{\gamma(x)}{2}}$, we have

$$\int_1^\infty |\psi_k|\, d\mu \le \left(\int_1^\infty |\eta_k(x)|^2\, dx\right)^{1/2} \left(\int_1^\infty e^{-\gamma(x)}\, dx\right)^{1/2} < \infty,$$

which implies $\psi_k \in L^1((1,\infty), d\mu)$. This finishes the proof of the proposition. □

Remark 7.11 If α and α' extend continuously up to 0, Hypotheses **H2**, **H3** and **H4** reduce to their counterparts at infinity.

7.5 QSD and Yaglom Limit

In this section, we study the existence of a QSD. One of the main technical problems here is that the drift may explode at 0. This could cause that ψ_1, a possible candidate to induce a QSD, is not integrable. For this reason, we shall work under the following assumption.

Definition 7.12 We say that Hypothesis **H0** is verified if **H1**, **H2** hold and $\psi_1 \in L^1(d\mu)$ (which is the case, for example, under **H3** or **H4**).

When $\psi_1 \in L^1(d\mu)$, a natural candidate for being a QSD is the normalized measure $\frac{\psi_1}{\langle \psi_1, 1\rangle_\mu} d\mu$, which turns out to be the conditional limit distribution. This is suggested by Theorem 7.3 and the main result of this section is the following theorem.

Theorem 7.13 *Assume that Hypothesis* **H0** *holds. Then*

$$d\nu_1 = \frac{\psi_1}{\langle \psi_1, 1\rangle_\mu} d\mu$$

is a QSD, that is, for every $t \ge 0$ and any Borel subset A of $(0,\infty)$,

$$\mathbb{P}_{\nu_1}(X_t \in A | t < T) = \nu_1(A).$$

Also for any $x > 0$ and any Borel subset A of $(0,\infty)$,

$$\begin{aligned} \lim_{t\to\infty} e^{\lambda_1 t} \mathbb{P}_x(t < T) &= \psi_1(x) \langle \psi_1, 1\rangle_\mu, \\ \lim_{t\to\infty} e^{\lambda_1 t} \mathbb{P}_x(X_t \in A, t < T) &= \nu_1(A) \psi_1(x) \langle \psi_1, 1\rangle_\mu. \end{aligned} \tag{7.10}$$

This implies, since $\psi_1 > 0$ on $(0,\infty)$, that

$$\lim_{t\to\infty} \mathbb{P}_x(X_t \in A | t < T) = \nu_1(A),$$

and the probability measure ν_1 is the Yaglom limit distribution. Moreover, for any probability measure ρ with compact support in $(0,\infty)$ we have

$$\lim_{t\to\infty} e^{\lambda_1 t}\mathbb{P}_\rho(t < T) = \langle\psi_1, 1\rangle_\mu \int \psi_1(x)\, d\rho(x),$$

$$\lim_{t\to\infty} e^{\lambda_1 t}\mathbb{P}_\rho(X_t \in A, t < T) = \nu_1(A)\langle\psi_1, 1\rangle_\mu \int \psi_1(x)\, d\rho(x), \tag{7.11}$$

$$\lim_{t\to\infty} \mathbb{P}_\rho(X_t \in A | t < T) = \nu_1(A).$$

Proof Thanks to the symmetry of the semigroup, we have for all f in $L^2(d\mu)$,

$$\int P_t f \psi_1\, d\mu = \int f P_t \psi_1\, d\mu = e^{-\lambda_1 t} \int f \psi_1\, d\mu.$$

Since $\psi_1 \in L^1(d\mu)$, this equality extends to all bounded f. In particular, we may use it with $f = \mathbb{1}_{(0,\infty)}$ and with $f = \mathbb{1}_A$. Noticing that

$$\int P_t(\mathbb{1}_{(0,\infty)})\psi_1\, d\mu = \mathbb{P}_{\nu_1}(t < T)\langle\psi_1, 1\rangle_\mu$$

and $\int P_t \mathbb{1}_A \psi_1\, d\mu = \mathbb{P}_{\nu_1}(X_t \in A, t < T)\langle\psi_1, 1\rangle_\mu$, we have shown that ν_1 is a QSD.

The rest of the proof is divided into two cases. First, we assume that μ is a bounded measure. Thanks to Theorem 7.1, we know that for any $x > 0$, any set $A \subset (0,\infty)$ such that $\mathbb{1}_A \in L^2(d\mu)$ and for any $t > 1$,

$$\begin{aligned}
\mathbb{P}_x(X_t \in A, t < T) &= \int \mathbb{P}_y(X_{t-1} \in A, t < T - 1) r(1, x, y)\, d\mu(y) \\
&= \int P_{t-1}(\mathbb{1}_A)(y) r(1, x, y)\, d\mu(y) \\
&= \int \mathbb{1}_A(y)\big(P_{t-1} r(1, x, \cdot)\big)(y)\, d\mu(y).
\end{aligned}$$

Since both $\mathbb{1}_A$ and $r(1, x, \cdot)$ are in $L^2(d\mu)$ and since **H2** is satisfied, we obtain using Theorem 7.3

$$\lim_{t\to\infty} e^{\lambda_1(t-1)}\mathbb{P}_x(X_t \in A, t < T) = \langle\mathbb{1}_A, \psi_1\rangle_\mu \big\langle r(1, x, \cdot), \psi_1\big\rangle_\mu. \tag{7.12}$$

Since

$$\int r(1, x, y)\psi_1(y)\, d\mu(y) = (P_1\psi_1)(x) = e^{-\lambda_1}\psi_1(x),$$

we get that ν_1 is the Yaglom limit.

If μ is not bounded (i.e. $\mathbb{1}_{(0,\infty)} \notin L^2(d\mu)$) we need an additional result to obtain the Yaglom limit. This lemma is a consequence of Harnack's Inequality and its proof can be found in Cattiaux et al. (2009). □

Lemma 7.14 *Assume $\psi_1 \in L^1(d\mu)$ then for all $x > 0$, there exists a locally bounded function $\Theta(x)$ such that for all $y > 0$ and all $t > 1$,*

$$r(t, x, y) \le \Theta(x) e^{-\lambda_1 t} \psi_1(y). \tag{7.13}$$

If (7.13) holds then $e^{\lambda_1 t} r(t, x, \cdot) \in L^1(d\mu)$, and this function is dominated by $\Theta(x)\psi_1$. Theorem 7.1 shows that $r(1, x, \cdot) \in L^2(d\mu)$, and then we can apply Theorem 7.3 to $r(t, x, \cdot) = P_{t-1} r(1, x, \cdot)$ to deduce that $\lim_{t\to\infty} e^{\lambda_1 t} r(t, x, \cdot)$ exists in $L^2(d\mu)$ and is equal to

$$e^{\lambda_1} \langle r(1, x, \cdot), \psi_1 \rangle_\mu \psi_1(\cdot) = \psi_1(x)\psi_1(\cdot).$$

Recall that convergence in L^2 implies almost sure convergence along subsequences. Therefore, for any sequence $t_n \to \infty$ there exists a subsequence t_n' such that

$$\lim_{n\to\infty} e^{\lambda_1 t_n'} r(t_n', x, y) = \psi_1(x)\psi_1(y) \quad \text{for } \mu\text{-almost all } y > 0.$$

Since

$$\mathbb{P}_x(t_n' < T) = \int_0^\infty r(t_n', x, y)\, d\mu(y),$$

Lebesgue's Bounded Convergence Theorem yields

$$\lim_{n\to\infty} e^{\lambda_1 t_n'} \mathbb{P}_x(t_n' < T) = \psi_1(x) \int_0^\infty \psi_1(y)\, d\mu(y).$$

That is, (7.12) holds with $A = (0, \infty)$ for the sequence t_n'. Since the limit does not depend on the subsequence, $\lim_{t\to\infty} e^{\lambda_1 t} \mathbb{P}_x(t < T)$ exists and is equal to the previous limit, hence (7.12) is still true. The rest of this part follows as before.

For the last part of the theorem, that is, for passing from the initial Dirac measures at every fixed $x > 0$ to the compactly supported case, we just use that $\Theta(\bullet)$ is bounded on compact sets included in $(0, \infty)$. This finishes the proof of Theorem 7.13. □

The positive real number λ_1 is the natural killing rate of the process. Indeed, the limit (7.10) obtained in Theorem 7.13 shows for any $x > 0$ and any $t > 0$,

$$\lim_{s\to\infty} \frac{\mathbb{P}_x(t + s < T)}{\mathbb{P}_x(s < T)} = e^{-\lambda_1 t}.$$

Let us also remark that

$$\mathbb{P}_{\nu_1}(t < T) = e^{-\lambda_1 t}.$$

7.5.1 Speed of Convergence to the Yaglom Limit

We now study the speed of convergence to the Yaglom limit. Informally, this speed of convergence is dominated by the difference between the two first eigenvalues as it is suggested by the expansion (7.4) in Theorem 7.3. In order to control the other terms in that expansion, we need the following result.

Lemma 7.15 *Under conditions* **H2** *and* **H4**, *the operator* P_1 *is bounded from* $L^\infty(\mu)$ *to* $L^2(d\mu)$. *Moreover, for any compact subset* K *of* $(0,\infty)$, *there is a constant* C_K *such that for any function* $f \in L^1(d\mu)$ *with support in* K *we have*

$$\|P_1 f\|_{L^2(d\mu)} \le C_K \|f\|_{L^1(d\mu)}.$$

Proof Let $g \in L^\infty(\mu)$, since

$$|P_1 g| \le P_1|g| \le \|g\|_{L^\infty(\mu)},$$

$$\int_1^\infty |P_1 g|^2\, d\mu \le \|g\|^2_{L^\infty(\mu)} \int_1^\infty e^{-\gamma(x)}\, dx,$$

where the last integral is finite by **H4**. We now recall that (see (7.5))

$$P_1 g(x) = e^{\frac{\gamma(x)}{2}} \tilde{P}_1\big(e^{-\frac{\gamma}{2}} g\big)(x).$$

It follows from Lemma 7.10 that uniformly in $x \in (0,1]$ we have using Hypothesis **H4**

$$\big|\tilde{P}_1\big(e^{-\frac{\gamma}{2}} g\big)(x)\big| \le \mathcal{O}(1)\|g\|_{L^\infty(\mu)} \int_0^\infty e^{-\frac{\gamma(y)}{2}} e^{-\frac{y^2}{4}}\, y\, dy \le \mathcal{O}(1)\|g\|_{L^\infty(\mu)}.$$

This implies

$$\int_0^1 |P_1 g|^2\, d\mu = \int_0^1 \big|\tilde{P}_1\big(e^{-\frac{\gamma}{2}} g\big)(x)\big|^2\, dx \le \mathcal{O}(1)\|g\|^2_{L^\infty(\mu)},$$

and the first part of the lemma follows. For the second part, we have from the Gaussian bound of Lemma 7.10 that for any $x > 0$ and for any f integrable and with support in K,

$$\begin{aligned}
\big|\tilde{P}_1\big(e^{-\frac{\gamma}{2}} f\big)(x)\big| &\le \mathcal{O}(1) \int_K e^{-\frac{\gamma(y)}{2}} e^{-\frac{(x-y)^2}{2}} \big|f(y)\big|\, dy \\
&\le \mathcal{O}(1) \sup_{z\in K} e^{\frac{\gamma(z)}{2}} \sup_{z\in K} e^{-\frac{(x-z)^2}{2}} \int_K e^{-\gamma(y)} \big|f(y)\big|\, dy \\
&\le \mathcal{O}(1) e^{-\frac{x^2}{4}} \int_K e^{-\gamma(y)} \big|f(y)\big|\, dy
\end{aligned}$$

since K is compact. This implies

$$\int_0^\infty |P_1 f|^2 \, d\mu = \int_0^\infty \left|\tilde{P}_1\left(e^{-\frac{\gamma}{2}} f\right)(x)\right|^2 dx \le \mathcal{O}(1) \|f\|^2_{L^1(d\mu)}. \qquad \square$$

We can now use the spectral decomposition of $r(1, x, \cdot)$ to obtain the desired speed of convergence.

Theorem 7.16 *Under conditions* **H2** *and* **H4**, *for all* $x > 0$ *and any measurable subset* A *of* $(0, \infty)$, *we have*

$$\lim_{t\to\infty} e^{(\lambda_2-\lambda_1)t}\left(\mathbb{P}_x(X_t \in A | t < T) - \nu_1(A)\right)$$
$$= \frac{\psi_2(x)}{\psi_1(x)}\left(\frac{\langle 1, \psi_1\rangle_\mu \langle \mathbb{1}_A, \psi_2\rangle_\mu - \langle 1, \psi_2\rangle_\mu \langle \mathbb{1}_A, \psi_1\rangle_\mu}{(\langle 1, \psi_1\rangle_\mu)^2}\right). \tag{7.14}$$

Proof Let h be a nonnegative bounded function with compact support in $(0, \infty)$. By using the semigroup property, Lemma 7.15 and the spectral decomposition for compact self adjoint semigroups (see Theorem 7.3), we have for any $t > 2$,

$$\int \mathbb{P}_x(X_t \in A, t < T) h(x) \, dx$$
$$= \langle he^\gamma, P_t \mathbb{1}_A\rangle_\mu = \langle P_1(he^\gamma), P_{(t-2)} P_1 \mathbb{1}_A\rangle_\mu$$
$$= e^{-\lambda_1(t-2)} \langle P_1(he^\gamma), \psi_1\rangle_\mu \langle \psi_1, P_1\mathbb{1}_A\rangle_\mu + e^{-\lambda_2(t-2)} \langle P_1(he^\gamma), \psi_2\rangle_\mu \langle \psi_2, P_1\mathbb{1}_A\rangle_\mu$$
$$+ R(h, A, t)$$

with

$$\left|R(h, A, t)\right| \le \sum_{i\ge 3} e^{-\lambda_i(t-2)} \left|\langle P_1(he^\gamma), \psi_i\rangle_\mu \langle \psi_i, P_1\mathbb{1}_A\rangle_\mu\right|$$
$$\le e^{-\lambda_3(t-2)} \left\| P_1(he^\gamma)\right\|_{L^2(d\mu)} \|P_1\mathbb{1}_A\|_{L^2(d\mu)},$$

because $\lambda_1 < \lambda_2 < \lambda_3 < \cdots$, the Cauchy–Schwarz Inequality and Parseval's Identity. Note that since P_1 is symmetric with respect to the scalar product $\langle , \rangle_\mu$, we have $\langle P_1(he^\gamma), \psi_1\rangle_\mu = e^{-\lambda_1}\langle he^\gamma, \psi_1\rangle_\mu$ and similarly for ψ_2. We have also $\langle \psi_1, P_1\mathbb{1}_A\rangle_\mu = e^{-\lambda_1}\langle \psi_1, \mathbb{1}_A\rangle_\mu$ and similarly for ψ_2. It follows immediately from Lemma 7.15 that for any fixed compact subset K of $(0, \infty)$, any A and any h satisfying the hypothesis of the proposition with support contained in K,

$$\left|R(h, A, t)\right| \le \mathcal{O}(1) e^{-\lambda_3(t-2)} \|h\|_{L^1(d\mu)} \le \mathcal{O}(1) e^{-\lambda_3(t-2)} \|h\|_{L^1(dx)},$$

since h has compact support in $(0, \infty)$. Therefore, letting h tend to a Dirac mass, we obtain that for any compact subset K of $(0, \infty)$, there is a constant D_K such that for any $x \in K$, for any measurable subset A of $(0, \infty)$, and for any $t > 2$, we have

$$\left|\mathbb{P}_x(X_t \in A, t < T) - e^{\gamma(x)}\psi_1(x)\langle \psi_1, \mathbb{1}_A\rangle_\mu e^{-\lambda_1 t} - e^{\gamma(x)}\psi_2(x)\langle \psi_2, \mathbb{1}_A\rangle_\mu e^{-\lambda_2 t}\right|$$

is bounded from above by $D_K e^{-\lambda_3 t}$. The proposition follows at once from

$$\mathbb{P}_x(X_t \in A | t < T) = \frac{\mathbb{P}_x(X_t \in A, t < T)}{\mathbb{P}_x(X_t \in (0,\infty), t < T)}. \qquad \square$$

7.6 The Q-Process

We describe in this section the law of the process conditioned to be never extinct, usually called the Q-process. This is similar to what we have proved in Theorem 6.26(iv) and we give here only the main points of the proof.

Proposition 7.17 *Assume* **H0** *holds. For all $x > 0$ and $s \geq 0$, we have for all $s \geq 0$ and all $A \in \mathcal{B}(\mathcal{C}([0,s]))$*

$$\lim_{t\to\infty} \mathbb{P}_x(X \in A | t < T) = Q_x(A),$$

where Q_x is the law of a diffusion process on $(0,\infty)$ with transition probability densities (w.r.t. the Lebesgue measure) given by

$$q(s,x,y) = e^{\lambda_1 s} \frac{\psi_1(y)}{\psi_1(x)} r(s,x,y) e^{-\gamma(y)},$$

that is, Q_x is locally absolutely continuous w.r.t. $\mathbb{P}_x$ and

$$Q_x(X \in A) = \mathbb{E}_x\left(\mathbb{1}_A(X) e^{\lambda_1 s} \frac{\psi_1(X_s)}{\psi_1(x)}, s < T\right).$$

Proof First check, thanks to Fubini's Theorem and $\kappa(0^+) < \infty$ in Hypothesis **H1**, that $\Lambda(0^+) > -\infty$. We can thus slightly change the notation (for this proof only) and define Λ as $\Lambda(x) = \int_0^x e^{\gamma(y)}\,dy$. From standard diffusion theory, $(\Lambda(X_{t\wedge T}); t \geq 0)$ is a local martingale, from which it is easy to derive that for any $y \geq x \geq 0$, $\mathbb{P}_y(T_x < T) = \Lambda(y)/\Lambda(x)$.

Now define $v(t,x) = \frac{\mathbb{P}_x(t<T)}{\mathbb{P}_1(t<T)}$. The same ideas used in the proof of (6.29) shows that for any $x \geq 0$, $v(t,x) \leq \frac{\Lambda(x)}{\Lambda(1)} + 1$. Now, thanks to Theorem 7.13, for all x it holds $e^{\lambda_1 t}\mathbb{P}_x(t < T) \to \psi_1(x)\langle 1, \psi_1\rangle_\mu$ as $t \to \infty$, and

$$\lim_{t\to\infty} v(t,x) = \frac{\psi_1(x)}{\psi_1(1)}.$$

Using the Markov property, it is easily seen that for large t,

$$\mathbb{P}_x(X \in A | t < T) = \mathbb{E}_x\left[\mathbb{1}_A(X) v(t-s, X_s), s < T\right] \frac{\mathbb{P}_1(t-s < T)}{\mathbb{P}_x(t < T)}.$$

The random variable in the above expectation is positive and bounded from above by $1+\Lambda(X_s)/\Lambda(1)$, which is integrable (see below). So we obtain the desired result using the Dominated Convergence Theorem.

To see that $\mathbb{E}_x(\Lambda(X_s)\mathbb{1}_{s<T})$ is finite, it is enough to use Itô's formula with the harmonic function Λ up to time $T \wedge T_M$. Since Λ is nonnegative, it follows that $\mathbb{E}_x(\Lambda(X_s)\mathbb{1}_{s<T\wedge T_M}) \le \Lambda(x)$ for all $M > 0$. The Monotone Convergence Theorem yields $\mathbb{E}_x(\Lambda(X_s)\mathbb{1}_{s<T}) \le \Lambda(x)$ (here we have used **H1**). □

Recall that ν_1 is the Yaglom limit.

Corollary 7.18 *Assume* **H0**. *Then for any Borel subset $B \subseteq (0,\infty)$ and any x,*

$$\lim_{s\to\infty} Q_x(X_s \in B) = \int_B \psi_1^2(y)\,d\mu(y) = \langle\psi_1, 1\rangle_\mu \int_B \psi_1(y)\,d\nu_1(y).$$

Proof We know from the proof of Theorem 7.13 that $e^{\lambda_1 s} r(s,x,\cdot)$ converges to $\psi_1(x)\psi_1(\cdot)$ in $L^2(d\mu)$ as $s \to \infty$. Hence, since $\mathbb{1}_B\psi_1 \in L^2(d\mu)$,

$$\psi_1(x)Q_x(X_s \in B) = \int \mathbb{1}_B(y)\psi_1(y)e^{\lambda_1 s} r(s,x,y)\,d\mu(y) \to \psi_1(x)\int_B \psi_1^2(y)\,d\mu(y)$$

as $s \to \infty$. We remind the reader that $d\nu_1 = \psi_1\,d\mu/\langle\psi_1, 1\rangle_\mu$. □

Remark 7.19 The stationary measure of the Q-process is absolutely continuous w.r.t. ν_1, with Radon–Nikodym derivative $\langle\psi_1, 1\rangle_\mu\psi_1$, which, thanks to Proposition 7.6, is nondecreasing.

7.7 Uniqueness of QSD

In this section, we study the uniqueness of the QSD. We relate this problem to the behavior at ∞ of the process X. In the problem of uniqueness, it is natural to study the notion of quasi-limiting distribution (q.l.d.).

Definition 7.20 A probability measure π supported on $(0,\infty)$ is a q.l.d. if there exists a probability measure ν such that the following limit exists in distribution

$$\lim_{t\to\infty} \mathbb{P}_\nu(X_t \in \bullet | t < T) = \pi(\bullet).$$

We say also that ν is attracted to π, or is in the domain of attraction of π, for the conditional evolution.

Obviously every QSD is a q.l.d. because such measures are fixed points for the conditional evolution. We prove that the reciprocal is also true, so both concepts coincide.

Lemma 7.21 *Let π a probability measure supported on* $(0,\infty)$. *If π is a q.l.d. then π is a QSD. In particular, there exists $\theta \geq 0$ such that for all $s > 0$*

$$\mathbb{P}_\pi(s < T) = e^{-\theta s}.$$

Proof By hypothesis, there exists a probability measure ν such that

$$\lim_{t\to\infty} \mathbb{P}_\nu(X_t \in \bullet | t < T) = \pi(\bullet)$$

in distribution. That is, for all continuous and bounded functions f, we have

$$\lim_{t\to\infty} \frac{\mathbb{P}_\nu(f(X_t), t < T)}{\mathbb{P}_\nu(t < T)} = \int f(x)\, d\pi(x).$$

If we take $f(x) = \mathbb{P}_x(X_s \in A, s < T)$ then, since $f(x) = \int_A r(s,x,y)\, d\mu(y)$, an application of Harnack's Inequality and of the Dominated Convergence Theorem ensures that f is continuous in $(0,\infty)$.

First, take $A = (0,\infty)$, so that $f(x) = \mathbb{P}_x(s < T)$. Then, we obtain for all $s \geq 0$

$$\lim_{t\to\infty} \frac{\mathbb{P}_\nu(t + s < T)}{\mathbb{P}_\nu(t < T)} = \mathbb{P}_\pi(s < T).$$

The left-hand side is easily seen to be exponential in s, and then there exists $\theta \geq 0$ such that

$$\mathbb{P}_\pi(s < T) = e^{-\theta s}.$$

Second, take $f(x) = \mathbb{P}_x(X_s \in A, s < T)$ to conclude that

$$\begin{aligned}
\mathbb{P}_\pi(X_s \in A, s < T) &= \lim_{t\to\infty} \mathbb{P}_\nu\big(f(X_t) | t < T\big)\\
&= \lim_{t\to\infty} \mathbb{P}_\nu(X_{t+s} \in A | t + s < T) \frac{\mathbb{P}_\nu(t + s < T)}{\mathbb{P}_\nu(t < T)}\\
&= e^{-\theta s}\pi(A),
\end{aligned}$$

and then π is a QSD. □

Under Hypothesis **H0** (see Theorem 7.13), the measure $d\nu_1 = \frac{\psi_1}{\langle \psi_1, 1\rangle_\mu}\, d\mu$ is the Yaglom limit, which in addition is a q.l.d. attracting all initial distributions with compact support on $(0,\infty)$. Here again, our assumptions on the behavior of α at infinity will allow us to characterize the domain of attraction of the QSD ν_1 associated to ψ_1. This turns out to be entirely different from the cases studied in the previous chapter.

Definition 7.22 We say that the diffusion process X *comes down from infinity* if there is $y > 0$ and a time $t > 0$ such that

$$\lim_{x\uparrow\infty} \mathbb{P}_x(T_y < t) > 0.$$

This terminology is equivalent to the property that ∞ is an entrance boundary for X (for instance, see Revuz and Yor 1999, p. 283).

Let us introduce the following condition

$$\mathbf{H5}: \quad \int_1^\infty e^{\gamma(y)} \int_y^\infty e^{-\gamma(z)}\, dz\, dy < \infty.$$

Tonelli's Theorem ensures that **H5** is equivalent to

$$\int_1^\infty e^{-\gamma(y)} \int_1^y e^{\gamma(z)}\, dz\, dy < \infty. \tag{7.15}$$

If **H5** holds then for any $y \geq 1$, $\int_y^\infty e^{-\gamma(z)}\, dz < \infty$. Applying the Cauchy–Schwarz Inequality, we get $(x-1)^2 = (\int_1^x e^{\frac{\gamma}{2}} e^{-\frac{\gamma}{2}}\, dz)^2 \leq \int_1^x e^{\gamma}\, dz \int_1^x e^{-\gamma}\, dz$, and therefore, **H5** implies that $\Lambda(\infty) = \infty$.

Remark 7.23 It is not obvious when Condition **H5** holds. In this direction, the following explicit conditions on α, all together, are sufficient for **H5** to hold:

- $\alpha(x) \geq q_0 > 0$ for all $x \geq x_0$;
- $\limsup_{x\to\infty} \frac{\alpha'(x)}{2\alpha^2(x)} < 1$;
- $\int_{x_0}^\infty \frac{1}{\alpha(x)}\, dx < \infty$.

Now we state the main result of this section.

Theorem 7.24 *Assume* **H0** *holds. Then the following are equivalent:*

(i) *X comes down from infinity;*
(ii) **H5** *holds;*
(iii) *ν_1 attracts all initial distributions ν supported in $(0,\infty)$, that is,*

$$\lim_{t\to\infty} \mathbb{P}_\nu(X_t \in \bullet | t < T) = \nu_1(\bullet).$$

In particular, any of these three conditions implies there is a unique QSD.

The proof of Theorem 7.24 requires some previous results.

Proposition 7.25 *Assume* **H1** *holds. If there is a unique QSD that attracts all initial distributions supported in $(0,\infty)$, then X comes down from infinity.*

Proof Let π be the unique QSD that attracts all distributions. We know that $\mathbb{P}_\pi(t < T) = e^{-\theta t}$, for some $\theta \geq 0$. Since absorption is certain, $\theta > 0$. For the rest of the proof, let ν be any initial distribution supported on $(0,\infty)$, which by hypothesis is in the domain of attraction of π, that is, for any bounded continuous f,

$$\lim_{t\to\infty} \int_0^\infty \mathbb{P}_\nu(X_t \in dx | t < T) f(x) = \int_0^\infty f(x)\, d\pi(x).$$

We now prove that for any $\lambda < \theta$, $\mathbb{E}_\nu(e^{\lambda T}) < \infty$. As in Lemma 7.21, we have for any s

$$\lim_{t\to\infty} \frac{\mathbb{P}_\nu(t+s<T)}{\mathbb{P}_\nu(t<T)} = e^{-\theta s}.$$

Now pick $\lambda \in (0,\theta)$ and $\varepsilon > 0$ such that $(1+\varepsilon)e^{\lambda-\theta} < 1$. An elementary induction shows that there is t_0 such that for any $t > t_0$, and any integer n,

$$\frac{\mathbb{P}_\nu(t+n<T)}{\mathbb{P}_\nu(t<T)} \le (1+\varepsilon)^n e^{-\theta n}.$$

Breaking down the integral $\int_{t_0}^\infty \mathbb{P}_\nu(s<T)e^{\lambda s}\,ds$ over intervals of the form $(n,n+1]$ and using the previous inequality, it is easily seen that this integral converges. This proves that $\mathbb{E}_\nu(e^{\lambda T}) < \infty$ for any initial distribution ν.

Now fix $\lambda = \theta/2$ and for any $x \ge 0$, let $g(x) = \mathbb{E}_x(e^{\lambda T}) < \infty$. We want to show that g is bounded, which trivially entails that X comes down from infinity. Thanks to the previous step, for any nonnegative random variable Y with law ν,

$$\mathbb{E}\big(g(Y)\big) = \mathbb{E}_\nu\big(e^{\lambda T}\big) < \infty.$$

Since Y can be any random variable, this implies that g is bounded. Indeed, observe that g is increasing and $g(0) = 1$, so that $a := 1/g(\infty)$ is well defined in $[0, 1)$. Then check that

$$d\nu(x) = \frac{g'(x)}{(1-a)g(x)^2}\,dx$$

is a probability density on $(0,\infty)$. Conclude computing $\int g\,d\nu$. □

Proposition 7.26 *The following are equivalent*

(i) *X comes down from infinity*;
(ii) **H5** *holds*;
(iii) *For any $a > 0$ there exists $y_a > 0$ such that* $\sup_{x>y_a} \mathbb{E}_x[e^{aT_{y_a}}] < \infty$.

Proof Since (i) is equivalent to ∞ being an entrance boundary and (ii) is equivalent to (7.15), we must show that "∞ is an entrance boundary" and (7.15) are equivalent. This will follow from Kallenberg (1997, Theorem 20.12(iii)). For that purpose, consider $Y_t = \Lambda(X_t)$. Under each one of the conditions (i) or (ii), we have $\Lambda(\infty) = \infty$. It is direct to prove that Y is in natural scale on the interval $(\Lambda(0),\infty)$, that is, for $\Lambda(0) < a \le y \le b < \infty = \Lambda(\infty)$

$$\mathbb{P}_y\big(T_a^Y < T_b^Y\big) = \frac{b-y}{b-a},$$

where T_a^Y is the hitting time of a for the diffusion Y. Then, ∞ is an entrance boundary for Y if and only if

$$\int_0^\infty y\,dm(y) < \infty,$$

where m is the speed measure of Y, which is given by

$$dm(y) = \frac{2dy}{(\Lambda'(\Lambda^{-1}(y))^2},$$

see Karatzas and Shreve (1988, formula (5.51)), because Y satisfies the SDE

$$dY_t = \Lambda'\big(\Lambda^{-1}(Y_t)\big)\, dB_t.$$

After a change of variables, we obtain

$$\int_0^\infty y\, dm(y) = \int_1^\infty e^{-\gamma(y)} \int_1^x e^{\gamma(z)}\, dz\, dx.$$

Therefore, we have shown the equivalence between (i) and (ii).

We continue the proof with (ii) $\Rightarrow$ (iii). Let $a > 0$, and pick x_a large enough so that

$$\int_{x_a}^\infty e^{\gamma(x)} \int_x^\infty e^{-\gamma(z)}\, dz\, dx \le \frac{1}{2a}.$$

Let J be the nonnegative increasing function defined on $[x_a, \infty)$ by

$$J(x) = \int_{x_a}^x e^{\gamma(y)} \int_y^\infty e^{-\gamma(z)}\, dz\, dy.$$

Then check that $J'' = 2\alpha J' - 1$, so that $\mathcal{L}J = -1/2$. Set now $y_a = 1 + x_a$, and consider a large $M > x$. Itô's formula gives, for $S = t \wedge T_M \wedge T_{y_a}$,

$$\mathbb{E}_x\big(e^{aS} J(X_S)\big) = J(x) + \mathbb{E}_x\left(\int_0^S e^{as}\big(aJ(X_s) + \mathcal{L}J(X_s)\big)\, ds\right).$$

From $\mathcal{L}J = -1/2$, and $J(X_s) < J(\infty) \le 1/(2a)$ for any $s \le T_{y_a}$, we get

$$\mathbb{E}_x\big[e^{aS} J(X_S)\big] \le J(x).$$

But J is increasing, hence for $x \ge y_a$ one obtains the inequality $1/(2a) > J(x) \ge J(y_a) > 0$. It follows that $\mathbb{E}_x(e^{a(t\wedge T_M \wedge T_{y_a})}) \le 1/(2aJ(y_a))$, and finally $\mathbb{E}_x(e^{aT_{y_a}}) \le 1/(2aJ(y_a))$ because of the Monotone Convergence Theorem. So (iii) holds.

Finally, it is clear that (iii) $\Rightarrow$ (i). □

Proposition 7.27 *Assume* **H0** *holds. If there exists* x_0 *such that* $\sup_{x\ge x_0} \mathbb{E}_x(e^{\lambda_1 T_{x_0}}) < \infty$, *then* ν_1 *attracts all initial distributions supported on* $(0, \infty)$.

The proof of this result requires the following control near 0 and ∞.

Lemma 7.28 *Assume* **H0** *holds, and* $\sup_{x\ge x_0} \mathbb{E}_x(e^{\lambda_1 T_{x_0}}) < \infty$. *For* $h \in L^1(d\mu)$ *strictly positive in* $(0, \infty)$, *we have*

$$\lim_{\epsilon\downarrow 0}\limsup_{t\to\infty}\frac{\int_0^\epsilon h(x)\mathbb{P}_x(t<T)\,d\mu(x)}{\int h(x)\mathbb{P}_x(t<T)\,d\mu(x)}=0, \tag{7.16}$$

$$\lim_{M\uparrow\infty}\limsup_{t\to\infty}\frac{\int_M^\infty h(x)\mathbb{P}_x(t<T)\,d\mu(x)}{\int h(x)\mathbb{P}_x(t<T)\,d\mu(x)}=0. \tag{7.17}$$

Proof We start with (7.16). Using Harnack's Inequality, we have for $\epsilon<1$ and large t,

$$\frac{\int_0^\epsilon h(x)\mathbb{P}_x(t<T)\,d\mu(x)}{\int h(x)\mathbb{P}_x(t<T)\,d\mu(x)}\le\frac{\mathbb{P}_1(t<T)\int_0^\epsilon h(z)\,d\mu(z)}{Cr(t-1,1,1)\int_1^2 h(x)\,d\mu(x)\int_1^2 d\mu(y)},$$

then

$$\limsup_{t\to\infty}\frac{\int_0^\epsilon h(x)\mathbb{P}_x(t<T)\,d\mu(x)}{\int h(x)\mathbb{P}_x(t<T)\,d\mu(x)}\le\frac{e^{-\lambda_1}\langle\psi_1,1\rangle_\mu\int_0^\epsilon h(z)\,d\mu(z)}{C\psi_1(1)\int_1^2 h(x)\,d\mu(x)\int_1^2 d\mu(y)},$$

and the first assertion of the statement follows.

For the second limit, we set $A_0:=\sup_{x\ge x_0}\mathbb{E}_x(e^{\lambda_1 T_{x_0}})<\infty$. Then for large $M>x_0$, we have

$$\mathbb{P}_x(t<T)=\int_0^t\mathbb{P}_{x_0}(u<T)\mathbb{P}_x\big(T_{x_0}\in d(t-u)\big)+\mathbb{P}_x(t<T_{x_0}).$$

Using that $\lim_{u\to\infty}e^{\lambda_1 u}\mathbb{P}_{x_0}(u<T)=\psi_1(x_0)\langle\psi_1,1\rangle_\mu$, we obtain that $B_0:=\sup_{u\ge 0}e^{\lambda_1 u}\mathbb{P}_{x_0}(u<T)<\infty$. Then

$$\begin{aligned}\mathbb{P}_x(t<T)&\le B_0\int_0^t e^{-\lambda_1 u}\mathbb{P}_x\big(T_{x_0}\in d(t-u)\big)+\mathbb{P}_x(t<T_{x_0})\\&\le B_0e^{-\lambda_1 t}\mathbb{E}_x\big(e^{\lambda_1 T_{x_0}}\big)+e^{-\lambda_1 t}\mathbb{E}_x\big(e^{\lambda_1 T_{x_0}}\big)\le e^{-\lambda_1 t}A_0(B_0+1),\end{aligned}$$

and (7.17) follows immediately. □

Proof (Proposition 7.27) Let ν be any fixed probability distribution whose support is contained in $(0,\infty)$. We must show that the conditional evolution of ν converges to ν_1. We begin by claiming that ν can be assumed to have a strictly positive density h with respect to μ. Indeed, let

$$\ell(y)=\int r(1,x,y)\,d\nu(x).$$

Using Tonelli's Theorem, we have

$$\begin{aligned}\int\int r(1,x,y)\,d\nu(x)\,d\mu(y)&=\int\int r(1,x,y)\,d\mu(y)\,d\nu(x)\\&=\int\mathbb{P}_x(T>1)\,d\nu(x)\le 1,\end{aligned}$$

which implies that $\int r(1,x,y)\,d\nu(x)$ is finite dy-a.s.. Also ℓ is strictly positive by Harnack's Inequality. Finally, define $h=\ell/\int \ell\,d\mu$. Notice that for $d\rho=h\,d\mu$

$$\mathbb{P}_\nu(X_{t+1}\in\bullet|t+1<T)=\mathbb{P}_\rho(X_t\in\bullet|t<T),$$

showing the claim.

Consider $M>\epsilon>0$ and any Borel set A included in $(0,\infty)$. Then

$$\left|\frac{\int \mathbb{P}_x(X_t\in A,t<T)h(x)\,d\mu(x)}{\int \mathbb{P}_x(t<T)h(x)\,d\mu(x)}-\frac{\int_\epsilon^M \mathbb{P}_x(X_t\in A,t<T)h(x)\,d\mu(x)}{\int_\epsilon^M \mathbb{P}_x(t<T)h(x)\,d\mu(x)}\right|$$

is bounded by the sum of the following two terms

$$I1=\left|\frac{\int \mathbb{P}_x(X_t\in A,t<T)h(x)\,d\mu(x)}{\int \mathbb{P}_x(t<T)h(x)\,d\mu(x)}-\frac{\int_\epsilon^M \mathbb{P}_x(X_t\in A,t<T)h(x)\,d\mu(x)}{\int \mathbb{P}_x(t<T)h(x)\,d\mu(x)}\right|,$$

$$I2=\left|\frac{\int_\epsilon^M \mathbb{P}_x(X_t\in A,t<T)h(x)\,d\mu(x)}{\int \mathbb{P}_x(t<T)h(x)\,d\mu(x)}-\frac{\int_\epsilon^M \mathbb{P}_x(X_t\in A,t<T)h(x)\,d\mu(x)}{\int_\epsilon^M \mathbb{P}_x(t<T)h(x)\,d\mu(x)}\right|.$$

We have the bound

$$I1\vee I2\le\frac{\int_0^\epsilon \mathbb{P}_x(t<T)h(x)\,d\mu(x)+\int_M^\infty \mathbb{P}_x(t<T)h(x)\,d\mu(x)}{\int \mathbb{P}_x(t<T)h(x)\,d\mu(x)}.$$

Thus, from Lemma 7.28, we get

$$\lim_{\epsilon\downarrow 0,M\uparrow\infty}\ \limsup_{t\to\infty}\left|\frac{\int \mathbb{P}_x(X_t\in A,t<T)h(x)\,d\mu(x)}{\int \mathbb{P}_x(t<T)h(x)\,d\mu(x)}\right.$$
$$\left.-\frac{\int_\epsilon^M \mathbb{P}_x(X_t\in A,t<T)h(x)\,d\mu(x)}{\int_\epsilon^M \mathbb{P}_x(t<T)h(x)\,d\mu(x)}\right|=0.$$

On the other hand, using (7.11), we have

$$\lim_{t\to\infty}\frac{\int_\epsilon^M \mathbb{P}_x(X_t\in A,t<T)h(x)\,d\mu(x)}{\int_\epsilon^M \mathbb{P}_x(t<T)h(x)\,d\mu(x)}=\frac{\int_A \psi_1(z)\,d\mu(z)}{\int_{\mathbb{R}^+}\psi_1(z)\,d\mu(z)}=\nu_1(A),$$

independently of $M>\epsilon>0$, and the result follows. □

The following corollary of Proposition 7.26 describes how fast the process comes down from infinity.

Corollary 7.29 *Assume* **H0** *and* **H5**. *Then for all* $\lambda<\lambda_1$,

$$\sup_{x>0}\mathbb{E}_x\left[e^{\lambda T}\right]<\infty.$$

Proof We have seen in Sect. 7.5 (Theorem 7.13) that for all $x > 0$, $\lim_{t\to\infty} e^{\lambda_1 t} \times \mathbb{P}_x(t < T) = \psi_1(x)\langle\psi_1, 1\rangle_\mu < \infty$, i.e. $\mathbb{E}_x[e^{\lambda T}] < \infty$ for all $\lambda < \lambda_1$. Applying Proposition 7.26 with $a = \lambda$ and the strong Markov property, it follows that $\sup_{x>y_\lambda} \mathbb{E}_x[e^{\lambda T}] < \infty$. Furthermore, thanks to the uniqueness of the solution of (7.1), $X_t^x \le X_t^{y_\lambda}$ a.s. for all $t > 0$ and all $x < y_\lambda$, hence $\mathbb{E}_x[e^{\lambda T}] \le \mathbb{E}_{y_\lambda}[e^{\lambda T}]$ for those x, completing the proof. □

The previous corollary states that the killing time for the process starting from infinity has exponential moments up to order λ_1.

The uniqueness of the QSD for birth-and-death chains coming down from infinity was studied in Theorem 5.5 in Sect. 5.3.

7.8 Examples

7.8.1 A Biological Model

The following singular diffusion appears in some biological models (for details, see Cattiaux et al. 2009):

$$dV_t = \sqrt{\sigma V_t}\, dB_t + h(V_t)\, dt. \tag{7.18}$$

We assume that $h \in C^1([0,\infty))$ and $h(0) = 0$. The main hypothesis on h is

$$\textbf{HH:}\quad \begin{cases} \text{(i)} & \lim\limits_{x\to\infty} \dfrac{h(x)}{\sqrt{x}} = -\infty, \\ \text{(ii)} & \lim\limits_{x\to\infty} \dfrac{xh'(x)}{h(x)^2} = 0. \end{cases}$$

In biological models, **HH** holds for any subcritical branching diffusion, and any logistic diffusion. Concerning assumption (i), the fact that h goes to $-\infty$ indicates strong competition in large populations, resulting in negative growth rates (as in the logistic case). Assumption (ii) is more technical, but it is fulfilled for most classical biological models.

Theorem 7.30 *Let V be the solution of* (7.18). *If h satisfies assumption* **HH**, *then for all initial laws with bounded support, the law of V_t conditioned on $\{V_t \neq 0\}$ converges exponentially fast to a probability measure ν, called the Yaglom limit.*

The law Q_x of the process V starting from x and conditioned to be never extinct exists and defines the so-called Q-process. This process converges, as $t \to \infty$, in distribution, to its unique invariant probability measure. This probability measure is absolutely continuous w.r.t. ν with a nondecreasing Radon–Nikodym derivative.

If, in addition, the following integrability condition is satisfied

$$\int_1^\infty \frac{dx}{-h(x)} < \infty,$$

then V comes down from infinity and the convergence of the conditional one-dimensional distributions holds for all initial laws, so that the Yaglom limit ν is the unique QSD.

7.8.2 Shiryaev Martingale

We came to this model in a correspondence with professor Shiryaev. The question was about the problem of a QSD in the context of the process

$$dX_t = X_t\,dB_t + dt, \qquad X_0 = x_0 > 0,$$

called *Shiryaev martingale*. Here there are two different cases. First, where the process is absorbed at $a > 0$ starting from $x > a$, and the more interesting case where the process is absorbed at 0 and a, starting from $x \in (0, a)$. Let us first explain what happens in the former case.

I. *QSD for* (a, ∞)

It is better to state this problem for a diffusion of the type

$$dY_t = dB_t - \alpha(Y_t)\,dt, \qquad Y_0 = y_0,$$

which is obtained by defining $Y_t = \log(X_t)$, and we get

$$dY_t = dB_t + \left(e^{-Y_t} - 1/2\right)dt, \qquad y_0 = \log(x_0).$$

In particular $\alpha(y) = 1/2 - e^{-y}$, and the state space considered is $(\log(a), \infty)$. For simplicity, in some places we shall denote $b = \log(a)$. In this situation, the theory of the QSD for Y holds under the hypothesis of Chap. 6, that is, when α satisfies Hypothesis **H**,

$$\int_b^\infty \int_b^x e^{\gamma(\xi)}\,d\xi\, e^{-\gamma(x)}\,dx = \int_b^\infty \int_b^x e^{-\gamma(\xi)}\,d\xi\, e^{\gamma(x)}\,dx = \infty,$$

where $\gamma(x) = 2\int_b^x \alpha(\psi)\,d\psi$. This hypothesis is that infinity is a natural boundary for Y and it prevents the process to explode to and from ∞. Also **H1** is satisfied, that is,

$$\Lambda(\infty) = \int_b^\infty e^{\gamma(y)}\,dy = \infty.$$

The spectrum of the infinitesimal generator of the process Y absorbed at b in $L^2(d\mu)$ is included in $(-\infty, -\underline{\lambda}]$ and this top of the spectrum $\underline{\lambda} = \underline{\lambda}(b)$ is characterized by the supremum of λ for which there is a nonnegative solution $u = u_{b,\lambda}$ to the eigenvector problem

$$\mathcal{L}^* u = -\lambda u, \qquad u(b) = 0, \qquad u'(b) = 1.$$

Here $\mathcal{L}^* u = \frac{1}{2}u'' + (\alpha u)'$. We impose $u(b) = 0$ because we want to absorb Y at b, and the drift is regular at b. In what follows, we denote by S_z the hitting time of z for the process Y.

According to Theorem 6.26, we have

(i) $0 \leq \underline{\lambda} = -\lim_{t\to\infty} \frac{\log(\mathbb{P}_{y_0}(t<S_b))}{t}$, that is, $\underline{\lambda}$ is the exponential tail parameter for the distribution of S_b, which does not depend on y_0.

(ii) For all $\lambda \in (0, \underline{\lambda}]$, the solution u_λ is nonnegative and integrable, and it defines a QSD, that is, if $d\nu_\lambda(y) = u_\lambda(y)\,dy / \int_b^\infty u_\lambda(z)\,dz$ then

$$\begin{aligned} \mathbb{P}_{\nu_\lambda}(t < S_b) &= e^{-\lambda t}, \\ \mathbb{P}_{\nu_\lambda}(Y_t \in dy, t < S_b) &= e^{-\lambda t}\, d\nu_\lambda(y), \quad \text{and, of course,} \\ \mathbb{P}_{\nu_\lambda}(Y_t \in A | t < S_b) &= \nu_\lambda(A). \end{aligned}$$

One of the main difficult points is to prove that $\underline{\lambda}(b) > 0$. We have studied in Sect. 6.5 this function as a function of b. This function is obviously increasing in b and it satisfies

$$\lim_{b\to\infty} \underline{\lambda}(b) = (1/2)^2/2 = 1/8 > 0$$

because asymptotically $\alpha \approx \alpha(\infty) = 1/2$, and for the constant drift $\alpha(\infty)$ the result is well known to be $\alpha(\infty)^2/2$. So, for large enough b, we know that $\underline{\lambda}(b) > 0$. The question is if for all b one has $\underline{\lambda}(b) > 0$. Moreover, for large b one has that $\underline{\lambda}(b) = 1/8$ (see Proposition 6.59, at least when $e^{-b}(b+3) < e^{-1}$, which gives, for example, that for $b \geq 3$ one is certain that $\underline{\lambda}(b) > 0$).

The conclusion that $\underline{\lambda}(b) > 0$ can be achieved by comparison with a two-valued drift $\tilde{\alpha}(y) = D\mathbb{1}_{b\leq y\leq 3} + \frac{1}{4}\mathbb{1}_{y>3}$. If $D < 0$ is sufficiently negative then $\alpha \geq \tilde{\alpha}$ and then absorption is faster for the process with drift α than for the one with drift $\tilde{\alpha}$. Then $\underline{\lambda}_\alpha(b) \geq \underline{\lambda}_{\tilde{\alpha}}(b)$ and the latter can be estimated to be strictly positive. It is also clear that $\lim_{b\to-\infty} \underline{\lambda}(b) = 0$. Once it is known that $\underline{\lambda}(b) > 0$, then Theorem 6.26 shows:

(iii) For any $y > b$ the Yaglom limit exists, that is,

$$\lim_{t\to\infty} \mathbb{P}_y(Y_t \in A | t < S_b) = \underline{\nu}(A),$$

where $\underline{\nu}$ is the probability measure whose density with respect to dy is proportional to the extremal $u_{\underline{\lambda}}$.

(iv) For any $y > b$ and any $s > 0$,

$$\lim_{t\to\infty} \mathbb{P}_y(Y_s \in A | t < S_b) = e^{\underline{\lambda} s}\mathbb{E}_y\left(\frac{u_{\underline{\lambda}}(Y_s)}{u_{\underline{\lambda}}(y)}, s < S_b\right),$$

which means that over $[0, s]$ the conditioned process converges to an h-process defined by $u_{\underline{\lambda}}$.

In terms of the original process X, the absorption rate is the same (because $T_a = S_b$), and there is a one parameter family of QSDs whose densities, with respect to dx, are proportional to

$$g_{a,\lambda}(x) = \frac{u_\lambda(\log(x))}{x}, \qquad x > a,$$

where $g_{a,\lambda}(a) = 0$ and $g'_{a,\lambda}(a) > 0$ for $\lambda \in (0, \underline{\lambda}(\log(a))]$. Each of these functions $g = g_{a,\lambda}$ verifies the second order o.d.e.

$$\frac{1}{2}\left(x^2 g\right)'' = g' - \lambda g,$$

or $(\mathcal{L}_X)^* g = -\lambda g$.

This process shows an interesting feature. For large values of a, the absorption rate is constant $1/8$, and for small values of a this absorption rate tends to 0.

II. *QSD in* $(0, a)$

As before we make a change of variable to get a diffusion of the type

$$dY_t = dW_t - \alpha(Y_t)\, dt, \qquad Y_0 = y_0,$$

which is obtained by defining $Y_t = -\log(X_t), dW_t = -dB_t$ to get

$$dY_t = dW_t - \left(e^{Y_t} - 1/2\right) dt, \qquad y_0 = -\log(x_0).$$

In particular, $\alpha(y) = e^y - 1/2$, and the state space considered is $(-\log(a), \infty)$ (as before $b = -\log(a)$). Contrary to the previous case, Hypothesis **H** is not verified. Nevertheless, α satisfies Hypotheses **H1**, **H2**, **H3**, **H4**, that is, α satisfies **H0**. On the other hand, it also satisfies **H5** and the measure $d\mu = e^{-\gamma}\, dx$ is finite. We summarize the results proved in this chapter.

(i) $-\mathcal{L}$ (and $-\mathcal{L}^*$) has a purely discrete spectrum $0 < \lambda_1 < \lambda_2 < \cdots$ on $L^2(d\mu)$. Furthermore, each λ_i is associated to a unique (up to a multiplicative constant) eigenfunction ψ_i of class $\mathcal{C}^2((b, \infty))$, which also belongs to $L^1(d\mu)$. This function satisfies the following o.d.e. on $(0, \infty)$:

$$\frac{1}{2}\psi_i'' - \alpha\psi_i' = -\lambda_i \psi_i.$$

(ψ_k) is an orthonormal basis of $L^2(d\mu)$, ψ_1 can be chosen nonnegative and if so, this eigenfunction is strictly positive on (b, ∞). We also have the pointwise representation for the transition densities for $Y_{t\wedge S_b}$, uniformly on compact sets of $(0, \infty) \times (b, \infty) \times (b, \infty)$:

$$r(t, x, y) = \sum_{k=1}^{\infty} e^{-\lambda_k t} \psi_k(x)\psi_k(y).$$

(ii) Therefore, on compact sets of $(0,\infty)\times(b,\infty)\times(b,\infty)$, we have

$$\lim_{t\to\infty} e^{\lambda_1 t} r(t,x,y) = \psi_1(x)\psi_1(y).$$

(iii) The probability measure $d\nu_1 = \psi_1\, d\mu / \int_b^\infty \psi_1(y)\, d\mu(y)$ is a QSD, namely, for every $t \geq 0$ and any Borel subset A of $(b,+\infty)$,

$$\mathbb{P}_{\nu_1}(Y_t \in A | t < S_b) = \nu_1(A),$$

and for any $y > b$ and any Borel subset A of $(b,+\infty)$,

$$\lim_{t\to\infty} e^{\lambda_1 t}\mathbb{P}_y(t < S_b) = \psi_1(y)\int \psi_1(z)\, d\mu(z),$$

$$\lim_{t\to\infty} e^{\lambda_1 t}\mathbb{P}_y(Y_t \in A, t < S_b) = \nu_1(A)\psi_1(y)\int \psi_1(z)\, d\mu(z),$$

$$\lim_{t\to\infty} \mathbb{P}_y(Y_t \in A | t < S_b) = \nu_1(A).$$

The probability measure ν_1 is the Yaglom limit distribution.

(iv) ν_1 is the unique limiting conditional distribution, that is,

$$\lim_{t\to\infty} \mathbb{P}_\nu(Y_t \in A | t < S_b) = \nu_1(A),$$

for any Borel set A and any initial distribution ν supported on (b,∞). In particular, ν_1 is the unique QSD.

As we can see, there is no mention of the boundary conditions of the eigenfunctions ψ_i, which are assumed to be in the domain of the infinitesimal generator of Y in $L^2(d\mu)$. The drift is regular at b, and one can show that $\psi_1(b) = 0$, $\psi_1'(b) > 0$. The exact value of this derivative is related to the normalization $\int_b^\infty \psi_1^2(z)\, d\mu(z) = 1$. In terms of the original diffusion, the eigenvalue problem is

$$\frac{1}{2}\big(x^2 g_\lambda\big)'' = g_\lambda' - \lambda g_\lambda$$

with $g_\lambda(a) = 0$ and $g_\lambda'(a) = 1$. Is not difficult to prove that for all $\lambda \in [0,\lambda_1]$ the associated eigenfunctions g_λ are nonnegative and integrable. In particular, g_0 is proportional to the density of an invariant measure (!) for the sub-Markov process $X_{t\wedge T_a}$. This is impossible because the process is absorbed at a and then $X_{t\wedge T_a}$ is loosing mass. So, what is the problem? To understand it, let us come back to the transformed process Y. For $\lambda = 0$, we have that

$$\rho(x) = \int_b^x e^{\gamma(z)}\, dz$$

is integrable on $[b,\infty)$ with respect to μ, that is, $1/r = \int_b^\infty \rho(x) e^{-\gamma(x)}\, dx < \infty$ and ρ verifies $\mathcal{L}\rho = 1/2\rho'' - \alpha\rho' = 0$. But a problem arises here, ρ does not belong to $L^2(d\mu)$, thus it is not in the domain of the generator in this space. On the other hand, in order to prove that $r\rho\, d\mu$ is invariant, that is,

$$\mathbb{P}_{r\rho\, d\mu}(Y_t \in A, t < S_b) = r \int_A \rho\, d\mu,$$

one needs to use integration by parts for $h(y) = \rho(y)e^{-\gamma(y)}$ which satisfies the equation $1/2h'' + (\alpha h)' = 0$. This computation gives an extra positive term at ∞, representing the flux there and is given by

$$\left(h' + 2\alpha h\right)\Big|_{\infty} = h'(b) + 2\alpha(b)h(b) = h'(b) > 0.$$

The conclusion is that $r\rho\, d\mu$ is not invariant.

Recapitulating, to have an invariant measure, it is not enough to solve the o.d.e. $(x^2 g)'' = 2g'$ with Dirichlet conditions at a, using positivity and integrability, because in order that g be in the domain of the infinitesimal generator one typically needs extra hypotheses at 0 which, in general, are difficult to get, especially in the case where the coefficient of the diffusion term produces a singularity at 0. Finally, there is a unique $\lambda_1 > 0$ such that $(x^2 g)'' = 2g' - 2\lambda_1 g$ has a nonnegative, integrable solution, which is also a QSD. This value of λ_1 satisfies

$$\lim_{t\to\infty} e^{\lambda_1 t}\mathbb{P}_x(t < T_a) = g(x) \int_0^a g(z)\, dz,$$

for all $x \in (0, a)$. Of course, $\lambda_1 = \lambda_1(a)$ and similarly $g = g_a$, which we recall satisfies the normalization $\int_0^a g^2(x)\, dx = 1$.

Chapter 8
Dynamical Systems

In this chapter, we study the existence of a QSD for a dynamical system with a trap. The result for subshifts of finite type with a Markovian trap is given in Theorem 8.20, Sect. 8.2. In Sect. 8.3, we study dynamical systems with a trap in the context of differential systems with some expanding conditions. The Pianigiani–Yorke QSD is given in Theorem 8.32. The main tool we use, in both cases, is the spectral analysis of a Ruelle–Perron–Frobenius operator.

We recall (see Sect. 1.4) that a (discrete time) dynamical system is given by a map $f : \mathcal{X} \to \mathcal{X}$. The map f describes the evolution of the state during one unit of time, and the orbit of an initial condition is obtained by applying f recursively.

For a Borel subset $B \subseteq \mathcal{X}$, and an initial condition $x \in \mathcal{X}$, we denote by $T_B(x)$ the first time the orbit reaches B. We consider a fixed forbidden region called a trap and denoted by $\mathcal{X}^{\mathrm{tr}}$. The time to reach this trap is denoted by $T = T_{\mathcal{X}^{\mathrm{tr}}}$. The formalism of this chapter also covers the case where $f : \mathcal{X}_0 \to \mathcal{X}$ with $\mathcal{X}_0$ a strict subset of $\mathcal{X}$. In this case, we put $\mathcal{X}^{\mathrm{tr}} = \mathcal{X} \setminus \mathcal{X}_0$.

We denote by $\mathcal{X}_n = \{T > n\}$ the set of initial conditions whose orbit has not reached $\mathcal{X}^{\mathrm{tr}}$ up to time n. For a given probability measure ν_0, we are interested in quantities like

$$\nu_0(T > n) = \nu_0(\mathcal{X}_n),$$

and

$$\nu_0\big(f^{-n}(B) | T > n\big) = \frac{\nu_0(f^{-n}(B) \cap \mathcal{X}_n)}{\nu_0(\mathcal{X}_n)},$$

where B is a measurable subset of $\mathcal{X}$.

8.1 Some General Results

We start by deriving some simple, but useful lemmas.

P. Collet et al., *Quasi-Stationary Distributions*, Probability and Its Applications, DOI 10.1007/978-3-642-33131-2_8, © Springer-Verlag Berlin Heidelberg 2013

Lemma 8.1 *For any $n \geq 1$ we have*

$$\mathcal{X}_n = f^{-1}(\mathcal{X}_{n-1}) \cap \mathcal{X}_0 = \mathcal{X}_{n-1} \cap f^{-n}(\mathcal{X}_0) = \mathcal{X}_{n-1} \cap f^{-n+1}(\mathcal{X}_1).$$

For any $k \leq n$,

$$f^k(\mathcal{X}_n) \subseteq \mathcal{X}_{n-k}$$

and

$$f^k(\mathcal{X}_n \backslash \mathcal{X}_{n+1}) \subseteq \mathcal{X}_{n-k} \backslash \mathcal{X}_{n-k+1}.$$

Proof The first two statements follow immediately from the definition of $\mathcal{X}_n$. To prove the third result, assume there exists $x \in \mathcal{X}_n \backslash \mathcal{X}_{n+1}$ such that $f^k(x) \in \mathcal{X}_{n-k+1}$. Then for any $k+1 \leq j \leq n+1$ we have $f^j(x) = f^{j-k}(f^k(x)) \in \mathcal{X}_0$, and $f^j(x) \in \mathcal{X}_0$ for $0 \leq j \leq k \leq n$ since $x \in \mathcal{X}_n$. Therefore, $x \in \mathcal{X}_{n+1}$, and we have a contradiction. Together with the second assertion, this proves the lemma. □

Lemma 8.2 *Assume the measure ν satisfies*

$$\frac{\nu(f^{-1}(B) \cap \mathcal{X}_1)}{\nu(\mathcal{X}_1)} = \nu(B), \tag{8.1}$$

for any Borel set $B \subseteq \mathcal{X}_0$. Then ν is a QSD.

Proof The proof is recursive. Let $n \geq 1$, and assume that we have already proved that

$$\frac{\nu(f^{-n}(B) \cap \mathcal{X}_n)}{\nu(\mathcal{X}_n)} = \nu(B),$$

for any Borel set $B \subseteq \mathcal{X}_0$. We have

$$\mathcal{X}_{n+1} = \mathcal{X}_n \cap f^{-n-1}(\mathcal{X}_0) = \mathcal{X}_n \cap f^{-n}(\mathcal{X}_1).$$

Therefore, using the recursion hypothesis, we get

$$\begin{aligned}\frac{\nu(f^{-n-1}(B) \cap \mathcal{X}_{n+1})}{\nu(\mathcal{X}_{n+1})} &= \frac{\nu(f^{-n-1}(B) \cap f^{-n}(\mathcal{X}_1) \cap \mathcal{X}_n)}{\nu(f^{-n}(\mathcal{X}_1) \cap \mathcal{X}_n)} \\ &= \frac{\nu(f^{-n}(f^{-1}(B) \cap \mathcal{X}_1) \cap \mathcal{X}_n)}{\nu(f^{-n}(\mathcal{X}_1) \cap \mathcal{X}_n)} \\ &= \frac{\nu(f^{-1}(B) \cap \mathcal{X}_1)\nu(\mathcal{X}_n)}{\nu(\mathcal{X}_1)\nu(\mathcal{X}_n)} = \nu(B).\end{aligned}$$ □

We now recall a lemma about the exponential decay of the probability of survival in a QSD. The proof in the case of dynamical systems is particularly simple.

Lemma 8.3 *Let* ν *be a QSD for the map* f *and the trap* $\mathcal{X}^{\mathrm{tr}}$. *With the above notations, we have*

$$\nu(\mathcal{X}_n) = \nu(\mathcal{X}_1)^n = \nu(T > n).$$

Proof The proof is recursive. The lemma is trivially true for $n = 0$ and $n = 1$. Assume the result holds up to $n \geq 1$. We have from the definition of a QSD

$$\nu(\mathcal{X}_{n+1}) = \nu\big(\mathcal{X}_n \cap f^{-n}(\mathcal{X}_1)\big) = \nu(\mathcal{X}_n)\nu(\mathcal{X}_1),$$

and the result follows. □

We now give a general result on the set theoretic support of the QSD measures. We define the set $\tilde{\mathcal{X}}_0$ by

$$\tilde{\mathcal{X}}_0 = \bigcap_{n=0}^{\infty} f^n(\mathcal{X}_n).$$

It is easy to see that

$$\tilde{\mathcal{X}}_0 = \big\{x \in \mathcal{X}_0 : \forall n, \exists (y_0, \ldots, y_n) \in \mathcal{X}_0^{n+1},\ y_0 = x,\ f(y_j) = y_{j-1}\ \forall 1 \leq j \leq n\big\}.$$

Proposition 8.4 *If* ν *is a QSD, then*

$$\nu\big(\tilde{\mathcal{X}}_0^c \cap \mathcal{X}_0\big) = 0.$$

Proof If ν is a QSD, we have for any measurable subset B of $\mathcal{X}_0$ and for any integer $n \geq 1$

$$\nu(B)\nu(\mathcal{X}_n) = \nu\big(f^{-n}(B) \cap \mathcal{X}_n\big) \leq \nu\big(f^{-n}\big(B \cap f^n(\mathcal{X}_n)\big)\big). \tag{8.2}$$

We now observe that the sequence of sets $(f^n(\mathcal{X}_n))$ is decreasing since

$$f^n(\mathcal{X}_n) = f^{n-1}\big(f(\mathcal{X}_n)\big) \subseteq f^{n-1}(\mathcal{X}_{n-1}).$$

Therefore, using $B = \mathcal{X}_0 \cap (f^n(\mathcal{X}_n))^c$ in (8.2), we have

$$\nu\big(\mathcal{X}_0 \cap f^n(\mathcal{X}_n)^c\big) \leq \frac{1}{\nu(\mathcal{X}_n)} \nu\big(f^{-n}\big(\mathcal{X}_0 \cap f^n(\mathcal{X}_n)^c \cap f^n(\mathcal{X}_n)\big)\big) = 0.$$

The proposition follows immediately. □

The next proposition is a kind of reciprocal of the previous one (see Demers and Young 2006).

Proposition 8.5 *Assume there exists a sequence* (y_n) *in* $\mathcal{X}_0$ *satisfying for any* $n \geq 1$ $f(y_n) = y_{n-1}$, *and* $f(y_0) \notin \mathcal{X}_0$. *Then there exists a QSD.*

Proof We will, in fact, construct a continuum of QSDs Fix $\rho \in (0, 1)$, and define the atomic measure

$$\nu = \frac{1}{1-\rho} \sum_{n=0}^{\infty} \rho^n \delta_{y_n},$$

where δ_{y_n} is the Dirac measure in y_n.

For any measurable subset B of $\mathcal{X}_0$, we have

$$\nu\big(f^{-1}(B) \cap \mathcal{X}_1\big) = \frac{1}{1-\rho} \sum_{n=1}^{\infty} \rho^n \mathbb{1}_{f^{-1}(B)}(y_n) = \frac{1}{1-\rho} \sum_{n=1}^{\infty} \rho^n \mathbb{1}_B\big(f(y_n)\big)$$

$$= \frac{1}{1-\rho} \sum_{n=1}^{\infty} \rho^n \mathbb{1}_B(y_{n-1}) = \rho\nu(B).$$

The proposition follows immediately. □

We define the set

$$\mathcal{X}_\infty = \bigcap_{n=0}^{\infty} \mathcal{X}_n.$$

This is the set of initial conditions whose trajectory stays forever in $\mathcal{X}_0$. We have $f(\mathcal{X}_\infty) \subseteq \mathcal{X}_\infty$, and if ν is a QSD, $\nu(\mathcal{X}_\infty) = 0$.

For the existence of a sequence (y_n) as in the previous proposition, we can state the following result.

Proposition 8.6 *Assume $\mathcal{X}_0$ is closed and for any n, $\mathcal{X}_0 \backslash \mathcal{X}_n$ is contained in a compact subset of $\mathcal{X}_\infty^c$, and $\mathcal{X}_n \backslash \mathcal{X}_{n+1} \neq \emptyset$. Then there exists a QSD.*

Note that if for some integer n, $\mathcal{X}_n \backslash \mathcal{X}_{n+1} \neq \emptyset$ (this means that all orbits "die" in a uniformly bounded time) then from Lemma 8.3 it follows that there cannot exist any QSD.

Proof By Lemma 8.1, for any n we have

$$f^n(\mathcal{X}_n \backslash \mathcal{X}_{n+1}) \subseteq \mathcal{X}_0 \backslash \mathcal{X}_1$$

and

$$f^{n+1}(\mathcal{X}_{n+1} \backslash \mathcal{X}_{n+2}) = f^n\big(f(\mathcal{X}_{n+1} \backslash \mathcal{X}_{n+2})\big) \subseteq f^n(\mathcal{X}_n \backslash \mathcal{X}_{n+1}).$$

Moreover, by hypothesis, we have for any n

$$f^n(\mathcal{X}_n \backslash \mathcal{X}_{n+1}) \neq \emptyset.$$

Let (x_n^0) be a sequence of points such that for any n

$$x_n^0 \in f^n(\mathcal{X}_n \backslash \mathcal{X}_{n+1}).$$

Therefore, for each n we can find points $x_n^1, \dots, x_n^n$ with $x_n^k \in \mathcal{X}_k \backslash \mathcal{X}_{k+1}$ such that for each $1 \leq k \leq n$

$$f(x_n^k) = x_n^{k-1}.$$

Since for each k, $\mathcal{X}_k \backslash \mathcal{X}_{k+1}$ is contained in a compact set, we can apply Cantor's diagonal procedure to the double sequence (x_n^k). Namely, there exists a sequence (n_j) of integers such that $(x_{n_j}^0)$ converges to a point $z^0 \in \overline{\mathcal{X}_0 \backslash \mathcal{X}_1}$. We can now find a second sequence (j_ℓ) of integers with $j_1 = 1$ such that the sequence $(x_{n_{j_\ell}}^1)$ also converges to a point $z^1 \in \overline{\mathcal{X}_1 \backslash \mathcal{X}_2}$. Note that by continuity of f we have $f(z^1) = z^0$. Continuing this procedure, we construct a sequence (z^r) such that for any $r \geq 0$

$$z^r \in \overline{\mathcal{X}_r \backslash \mathcal{X}_{r+1}}$$

and

$$f(z^{r+1}) = z^r.$$

Note that since $\mathcal{X}_0$ is closed we have $(z^r) \subseteq \mathcal{X}_0$. We now claim that there is a finite integer r such that $f^r(z^0) \notin \mathcal{X}_0$. Assume the opposite, then $z^0 \in \mathcal{X}_n$ for all n, and in particular, $z^0 \in \mathcal{X}_\infty$. Since all the $\mathcal{X}_n$ are closed, $\mathcal{X}_\infty$ is also closed. On the other hand, by hypothesis, $\mathcal{X}_0 \backslash \mathcal{X}_1$ is contained in a compact set disjoint from $\mathcal{X}_\infty$, and we have a contradiction. □

8.1.1 Counterexamples

It is easy to construct various examples where QSDs do not exist. For example, if $\mathcal{X}_n = \emptyset$ for some n, the existence of a QSD would contradict Lemma 8.3. As an example of this situation, we consider for $\mathcal{X}$ the unit circle $\mathbb{T}$, and for f the rotation by an angle α irrational with respect to $2/\pi$. Let the trap be an open interval $I =]\vartheta_1, \vartheta_2[$ $(\vartheta_1 \neq \vartheta_2)$. We claim that the function T_I is bounded. Note that f is invertible, and its inverse f^{-1} is the rotation by the angle $-\alpha$. We claim that the countable sequence of open intervals $(f^{-n}(I))$ is an open covering of $\mathbb{T}$. This follows at once from the fact that every orbit is dense (see, for example, Robinson 2008). Since $(f^{-n}(I))$ is a covering of the compact set $\mathbb{T}$ by open sets, we can extract a finite cover. This implies that T_I is bounded, and, by Lemma 8.3, no QSD can exist.

8.2 Gibbs Measures and QSDs

In this section, we will construct QSDs which are analogs to the Gibbs states of statistical mechanics. We first introduce some notations and recall some definitions and results about Gibbs measures which will be useful later on.

8.2.1 Notations and Gibbs Measures

We will denote by $\mathscr{A}$ a finite alphabet. There is no loss of generality to assume that $\mathscr{A}$ is

$$\mathscr{A} = \{1, \ldots, |\mathscr{A}|\},$$

where $|\mathscr{A}|$ is the cardinal of the alphabet.

For $\underline{S} \in \mathscr{A}^{\mathbb{Z}}$ and for two integers $p \le q$ (p may be $-\infty$ and q may be $+\infty$), it is convenient to denote by $\underline{S}_p^q$ the sequence $(S_p, \ldots, S_q)$ (more generally, we denote by a_p^q a sequence $(a_p, \ldots, a_q)$ of elements of $\mathscr{A}$). If $\underline{S}$ and $\underline{S}'$ are two elements of $\mathscr{A}^{\mathbb{Z}}$, and if $p \le q$ and $r \le s$ are four integers with q and r finite (p may be $-\infty$ and s may be $+\infty$), we will define the concatenated sequence $\underline{S}_p^q \bullet \underline{S}'^s_r$ by

$$\left(\underline{S}_p^q \bullet \underline{S}'^s_r\right)_j = \begin{cases} S_j & \text{for } p \le j \le q, \\ S'_{j+r-q-1} & \text{for } q+1 \le j \le s-r+q+1. \end{cases}$$

For a finite sequence a_p^q (recall p and q are integers with $p \le q$), we denote by $C_{a_p^q}$ the associated cylinder set (or shortly, the cylinder) which is the subset of $\mathscr{A}^{\mathbb{Z}}$ given by

$$C_{a_p^q} = \left\{\underline{S} \in \mathscr{A}^{\mathbb{Z}} : S_j = a_j \text{ for } p \le j \le q\right\}.$$

We now define a metric d on $\mathscr{A}^{\mathbb{Z}}$ as follows. For $\underline{S}$ and $\underline{S}'$ in $\mathscr{A}^{\mathbb{Z}}$, let

$$\delta(\underline{S}, \underline{S}') = \min\left\{|k| : S_k \ne S'_k\right\}.$$

In words, $\delta(\underline{S}, \underline{S}')$ is the minimal distance to the origin of a position where the two sequences differ. We now choose a real number $\rho \in]0, 1[$ and define the distance d by

$$d(\underline{S}, \underline{S}') = e^{\delta(\underline{S}, \underline{S}')}.$$

It is left to the reader to verify that d is a metric (and even an ultrametric). Moreover, equipped with this metric, $\mathscr{A}^{\mathbb{Z}}$ is a compact set. The metric d depends on the number $\rho \in]0, 1[$. It is left to the reader to verify that the topology does not depend on ρ. In the sequel, we will fix $\rho = 1/2$ once and for all, but the qualitative results are independent of ρ. A cylinder set is open and closed in this metric. We will denote by $\mathcal{C}(\mathscr{A}^{\mathbb{Z}})$ the set of continuous functions on $\mathscr{A}^{\mathbb{Z}}$. This is a Banach space when equipped with the sup norm (see, for example, Dunford and Schwartz 1958). It is easy to verify that the characteristic function of a cylinder set is continuous.

We recall that a function U defined on $\mathscr{A}^{\mathbb{Z}}$ is α-Hölder ($\alpha \in]0, 1]$) if there exists a constant $C > 0$ such that for any $\underline{S}$ and $\underline{S}'$ we have

$$\left|\mathrm{U}(\underline{S}) - \mathrm{U}(\underline{S}')\right| \le C d(\underline{S}, \underline{S}')^{\alpha}.$$

In the sequel, we will denote by $\mathcal{H}_\alpha(\mathscr{A}^{\mathbb{Z}})$ the set of α-Hölder functions. This is a Banach space when equipped with the norm (see Dunford and Schwartz 1958)

$$\|\mathrm{U}\|_{\mathcal{H}_\alpha(\mathscr{A}^{\mathbb{Z}})} = \sup_{\underline{S}\neq\underline{S}'} \frac{|\mathrm{U}(\underline{S}) - \mathrm{U}(\underline{S}')|}{d(\underline{S}, \underline{S}')^\alpha} + \sup_{\underline{S}} \big|\mathrm{U}(\underline{S})\big|.$$

More generally, we will use the notation $\mathcal{C}(F)$ and $\mathcal{H}_\alpha(F)$ for the set of continuous functions and for the set of α-Hölder continuous functions of a metric space F. If F is compact, these spaces are Banach spaces for the sup norm and for the $\mathcal{H}_\alpha$ norm, respectively.

Let M be a matrix of size $|\mathscr{A}| \times |\mathscr{A}|$ with entries equal to zero or one. We will denote by $\mathcal{X}^M$ the subset of $\underline{S} \in \mathscr{A}^{\mathbb{Z}}$ defined by

$$\mathcal{X}^M = \big\{\underline{S} \in \mathscr{A}^{\mathbb{Z}} : \forall j \in \mathbb{Z}, M(S_j, S_{j+1}) = 1\big\}.$$

In other words, the matrix M describes which letter can follow a given letter. One can also think that the matrix M describes very unsophisticated grammatical rules. It is left to the reader to verify that $\mathcal{X}^M$ is closed, and hence compact. A finite sequence a_p^q (with $q > p$) will be called M compatible (or M compatible) if

$$M(a_j, a_{j+1}) = 1$$

for all $p \le j \le q-1$. It follows immediately from this definition that if a_p^q is a finite sequence of elements of the alphabet $\mathscr{A}$, we have $C_{a_p^q} \cap \mathcal{X}^M \neq \emptyset$ if and only if a_p^q is M compatible. In the sequel, we will only consider cylinder sets defined by M compatible sequences.

In the sequel, we will always assume that the matrix M is irreducible and aperiodic, namely, there is an integer $k_M \ge 0$ such that M^{k_M} has all its entries nonzero. This has the following important consequence.

Lemma 8.7 *Let $p \le q$ and $r \le s$ with $r \ge k_M + q$. Then for any pair of M compatible sequences a_p^q and b_r^s, we can find an M compatible sequence c_r^s such that*

$$c_p^q = a_p^q, \quad \textit{and} \quad c_r^s = b_r^s.$$

The proof is left to the reader.

The shift map (shift, for short) $\sigma : \mathscr{A}^{\mathbb{Z}} \to \mathscr{A}^{\mathbb{Z}}$ is defined on by

$$\forall j \in \mathbb{Z}: \quad (\sigma \underline{S})_j = S_{j+1}.$$

It is left to the reader to verify that the shift map is continuous and the subset $\mathcal{X}^M$ is invariant. Moreover, σ is invertible on $\mathcal{X}^M$, and its inverse is the shift in the other direction. The dynamical system given by the metric space $\mathcal{X}^M$ and the shift is called a subshift of finite type (the subshift of finite type defined by the matrix M). It follows easily from the fact that M is irreducible and aperiodic that the subshift of finite type defined by M is topologically strongly mixing (see Denker et al. 1976 for the definition).

This dynamical system has, in general, many invariant measures. An interesting class of invariant measures are the Gibbs measures (Gibbs states) coming from statistical mechanics of one dimensional spin systems (see Bowen 1975 or Ruelle 1978). They are defined as follows. Let U be a (real) Hölder continuous function.

Definition 8.8 A measure μ on $\mathcal{X}^M$ is a Gibbs measure if

(i) μ is shift invariant;
(ii) There exist two constants P and $C > 1$ such that for any M compatible cylinder set $C_{a_r^q}$ and any $\underline{S} \in C_{a_r^q}$

$$\frac{1}{C} \le \mu(C_{a_r^q}) e^{P(q-r)} e^{-\Phi_r^q(\underline{S})} \le C$$

where

$$\Phi_r^q(\underline{S}) = \sum_{j=r}^{q} \mathrm{U}(\sigma^j \underline{S}).$$

The number P is called the pressure of the potential U by analogy with statistical mechanics of one dimensional discrete spin systems (we refer to Bowen 1975 and Ruelle 1978 for more details and properties of the Gibbs measures). The reader is invited to consider the case where the potential U depends only on the two symbols S_0 and S_1 and to make a connection with Markov chains on a finite state space.

In the sequel, we will assume that the potential $\mathrm{U}(\underline{S})$ depends only on the symbols S_j for $j \ge 0$. This is not a real restriction due to a theorem of Ruelle. We refer to Bowen (1975) and Ruelle (1978) for the details.

A main tool in the study of the Gibbs measures described above is the Ruelle–Perron–Frobenius operator $\mathscr{P}_M$. We will denote by

$$\mathcal{X}^{M,+} = \{\underline{S} \in \mathscr{A}_+^{\mathbb{Z}} : \forall j \in \mathbb{Z}_+, M(S_j, S_{j+1}) = 1\}.$$

The set of M compatible sequences a_0^∞. Note that this is a compact subset which is invariant by the shift, but σ is not invertible on this set. The dynamical system defined by $\mathcal{X}^{M,+}$ and σ is called a unilateral subshift of finite type. The operator $\mathscr{P}_M$ acting on a function $g \in \mathcal{C}(\mathcal{X}^{M,+})$ is defined by

$$(\mathscr{P}_M g)(\underline{S}) = \sum_{\substack{a \in \mathscr{A} \\ M(a, S_0) = 1}} g(a \bullet \underline{S}) e^{\mathrm{U}(a \bullet \underline{S})},$$

where $\underline{S} \in \mathcal{X}^{M,+}$. Note that the sum is over the symbols $a \in \mathscr{A}$ such that the sequence $a \bullet \underline{S} = (a, S_0, S_1, \ldots)$ is M compatible. It is left to the reader to verify that if the potential U is an α-Hölder function, then $\mathscr{P}_M$ is a bounded operator in $\mathcal{H}_\alpha$ (and in any $\mathcal{H}_\beta$ with $\beta \in]0, \alpha]$). A better notation for the Ruelle–Perron–Frobenius operator associated to the potential U should be $\mathscr{P}_{M,\mathrm{U}}$. However, in the sequel, we will fix the potential once and for all and vary only the matrix M. This is why we alleviated the notation by not mentioning explicitly the potential. The following formula is an easy consequence of the definition

$$\mathscr{P}_M(hg \circ \sigma) = g\mathscr{P}_M(h). \tag{8.3}$$

The Gibbs measure for a Hölder potential is constructed out of two pieces which are given by the following theorem.

Theorem 8.9 *For a given potential* $\mathrm{U} \in \mathcal{H}_\alpha$ *on* $\mathcal{X}^{M,+}$ ($\alpha \in]0,1]$), *there exist a unique real number* P *and a unique probability measure* ζ *on* $\mathcal{X}^{M,+}$ *satisfying*

$$\mathscr{P}_M^\dagger \zeta = e^P \zeta$$

or in other words,

$$\int (\mathscr{P}_M g)\, d\zeta = e^P \int g\, d\zeta,$$

for any continuous function g *on* $\mathcal{X}^{M,+}$. *There is also a unique positive continuous function* φ *on* $\mathcal{X}^{M,+}$ *such that*

$$\mathscr{P}_M \varphi = e^P \varphi$$

and

$$\int \varphi\, d\zeta = 1.$$

The measure $d\mu = \varphi\, d\zeta$ *given on the cylinder set* $C_{a_0^p}$ *of* $\mathcal{X}^{M,+}$ *by*

$$\begin{aligned}\mu(C_{a_0^p}) &= e^{-(p+1)P} \int_{\mathcal{X}^{M,+}} \mathscr{P}_M^{p+1}(\varphi \mathbb{1}_{C_{a_0^p}})\, d\zeta \\ &= e^{-(p+1)P} \int_{\mathcal{X}^{M,+}} \varphi\big(a_0^p \bullet S_0^\infty\big) \prod_{j=0}^{p} e^{\mathrm{U}(a_j^p \bullet S_0^\infty)}\, d\zeta\big(S_0^\infty\big)\end{aligned}$$

is invariant with respect to the shift, it is also exponentially mixing on functions in $\mathcal{H}_\beta(\mathcal{X}^{M,+})$ *for* $\beta \in]0,\alpha[$. *This measure is a Gibbs measure and satisfies for any* $g \in \mathcal{C}(\mathcal{X}^{M,+})$

$$e^{-P} \int \frac{\mathscr{P}_M(g\varphi)}{\varphi}\, d\mu = \int g\, d\mu.$$

This theorem is often called the Ruelle–Perron–Frobenius Theorem. The measure ζ is often called a transverse measure. We refer to Bowen (1975) and Ruelle (1978) for the proofs and definition of mixing. Recall (see the previous references) that a probability measure on $\mathcal{X}^{M,+}$ is completely determined by its values on the cylinder sets $(C_{a_0^p})$.

It is easy to extend the above measure μ to a Gibbs measure on $\mathcal{X}^M$. Namely, for any $p < q$ with $p < 0$, we define for a_p^q M compatible

$$\mu(C_{a_p^q}) = \mu\big(\sigma^p C_{a_p^q}\big).$$

It is left to the reader to verify that this measure satisfies all the assumptions of a Gibbs measure, and is exponentially mixing (see Bowen 1975 and Ruelle 1978).

The Ruelle–Perron–Frobenius operator has some analogy with the transition probability for Markov chains. Let (Ω, f) be a dynamical system. We denote by $\mathscr{K}$ the Koopman operator defined by

$$\mathscr{K}(g) = g \circ f.$$

This operator defines the time evolution on the observables, and is analogous to the action of the Markov transition matrix on functions (see also the definition of the semigroup associated to a Markov process in (2.17) of Sect. 2.6). If one were to find an invariant measure, one would have to look at the adjoint of this operator. Assume now that a measure μ_0 is such that the image measure $\mu_0 \circ f^{-1}$ (f^{-1} denotes the set theoretic inverse) is absolutely continuous with respect to μ_0 with density h_0. In this case, we may look for an invariant measure absolutely continuous with respect to μ_0. For any (square integrable) functions g_1 and g_2, we have by an easy computation, assuming that the equation $f(y) = x$ has only countably many solutions for any x,

$$\int g_2 \mathscr{K}(g_1)\, d\mu = \int g_2 g_1 \circ f\, d\mu \int \mathscr{L}(g_2) g_1\, d\mu$$

where

$$\mathscr{L}(g)(x) = \sum_{y,\, f(y)=x} h_0(y) g(y).$$

A nonnegative eigenfunction of eigenvalue 1 of $\mathscr{L}$ will be the density with respect to μ_0 of an invariant measure.

8.2.2 Gibbs QSD

We now consider a matrix M' with the same dimension as M and with entries also equal to zero or one. We denote by P', ζ' and φ' the elements given by Theorem 8.9 for the operator $\mathscr{P}_{M'}$ acting on continuous functions with domain $\mathcal{X}^{M',+}$.

From now on, we assume that M "dominates" M' in the following sense

$$M(i, j) = 0 \quad \Longrightarrow \quad M'(i, j) = 0.$$

In other words, M' allows fewer transitions than M. We will also assume that M' is irreducible and aperiodic, namely, there exists an integer $k_{M'}$ such that the matrix

$M'^{k_{M'}}$ has all its entries nonzero. Obviously, this implies that M itself is irreducible and aperiodic.

The results below can probably be extended to more general cases. We now define the trap $\mathcal{X}^{\mathrm{tr}}$ the subset of $\mathcal{X}^M$ given by

$$\mathcal{X}^{\mathrm{tr}} = \{\underline{S} \in \mathcal{X}^M : M'(S_0, S_1) = 0\}.$$

Recall that if $\underline{S} \in \mathcal{X}^M$, we have $M(S_0, S_1) = 1$, so $\mathcal{X}^{\mathrm{tr}}$ is the subset of sequences in $\mathcal{X}^M$ such that the pair of symbols in position zero and one is prohibited by the matrix M' although allowed by M. We now want to explore the existence of a Yaglom limit for this trap and initial Gibbs measure μ associated to the α-Hölder potential U. We will first work with the unilateral shift on $\mathcal{X}^{M,+}$. Recall that we assumed the potential U is defined on this set. The Ruelle–Perron–Frobenius operator $\mathscr{P}_{M'}$ will, of course, play a role, and we can apply Theorem 8.9 to this operator. There is, however, another useful operator $\mathscr{P}_{M',M}$ whose action on a function $g \in C(\mathcal{X}^{M,+})$ is given by

$$\mathscr{P}_{M',M}\, g(\underline{S}) = \sum_{\substack{a \in \mathscr{A} \\ M'(a,S_0)=1}} g(a \bullet \underline{S}) e^{\mathrm{U}(a \bullet \underline{S})}.$$

It is left to the reader to verify that this is a bounded operator on $C(\mathcal{X}^{M,+})$. Note also that $C(\mathcal{X}^{M',+})$ is invariant by $\mathscr{P}_{M',M}$, and on this space this operator coincides with $\mathscr{P}_{M'}$. Moreover, for any integer $p \geq 1$,

$$\mathscr{P}^p_{M',M}\, g(\underline{S}) = \sum_{\substack{a_0^{p-1}\ M'\text{compatible} \\ M'(a_{p-1},S_0)=1}} g\big(a_0^{p-1} \bullet \underline{S}\big) \prod_{j=0}^{p-1} e^{\mathrm{U}(a_j^{p-1} \bullet \underline{S})}. \tag{8.4}$$

We have, of course, with the notations of Sect. 1.4 $\mathcal{X}_0 = \mathcal{X}^{M,+} \backslash \mathcal{X}^{\mathrm{tr}}$, and more generally,

$$\mathcal{X}_n = \{\underline{S} \in \mathcal{X}^{M,+} : M'(S_j, S_{j+1}) = 1, \text{for } 0 \leq j \leq n\}.$$

It follows immediately from (8.3) and the last statement of Theorem 8.9 that if $g \in C(\mathcal{X}^{M,+})$, we have

$$\int \mathbb{1}_{\mathcal{X}_n} g \circ \sigma^n\, d\mu = \int \prod_{j=0}^{n} \mathbb{1}_{\mathcal{X}_0} \circ \sigma^j g \circ \sigma^n\, d\mu = e^{-nP} \int_{\mathcal{X}_0} \big(\mathscr{P}^n_{M',M}\varphi\big) g\, d\zeta. \tag{8.5}$$

We see with this formula that it is interesting to investigate the spectral properties of the operator $\mathscr{P}_{M',M}$. We will first derive two useful lemmas.

Lemma 8.10 *Assume the potential* U *is α-Hölder for some $\alpha \in]0, 1]$. Then there exists a constant $D > 0$ such that for any $\underline{x}, \underline{x}' \in \mathcal{X}^{M,+}$, for any integer $p \geq 1$,* *and for any finite sequence* a_0^{p-1}, M *compatible and satisfying* $M(a_{p-1}, \varkappa_0) = M(a_{p-1}, \varkappa'_0) = 1$, *we have*

$$\frac{\prod_{j=0}^{p-1} e^{\mathrm{U}(a_j^{p-1} \bullet \underline{\varkappa})}}{\prod_{j=0}^{p-1} e^{\mathrm{U}(a_j^{p-1} \bullet \underline{\varkappa}')}} \le e^{Dd(\underline{\varkappa}, \underline{\varkappa}')^\alpha}.$$

The proof is left to the reader.

Lemma 8.11 *Assume the matrix* M' *is irreducible and aperiodic and denote by* $k_{M'}$ *the smallest integer such that the matrix* $M'^{k_{M'}}$ *has all its entries nonzero. Assume the potential* U *is* α*-Hölder for some* $\alpha \in]0, 1]$. *Then there exists a constant* $C > 0$ *such that for any* $\underline{S}, \underline{S}' \in \mathcal{X}^{M,+}$ *and any integer* p *we have*

$$\frac{\mathscr{P}^p_{M',M} 1(\underline{S})}{\mathscr{P}^p_{M',M} 1(\underline{S}')} \le e^{Cd(\underline{S}, \underline{S}')^\alpha}.$$

Proof We first consider the case where $d(\underline{S}, \underline{S}') < 1$. Then it follows from the definition of the distance d that $S_0 = S'_0$. Let a_0^{p-1} be M' compatible and such that $M'(a_{p-1}, S_0) = 1$. Then $M'(a_{p-1}, S'_0) = 1$, and, by Lemma 8.10, we have

$$\prod_{j=0}^{p-1} e^{\mathrm{U}(a_j^{p-1} \bullet \underline{S})} \le e^{Dd(\underline{S}, \underline{S}')^\alpha} \prod_{j=0}^{p-1} e^{\mathrm{U}(a_j^{p-1} \bullet \underline{S}')}.$$

Therefore, the lemma follows in this case if we take $C > D$ by summing over a_0^{p-1} and using formula (8.4). If $S_0 \ne S'_0$ and $p > k_{M'}$, we observe that for any a_0^{p-1}, M' compatible and such that $M'(a_{p_1}, S_0) = 1$, it follows from Lemma 8.7 that we can find a finite sequence $b_0^{k_{M'}}$ which is M' compatible, with $b_0 = a_{p-k_{M'}}$ and $M'(b_{k_{M'}}, S'_0) = 1$. Denote by a'^{p-1}_0 the finite sequence $a_0^{p-1-k_{M'}} \bullet b_0^{k_{M'}}$. It follows from Lemma 8.10 that

$$\begin{aligned}
\prod_{j=0}^{p-1} e^{\mathrm{U}(a_j^{p-1} \bullet \underline{S})} &= \prod_{j=0}^{p-1-k_{M'}} e^{\mathrm{U}(a_j^{p-1} \bullet \underline{S})} \prod_{p-k_{M'}}^{p-1} e^{\mathrm{U}(a_j^{p-1} \bullet \underline{S})} \\
&\le \prod_{j=0}^{p-1-k_{M'}} e^{\mathrm{U}(a_j^{p-1} \bullet \underline{S})} e^{k_{M'} \|\mathrm{U}\|_{\mathcal{H}_\alpha(\mathcal{X}^{M,+})}} \\
&\le e^{2k_{M'} \|\mathrm{U}\|_{\mathcal{H}_\alpha(\mathcal{X}^{M,+})}} e^D \prod_{j=0}^{p-1} e^{\mathrm{U}(a'^{p-1}_j \bullet \underline{S}')}.
\end{aligned}$$

Summing over $a_0^{p-1-k_{M'}} = a'^{p-1-k_{M'}}_0$ and then over $a_{p-k_{M'}}^{p-1}$ and $a'^{p-1}_{p-k_{M'}}$, we get

$$\mathscr{P}^p_{M',M}\mathbf{1}(\underline{S}) \le e^{2k_{M'}\|\mathrm{U}\|_{\mathcal{H}_\alpha(\mathcal{X}^{M,+})}} e^D |\mathscr{A}|^{k_{M'}} \mathscr{P}^p_{M',M}\mathbf{1}(\underline{S}').$$

Since $d(\underline{S}, \underline{S}') = 1$, the result follows in this case by taking

$$C > D + 2k_{M'}\|\mathrm{U}\|_{\mathcal{H}_\alpha(\mathcal{X}^{M,+})} + k_{M'} \log |\mathscr{A}|.$$

Finally, it remains to prove the result in the case $p \le k_{M'}$. Since M' is irreducible and aperiodic, for any $S_0 \in \mathscr{A}$ there is an $a \in \mathscr{A}$ such that $M'(a, S_0) = 1$. It is the easy to verify that there exists a constant $C > 1$ such that

$$\frac{1}{C} \le \inf_{0\le p\le k_{M'}} \inf_{\underline{S}\in\mathcal{X}^{M,+})} \mathscr{P}^p_{M',M}\mathbf{1}(\underline{S}) \le \sup_{0\le p\le k_{M'}} \sup_{\underline{S}\in\mathcal{X}^{M,+})} \mathscr{P}^p_{M',M}\mathbf{1}(\underline{S}) \le C.$$

This finishes the proof of the lemma. □

The following lemma will be used several times later on.

Lemma 8.12 *For any $g \in \mathcal{C}^+(\mathcal{X}^{M,+})$, with g not identically zero, there exists an integer p_g such that for any integer $p \ge p_g$ we have*

$$\inf_{\underline{S}\in\mathcal{X}^{M,+}} \mathscr{P}^p_{M',M} g > 0.$$

Proof Let $\underline{S}_0$ be such that $g(\underline{S}_0) > 0$. Since g is continuous, there is a number $1 > \epsilon > 0$ such that for any $\underline{S}'$ with $d(\underline{S}_0, \underline{S}') \le \epsilon$ we have $g(\underline{S}') \ge g(\underline{S}_0)/2$. Let p be an integer such that $p > k_{M'} - 2\log\epsilon^{-1}$. Using Lemma 8.7 and (8.4), it follows easily that in the formula

$$\mathscr{P}^p_{M',M} g(\underline{S}) = \sum_{\substack{a_0^{p-1}\, M'\text{compatible} \\ M'(a_{p-1}, S_0)=1}} g\big(a_0^{p-1} \bullet \underline{S}\big) \prod_{j=0}^{p-1} e^{\mathrm{U}(a_j^{p-1}\bullet \underline{S})}$$

there exists a sequence a_0^{p-1}, M' compatible with $M'(a_{p-1}, S_0) = 1$ and such that

$$d\big(a_0^{p-1} \bullet \underline{S}, \underline{S}_0\big) < \epsilon.$$

Indeed, one takes $a_0^{p-1-k_{M'}} = (\underline{S}_0)_0^{p-1-k_{M'}}$ and uses Lemma 8.7 to complete the sequence. We conclude that there exists a number $c > 0$ such that for any $\underline{S}$ we have

$$\mathscr{P}^p_{M',M} g(\underline{S}) \ge c g(\underline{S}_0).$$

□

As observed above, $\mathscr{P}_{M',M}$ is a bounded linear operator in the Banach space $\mathcal{C}(\mathcal{X}^{M,+})$. Denote by $\mathcal{C}^+(\mathcal{X}^{M,+})$ the set of continuous nonnegative functions on

$\mathcal{X}^{M,+}$. It is easy to verify that $\mathcal{C}^+(\mathcal{X}^{M,+})$ is a cone with nonempty interior (the ball of radius $1/2$ centered around the constant function $\mathbf{1}$ is contained in $\mathcal{C}^+(\mathcal{X}^{M,+})$). It is also easy to verify that the bounded linear operator $\mathscr{P}_{M',M}$ maps the cone $\mathcal{C}^+(\mathcal{X}^{M,+})$ into itself. Therefore, by a theorem of Krein (see Oikhberg and Troitsky 2005), there is an element $\zeta_{M',M}$ of the dual Banach space $\mathcal{C}(\mathcal{X}^{M,+})^\dagger$ which is an eigenvector of the adjoint operator $\mathscr{P}^\dagger_{M',M}$. Moreover, $\zeta_{M',M}$ is nonnegative on $\mathcal{C}(\mathcal{X}^{M,+})$, and since $\mathcal{X}^{M,+}$ is compact, it follows that $\zeta_{M',M}$ is a positive measure (in particular, $\zeta_{M',M}(\mathcal{X}^{M,+}) > 0$; see, for example, Dunford and Schwartz 1958). Let λ be the associated eigenvalue, namely, for any function $g \in \mathcal{C}(\mathcal{X}^{M,+})$, we have

$$\int_{\mathcal{X}^{M,+}} \mathscr{P}_{M',M}\, g\, d\zeta_{M',M} = \lambda \int_{\mathcal{X}^{M,+}} g\, d\zeta_{M',M}. \tag{8.6}$$

We have, since the constant function $\mathbf{1}$ belongs to $\mathcal{C}(\mathcal{X}^{M,+})$,

$$\int_{\mathcal{X}^{M,+}} (\mathscr{P}_{M',M}\mathbf{1})\, d\zeta_{M',M} = \lambda \int_{\mathcal{X}^{M,+}} \mathbf{1}\, d\zeta_{M',M} = \lambda \zeta_{M',M}\big(\mathcal{X}^{M,+}\big). \tag{8.7}$$

It is easy to verify from the definition of $\mathscr{P}_{M',M}$ and the irreducibility of M' that there exists a constant $c > 0$ such that

$$\inf_{\underline{S} \in \mathcal{X}^{M,+}} (\mathscr{P}_{M',M}\mathbf{1})(\underline{S}) \geq c.$$

Using the identity (8.7), this implies $\lambda > 0$. We define P^* by

$$P^* = \log \lambda.$$

It is now convenient to introduce the operator $\tilde{\mathscr{P}}_{M',M}$ defined by

$$\tilde{\mathscr{P}}_{M',M} = e^{-P^*} \mathscr{P}_{M',M} \tag{8.8}$$

which satisfies

$$\int (\tilde{\mathscr{P}}_{M',M}\, g)\, d\zeta_{M',M} = \int g\, d\zeta_{M',M}$$

for any $g \in \mathcal{C}(\mathcal{X}^{M,+})$.

Lemma 8.13 *There exists a constant $C > 0$ such that for any integer n*

$$e^{-C} \leq \tilde{\mathscr{P}}^n_{M',M}\mathbf{1} \leq e^C.$$

Proof The proof follows at once from Lemma 8.11 and the identity

$$\int_{\mathcal{X}^{M,+}} \tilde{\mathscr{P}}^n_{M',M}\mathbf{1}\, d\zeta_{M',M} = \zeta_{M',M}\big(\mathcal{X}^{M,+}\big),$$

which is an immediate consequence of (8.6). □

In order to derive some spectral properties of the operator $\mathscr{P}_{M',M}$, we first recall the definition of a quasi-compact operator.

Definition 8.14 An operator $\mathcal{L}$ in a Banach space $\mathscr{B}$ is quasi-compact (quasi-strongly completely continuous in the older terminology) if there exists a constant $C > 0$ and a sequence $(\mathcal{K}_p)$ of compact operators such that $\|\mathcal{L}^n\| \le C$ for any integer n, and

$$\lim_{n\to\infty} \left\| \mathcal{L}^n - \mathcal{K}_n \right\| = 0.$$

Theorem 8.15 *Assume the matrix M' is irreducible and aperiodic and the potential* U *is α-Hölder for some $\alpha \in]0, 1]$. Then, for any $\beta \in]0, \alpha[$, the operator $\mathscr{P}_{M',M}$ is quasi-compact in the Banach space $\mathcal{H}_\beta(\mathcal{X}^{M,+})$.*

Proof For each integer $p \ge 1$ and for each finite sequence a_0^{p-1} which is M' compatible, we choose once and for all a sequence $\underline{\varkappa}(a_0^{p-1}) = (\varkappa_0, \varkappa_1, \varkappa_2, \dots) \in \mathcal{X}^{M,+}$ such that $M'(a_{p-1}, \varkappa_0) = 1$, and we define a bounded linear operator Δ_p on $\mathcal{C}(\mathcal{X}^{M,+})$ by

$$\Delta_p g(\underline{S}) = e^{-pP^*} \sum_{\substack{a_0^{p-1}\ M'\text{compatible} \\ M'(a_{p-1}, S_0)=1}} g\big(a_0^{p-1} \bullet \underline{\varkappa}(a_0^{p-1})\big) \prod_{j=0}^{p-1} e^{\mathrm{U}(a_j^{p-1} \bullet \underline{S})}.$$

We now estimate the norm of the operator $\tilde{\mathscr{P}}^p_{M',M} - \Delta_p$ in $\mathcal{H}_\beta(\mathcal{X}^{M,+})$. For any $g \in \mathcal{H}_\beta(\mathcal{X}^{M,+})$, using $d(a_0^{p-1} \bullet \underline{S}, a_0^{p-1} \bullet \underline{\varkappa}(a_0^{p-1})) \le 2^{-p}$, we obtain

$$\begin{aligned}
&\sup_{\underline{S}\in(\mathcal{X}^{M,+})} \left| \tilde{\mathscr{P}}^p_{M',M} g(\underline{S}) - \Delta_p g(\underline{S}) \right| \\
&\quad \le \sup_{\underline{S}\in(\mathcal{X}^{M,+})} e^{-pP^*} \sum_{\substack{a_0^{p-1}\ M'\text{compatible} \\ M'(a_{p-1}, S_0)=1}} \prod_{j=0}^{p-1} e^{\mathrm{U}(a_j^{p-1} \bullet \underline{S})} \big| g\big(a_0^{p-1} \bullet \underline{S}\big) \\
&\qquad - g\big(a_0^{p-1} \bullet \underline{\varkappa}(a_0^{p-1})\big)\big| \\
&\quad \le \|g\|_{\mathcal{H}_\beta(\mathcal{X}^{M,+})} 2^{-p\beta} \sup_{\underline{S}\in(\mathcal{X}^{M,+})} e^{-pP^*} \sum_{\substack{a_0^{p-1}\ M'\text{compatible} \\ M'(a_{p-1}, S_0)=1}} \prod_{j=0}^{p-1} e^{\mathrm{U}(a_j^{p-1} \bullet \underline{S})} \\
&\quad = \|g\|_{\mathcal{H}_\beta(\mathcal{X}^{M,+})} 2^{-p\beta} \sup_{\underline{S}\in(\mathcal{X}^{M,+})} \tilde{\mathscr{P}}^p_{M',M} \mathbf{1}(\underline{S}).
\end{aligned}$$

Therefore, using Lemma 8.13,

$$\sup_{\underline{S}\in(\mathcal{X}^{M,+})} \left|\tilde{\mathscr{P}}^{p}_{M',M} g(\underline{S}) - \Delta_p g(\underline{S})\right| \le e^C \|g\|_{\mathcal{H}_\beta(\mathcal{X}^{M,+})} 2^{-p\beta}.$$

Similarly, if $p = 2q$ and $d(\underline{S}, \underline{S}') \ge 2^{-q}$, we have from the previous estimate

$$\begin{aligned}&\left|\left(\tilde{\mathscr{P}}^{2q}_{M',M} g(\underline{S}) - \Delta_{2q} g(\underline{S})\right) - \left(\tilde{\mathscr{P}}^{2q}_{M',M} g(\underline{S}') - \Delta_{2q} g(\underline{S}')\right)\right|\\ &\quad \le 2e^C \|g\|_{\mathcal{H}_\beta(\mathcal{X}^{M,+})} 2^{-p\beta} \le 2e^C \|g\|_{\mathcal{H}_\beta(\mathcal{X}^{M,+})} 2^{-q\beta} d(\underline{S}, \underline{S}')^\beta.\end{aligned}$$

We now assume $d(\underline{S}, \underline{S}') < 2^{-q}$. We have

$$\begin{aligned}&\left|\left(\tilde{\mathscr{P}}^{2q}_{M',M} g(\underline{S}) - \Delta_{2q} g(\underline{S})\right) - \left(\tilde{\mathscr{P}}^{2q}_{M',M} g(\underline{S}') - \Delta_{2q} g(\underline{S}')\right)\right|\\ &\quad \le |I_1| + |I_2|,\end{aligned}$$

where

$$I_1 = \tilde{\mathscr{P}}^{2q}_{M',M} g(\underline{S}) - \tilde{\mathscr{P}}^{2q}_{M',M} g(\underline{S}') \quad \text{and} \quad I_2 = \Delta_{2q} g(\underline{S}) - \Delta_{2q} g(\underline{S}').$$

We observe that if $d(\underline{S}, \underline{S}') < 1$ then $S_0 = S'_0$. Therefore,

$$\begin{aligned}I_1 &= e^{-pP^*} \sum_{\substack{a_0^{p-1}\ M'\text{compatible}\\ M'(a_{p-1},S_0)=1}} \Bigg(g\left(a_0^{p-1} \bullet \underline{S}\right) \prod_{j=0}^{p-1} e^{\mathrm{U}(a_j^{p-1}\bullet\underline{S})}\\ &\qquad - g\left(a_0^{p-1} \bullet \underline{S}'\right) \prod_{j=0}^{p-1} e^{\mathrm{U}(a_j^{p-1}\bullet\underline{S}')}\Bigg)\\ &= I_{1,1} + I_{1,2},\end{aligned}$$

where

$$I_{1,1} = e^{-pP^*} \sum_{\substack{a_0^{p-1}\ M'\text{compatible}\\ M'(a_{p-1},S_0)=1}} \left(g\left(a_0^{p-1} \bullet \underline{S}\right) - g\left(a_0^{p-1} \bullet \underline{S}'\right)\right) \prod_{j=0}^{p-1} e^{\mathrm{U}(a_j^{p-1}\bullet\underline{S})}$$

and

$$I_{1,2} = e^{-pP^*} \sum_{\substack{a_0^{p-1}\ M'\text{compatible}\\ M'(a_{p-1},S_0)=1}} g\left(a_0^{p-1} \bullet \underline{S}'\right) \left(\prod_{j=0}^{p-1} e^{\mathrm{U}(a_j^{p-1}\bullet\underline{S})} - \prod_{j=0}^{p-1} e^{\mathrm{U}(a_j^{p-1}\bullet\underline{S}')}\right).$$

For $I_{1,1}$, we observe that $d(a_0^{p-1} \bullet \underline{S}, a_0^{p-1} \bullet \underline{S}') \le 2^{-p} d(\underline{S}, \underline{S}')$; therefore, using the estimate (8.13), we obtain

$$|I_{1,1}| \le 2^{-p\beta} d\big(\underline{S}, \underline{S}'\big)^{\beta} \|g\|_{\mathcal{H}_\beta(\mathcal{X}^{M,+})} e^{-pP^*} \sum_{\substack{a_0^{p-1}\ M'\text{compatible} \\ M'(a_{p-1}, S_0)=1}} \prod_{j=0}^{p-1} e^{\mathrm{U}(a_j^{p-1} \bullet \underline{S})}$$

$$= 2^{-p\beta} d\big(\underline{S}, \underline{S}'\big)^{\beta} \|g\|_{\mathcal{H}_\beta(\mathcal{X}^{M,+})} \tilde{\mathscr{P}}^p_{M',M} \mathbf{1}(\underline{S}) \le 2^{-p\beta} e^C d\big(\underline{S}, \underline{S}'\big)^{\beta} \|g\|_{\mathcal{H}_\beta(\mathcal{X}^{M,+})}.$$

For $I_{1,2}$, we have

$$|I_{1,2}| \le \|g\|_{\mathcal{H}_\beta(\mathcal{X}^{M,+})} e^{-pP^*} \sum_{\substack{a_0^{p-1}\ M'\text{compatible} \\ M'(a_{p-1}, S_0)=1}} \left| \prod_{j=0}^{p-1} e^{\mathrm{U}(a_j^{p-1} \bullet \underline{S})} - \prod_{j=0}^{p-1} e^{\mathrm{U}(a_j^{p-1} \bullet \underline{S}')} \right|$$

$$= \|g\|_{\mathcal{H}_\beta(\mathcal{X}^{M,+})} e^{-pP^*} \sum_{\substack{a_0^{p-1}\ M'\text{compatible} \\ M'(a_{p-1}, S_0)=1}} \prod_{j=0}^{p-1} e^{\mathrm{U}(a_j^{p-1} \bullet \underline{S})}$$

$$\times \left| 1 - \prod_{j=0}^{p-1} e^{(\mathrm{U}(a_j^{p-1} \bullet \underline{S}') - \mathrm{U}(a_j^{p-1} \bullet \underline{S}))} \right|.$$

Using Lemma 8.10 and the estimate (8.13), we obtain

$$|I_{1,2}| \le De^D d\big(\underline{S}, \underline{S}'\big)^{\alpha} \|g\|_{\mathcal{H}_\beta(\mathcal{X}^{M,+})} \tilde{\mathscr{P}}^p_{M',M} \mathbf{1}(\underline{S})$$
$$\le De^{D+C} 2^{-q(\alpha-\beta)} d\big(\underline{S}, \underline{S}'\big)^{\beta} \|g\|_{\mathcal{H}_\beta(\mathcal{X}^{M,+})}.$$

Finally, since $S_0 = S_0'$, we have

$$I_2 = e^{-pP^*} \sum_{\substack{a_0^{p-1}\ M'\text{compatible} \\ M'(a_{p-1}, S_0)=1}} g\big(a_0^{p-1} \bullet \underline{\varkappa}(a_0^{p-1})\big) \left(\prod_{j=0}^{p-1} e^{\mathrm{U}(a_j^{p-1} \bullet \underline{S})} - \prod_{j=0}^{p-1} e^{\mathrm{U}(a_j^{p-1} \bullet \underline{S}')} \right),$$

and we can estimate this term the same way as we estimated $I_{1,2}$. We obtain

$$|I_2| \le De^{D+C} 2^{-q(\alpha-\beta)} d\big(\underline{S}, \underline{S}'\big)^{\beta} \|g\|_{\mathcal{H}_\beta(\mathcal{X}^{M,+})}.$$

Regrouping all the estimates, we conclude that there exists a constant $C' > 0$ such that for any integer p

$$\big\| \tilde{\mathscr{P}}^p_{M',M} - \Delta_p \big\|_{\mathcal{H}_\beta} \le C' 2^{-p(\alpha-\beta)/2}. \tag{8.9}$$

We now estimate the norm of the operator Δ_p in $\mathcal{H}_\beta(\mathcal{X}^{M,+})$. We have using the estimate (8.13)

$$\left|\Delta_p g(\underline{S})\right| \le e^{-pP^*} \|g\|_{\mathcal{H}_\beta(\mathcal{X}^{M,+})} \sum_{\substack{a_0^{p-1}\, M'\text{compatible} \\ M'(a_{p-1}, S_0)=1}} \prod_{j=0}^{p-1} e^{\mathrm{U}(a_j^{p-1} \bullet \underline{S})}$$

$$= \|g\|_{\mathcal{H}_\beta(\mathcal{X}^{M,+})} \tilde{\mathscr{P}}^p_{M',M} \mathbf{1}(\underline{S}) \le e^C \|g\|_{\mathcal{H}_\beta(\mathcal{X}^{M,+})}.$$

If $S_0 \ne S_0'$, we obtain

$$\left|\Delta_p g(\underline{S}) - \Delta_p g(\underline{S}')\right| \le 2e^C \|g\|_{\mathcal{H}_\beta(\mathcal{X}^{M,+})} d(\underline{S}, \underline{S}').$$

Assume now $S_0 = S_0'$. We have

$$\Delta_p g(\underline{S}) - \Delta_p g(\underline{S}')$$
$$= e^{-pP^*} \sum_{\substack{a_0^{p-1}\, M'\text{compatible} \\ M'(a_{p-1}, S_0)=1}} g\big(a_0^{p-1} \bullet \underline{\varkappa}(a_0^{p-1})\big) \left(\prod_{j=0}^{p-1} e^{\mathrm{U}(a_j^{p-1} \bullet \underline{S})} - \prod_{j=0}^{p-1} e^{\mathrm{U}(a_j^{p-1} \bullet \underline{S}')} \right),$$

and this quantity can be estimated in the same way as we estimated $I_{1,2}$. We obtain

$$\left|\Delta_p g(\underline{S}) - \Delta_p g(\underline{S}')\right| \le D e^{D+C} d(\underline{S}, \underline{S}')^\beta \|g\|_{\mathcal{H}_\beta(\mathcal{X}^{M,+})}.$$

We conclude that

$$\sup_p \|\Delta_p\|_{\mathcal{H}_\beta(\mathcal{X}^{M,+})} \le 2e^C + De^{D+C}.$$

Therefore, using (8.9), we obtain

$$\sup_p \left\| \tilde{\mathscr{P}}^p_{M',M} \right\|_{\mathcal{H}_\beta(\mathcal{X}^{M,+})} \le C' + 2e^C + De^{D+C}. \tag{8.10}$$

We now observe that the operator Δ_p is of finite rank and hence compact (see Dunford and Schwartz 1958). The estimates (8.9) and (8.10) imply that the operator $\mathscr{P}^p_{M',M}$ is quasi-compact in the Banach space $\mathcal{H}_\beta(\mathcal{X}^{M,+})$. □

For quasi-compact operators, there is an important result of Yosida and Kakutani known as the Uniform Ergodic Theorem concerning their spectral theory. We now recall this result (see Dunford and Schwartz 1958 or Yosida and Kakutani 1941).

Theorem 8.16 (Uniform Ergodic Theorem) *Let $\mathcal{L}$ be a quasi-compact operator. Then the eigenvalues of modulus one (if any) are of finite number and with finite* *multiplicity. Let s be their number. If $s > 0$ we will denote these eigenvalues of modulus one by $\lambda_1, \dots, \lambda_s$. The operator $\mathcal{L}$ can be written*

$$\mathcal{L} = \sum_{i=1}^{s} \lambda_i P_i + R$$

where the P_i are spectral projections associated to the eigenvalues λ_i (the sum is taken to be zero in the above formula if $s = 0$). They have finite dimensional range and satisfy $P_i^2 = P_i$ for $1 \leq i \leq s$, and $P_i P_j = 0$ for $i \neq j$. Moreover, the operator R has spectral radius $\rho < 1$ (namely, $\lim_{n\to\infty} \|R^n\|^{1/n} = \rho < 1$). This operator satisfies also $RP_i = P_i R = 0$ for $1 \leq i \leq s$. Therefore, outside the complex disk $|z| \leq \rho$, the spectrum of $\mathcal{L}$ consists only of the eigenvalues $\lambda_1, \dots, \lambda_s$.

We can now apply the uniform ergodic Theorem 8.16 to the operator $\tilde{\mathscr{P}}_{M',M}$. From the existence of the measure $\zeta_{M',M}$, we conclude that 1 belongs to the spectrum of $\tilde{\mathscr{P}}_{M',M}$ (recall that the spectrum of the adjoint of $\tilde{\mathscr{P}}_{M',M}$ is the complex conjugate of the spectrum).

We will now prove that the only eigenvalue of modulus one is the number 1, and this is an eigenvalue of finite multiplicity.

Lemma 8.17 *The operator $\tilde{\mathscr{P}}_{M',M}$ has in $\mathcal{H}_\beta(\mathcal{X}^{M,+})$ an eigenvalue 1 of multiplicity one, and the eigenvector φ^* can be chosen strictly positive on $\mathcal{X}^{M,+}$.*

Proof It follows from Lemma 8.13 that the functions g_n defined for any $n \geq 1$ by

$$g_n = \frac{1}{n} \sum_{j=0}^{n-1} \tilde{\mathscr{P}}^j_{M',M} \mathbf{1}$$

satisfy $g_n \geq e^{-C}$. Moreover, by the Uniform Ergodic Theorem 8.16 and Proposition 8.15, it follows that this sequence converges in $\mathcal{H}_\beta(\mathcal{X}^{M,+})$ to a function φ^* which should also satisfy $\varphi^* \geq e^{-C}$. Since

$$\tilde{\mathscr{P}}_{M',M} g_n = \frac{n+1}{n} g_{n+1} - \frac{1}{n+1},$$

we conclude by taking the limit that φ^* is an eigenvector of $\tilde{\mathscr{P}}_{M',M}$ of eigenvalue one which is obviously strictly positive. Assume there exits a $\widetilde{\varphi}$, an eigenvector of $\tilde{\mathscr{P}}_{M',M}$ of eigenvalue 1 and not proportional to φ^*. If $\widetilde{\varphi}$ is not real, its real and imaginary parts must be eigenvectors of $\tilde{\mathscr{P}}_{M',M}$ of eigenvalue 1. Therefore, it is enough to assume $\widetilde{\varphi}$ real. Let η be the smallest real number such that

$$g = \widetilde{\varphi} + \eta \varphi^* \geq 0.$$

In particular, by continuity of g and compactness of $\mathcal{X}^{M,+}$, there is an $\underline{S}_0 \in \mathcal{X}^{M,+}$ such that $g(\underline{S}_0) = 0$. However, since $g \geq 0$, we have, by Lemma 8.12, that for p

large enough

$$\tilde{\mathscr{P}}^{p}_{M',M} g > 0.$$

But this is a contradiction since

$$\tilde{\mathscr{P}}^{p}_{M',M} g = g$$

vanishes in $\underline{S}_0$. Therefore, the eigenvalue one of $\tilde{\mathscr{P}}_{M',M}$ is of multiplicity one. □

From now on, we will denote by φ^* the eigenvector of $\tilde{\mathscr{P}}_{M',M}$ of eigenvalue 1 satisfying

$$\int \varphi^* \, d\zeta = 1.$$

Lemma 8.18 *Let $g \in \mathcal{H}_\beta(\mathcal{X}^{M,+})$ be a nonnegative function which is not proportional to φ^* (otherwise the result is obvious). Then there exists a constant $\gamma \geq 0$ such that for any $\underline{S} \in \mathcal{X}^{M,+}$ we have*

$$\lim_{n\to\infty} \big| \big(\tilde{\mathscr{P}}^{n}_{M',M} g\big)(\underline{S}) - \gamma \varphi^*(\underline{S}) \big| = 0.$$

Proof We define recursively a sequence of nonnegative numbers (γ_n) and a sequence of nonnegative function (g_n) in $\mathcal{H}_\beta(\mathcal{X}_0^+)$ by

$$\gamma_n = \inf_{\underline{S} \in \mathcal{X}^{M,+}} \frac{\tilde{\mathscr{P}}^{n}_{M',M} g(\underline{S})}{\varphi^*(\underline{S})}$$

and

$$g_n = \tilde{\mathscr{P}}^{n}_{M',M} g - \gamma_n \varphi^*.$$

In particular, $g_n \geq 0$ and

$$\inf_{\underline{S} \in \mathcal{X}_0^+} g_n(\underline{S}) = 0.$$

From the relation

$$\begin{aligned} \tilde{\mathscr{P}}^{n+1}_{M',M} g &= \gamma_{n+1} \varphi^* + g_{n+1} = \tilde{\mathscr{P}}_{M',M}\big(\tilde{\mathscr{P}}^{n}_{M',M} g\big) \\ &= \tilde{\mathscr{P}}_{M',M}\big(\gamma_n \varphi^* + g_n\big) = \gamma_n \varphi^* + \tilde{\mathscr{P}}_{M',M}(g_n), \end{aligned}$$

we derive $\gamma_{n+1} \geq \gamma_n$ and

$$\tilde{\mathscr{P}}_{M',M}(g_n) \geq g_{n+1}.$$

Since the operator $\tilde{\mathscr{P}}_{M',M}$ is quasi-compact (see the definition and Proposition 8.27), we have

$$\sup_n \big\| \tilde{\mathscr{P}}^{n}_{M',M} g \big\|_{\mathcal{H}_\beta(\mathcal{X}^{M,+})} < \infty,$$

and for any n

$$\gamma_n \leq \frac{\|\tilde{\mathscr{P}}^n_{M',M} g\|_{\mathcal{H}_\beta(\mathcal{X}^{M,+})}}{\inf_{\underline{S}\in\mathcal{X}^{M,+}} \varphi^*(\underline{S})} \leq C \frac{\|g\|_{\mathcal{H}_\beta(\mathcal{X}^{M,+})}}{\inf_{\underline{S}\in\mathcal{X}^{M,+}} \varphi^*(\underline{S})} < \infty.$$

Therefore, the nondecreasing sequence (γ_n) converges to a nonnegative number γ. Assume that (g_n) does not converge to zero in $\mathcal{C}(\mathcal{X}_0^+)$. We are going to derive a contradiction. Since the sequence (g_n) is bounded in $\mathcal{H}_\beta(\mathcal{X}^{M,+})$ it follows from Ascoli's Theorem (see, for example, Dunford and Schwartz 1958) that this sequence is compact in $\mathcal{C}(\mathcal{X}^{M,+})$. Let g_* denote a nonzero accumulation point, namely,

$$g_* = \lim_{j\to\infty} g_{n_j}$$

for some infinite sequence of integers (n_j). Since $g_* \geq 0$ is not identically zero, it follows from Lemma 8.12 that there is an integer p such that

$$\tilde{\mathscr{P}}^p_{M',M} g_* > 0.$$

In particular, there is a number $\delta > 0$ such that

$$\tilde{\mathscr{P}}^p_{M',M} g_* = \delta\varphi^* + u$$

with $u \geq 0$. We have

$$\begin{aligned}\tilde{\mathscr{P}}^{n_j+p}_{M',M} g &= \gamma_{n_j+p}\varphi^* + g_{n_j+p} = \tilde{\mathscr{P}}^p_{M',M}\big(\tilde{\mathscr{P}}^{n_j}_{M',M} g\big)\\ &= \gamma_{n_j}\varphi^* + \tilde{\mathscr{P}}^p_{M',M} g_{n_j} \geq \left(\gamma_{n_j} + \inf_{\underline{S}} \frac{\tilde{\mathscr{P}}^p_{M',M} g_{n_j}(\underline{S})}{\varphi^*(\underline{S})}\right)\varphi^*.\end{aligned}$$

However,

$$\lim_{j\to\infty}\left(\gamma_{n_j} + \inf_{\underline{S}} \frac{\tilde{\mathscr{P}}^p_{M',M} g_{n_j}(\underline{S})}{\varphi^*(\underline{S})}\right) \geq \gamma + \delta,$$

and for large j we have a contradiction to the definition of $\gamma_{n_j+p} \leq \gamma$. Therefore, all the accumulation points of the sequence (g_n) are equal to zero and hence this sequence converges to zero. □

The following theorem establishes some spectral properties of the operator $\mathscr{P}_{M',M}$.

Theorem 8.19 *Assume the matrix M' is irreducible and aperiodic and the potential* U *is α-Hölder for some $\alpha \in]0,1]$. Then, for any $\beta \in]0,\alpha[$, the spectrum of the operator $\mathscr{P}_{M',M}$ in the Banach space $\mathcal{H}_\beta(\mathcal{X}^{M,+})$ consists of a positive eigenvalue e^{P^*} with eigenspace of dimension* 1 *and an eigenvector φ^* which can be chosen strictly positive. The function φ^* and the number P^* can be chosen independent* *of β. The corresponding linear form which is an eigenvector of the adjoint for the same eigenvalue can be chosen as a probability measure $\zeta_{M',M}$.*

Moreover, there exists $\rho^* \in]0, e^{P^*}[$ such that the rest of the spectrum of $\mathscr{P}_{M',M}$ is contained in the disk $\{z \in \mathbb{C} : \|z\| \leq \rho^*\}$.

Finally, we have the equalities $P^* = P'$ and $\varphi' = a\varphi^*|_{\mathcal{X}^{M',+}}$, where $a > 0$ is a normalizing constant.

Proof By the Uniform Ergodic Theorem 8.16, we know that there are only finitely many eigenvalues of modulus 1, and, by Lemma 8.17, we know that the eigenvalue 1 is simple. Assume ξ is a nonzero eigenvector of $\tilde{\mathscr{P}}_{M',M}$ in $\mathcal{H}_\beta(\mathcal{X}^{M,+})$ of eigenvalue λ of modulus 1 with $\lambda \neq 1$. We can write

$$\xi = \xi_1 - \xi_2 + i(\xi_3 - \xi_4)$$

with $\xi_i \geq 0$ and $\xi_i \in \mathcal{H}_\beta(\mathcal{X}^{M,+})$ for $i = 1, \dots, 4$. Moreover, since ξ is not identically zero, at least one of these functions must be not identically zero. Applying Lemma 8.18, we conclude that the sequence $(\tilde{\mathscr{P}}^n_{M',M}\xi(\underline{S}))$ converges for any $\underline{S}$. But this sequence is also equal to $(\lambda^n\xi(\underline{S}))$ which cannot converge, unless $\lambda = 1$. This proves that 1 is the unique eigenvalue of modulus 1. The stated spectral properties follow from the Uniform Ergodic Theorem 8.16 and formula (8.8).

The last statement of the theorem follows from the fact that $C(\mathcal{X}^{M',+})$ is invariant by $\mathscr{P}_{M',M}$, and on this space this operator coincides with $\mathscr{P}_{M'}$. □

We can now prove a result about the Yaglom limit of the measure μ and the existence of a QSD.

Theorem 8.20 *Under the above assumptions on M and M', for any Gibbs measure μ associated to an α-Hölder potential* U, *there exists a QSD ν given by*

$$d\nu = \frac{\varphi^*}{\int_{\mathcal{X}_0} \varphi^* \, d\zeta} \, d\zeta \tag{8.11}$$

where ζ is defined in Theorem 8.9. *Moreover, $\nu(T > n) = e^{n(P'-P)}$, that is, the exponential survival rate of ν is $P - P'$.*

For any function $g \in C(\mathcal{X}^M)$, we have

$$\lim_{n\to\infty} \frac{1}{\mu(\mathcal{X}_n)} \int_{\mathcal{X}_n} g \circ \sigma^n \, d\mu = \int_{\mathcal{X}_0} g \, d\nu. \tag{8.12}$$

Furthermore, there are two constants $C > 0$ and $\rho \in]0, 1[$ such that for any $g \in C(\mathcal{X}^M)$ we have

$$\left| \frac{1}{\mu(\mathcal{X}_n)} \int_{\mathcal{X}_n} g \circ \sigma^n \, d\mu - \int_{\mathcal{X}_0} g \, d\nu \right| \leq C\|g\|_{C(\mathcal{X}^M)}\rho^n. \tag{8.13}$$

In particular, ν is the unique QSD with continuous Radon–Nykodim derivative with respect to ζ.

Proof Let $g \in C(\mathcal{X}^M)$. From (8.5), we have

$$\frac{1}{\mu(\mathcal{X}_n)} \int_{\mathcal{X}_n} g \circ \sigma^n \, d\mu = \frac{\int_{\mathcal{X}_n} g \circ \sigma^n \, d\mu}{\int_{\mathcal{X}_n} d\mu} = \frac{\int_{\mathcal{X}_0} (\mathscr{P}^n_{M',M} \varphi) g \, d\zeta}{\int_{\mathcal{X}_0} (\mathscr{P}^n_{M',M} \varphi) \, d\zeta}.$$

From Theorem 8.19, we obtain that

$$\lim_{n \to \infty} \frac{1}{\mu(\mathcal{X}_n)} \int_{\mathcal{X}_n} g \circ \sigma^n \, d\mu = \int_{\mathcal{X}_0} g \, d\nu,$$

proving that ν is a QSD. Relations (8.12) and (8.13) also follow from this discussion. To finish the proof, we only need to identify the exponential survival rate of ν. This is obtained by noticing that

$$\begin{aligned} \nu(\mathcal{X}_1) \int_{\mathcal{X}_0} \varphi^* \, d\zeta &= \int_{\mathcal{X}_0} \mathbf{1}_{\mathcal{X}_0} \circ \sigma \varphi^* \, d\zeta = e^{-P} \int P_M\big(\mathbf{1}_{\mathcal{X}_0} \circ \sigma \mathbf{1}_{\mathcal{X}_0} \varphi^*\big) \, d\zeta \\ &= e^{-P} \int_{\mathcal{X}_0} P_M\big(\mathbf{1}_{\mathcal{X}_0} \varphi^*\big) \, d\zeta = e^{-P} \int_{\mathcal{X}_0} P_{M,M'}\big(\varphi^*\big) \, d\zeta \\ &= e^{P'-P} \int_{\mathcal{X}_0} \varphi^* \, d\zeta, \end{aligned}$$

from where the result follows. □

It also follows that the QSD ν constructed above is the Yaglom limit of all the initial distributions absolutely continuous with respect to the measure ν.

It also follows easily from Theorem 8.19 that for any Borel set D

$$\lim_{n \to \infty} \frac{\zeta(D \cap \mathcal{X}_n)}{\zeta(\mathcal{X}_n)} = \frac{\zeta_{M',M}(D \cap \mathcal{X}_\infty)}{\zeta_{M',M}(\mathcal{X}_\infty)}.$$

The shift on the invariant set $\mathcal{X}_\infty$ equipped with the Gibbs state of potential U is an analog of an h-process.

8.3 Pianigiani–Yorke QSD

We now consider a particular case where $\mathcal{X}_0 \subset f(\mathcal{X}_0)$, with strict inclusion. We start with a simple, but rather typical example. The phase space is the complex plane $\mathbb{C}$ and the map f is given by

$$f(z) = z^2.$$

For $R > 1$, let $\mathcal{A}_R$ be the annulus

$$\mathcal{A}_R = \big\{z : R^{-1} \le |z| \le R\big\}.$$

It is easy to verify that

$$f(\mathcal{A}_R) = \mathcal{A}_{R^2}.$$

Let $R_0 > 1$ be a fixed number, and let $\mathcal{X}_0 = \mathcal{A}_{R_0}$. A simple computation leads to

$$\mathcal{X}_n = \mathcal{A}_{R_0^{2^{-n}}}$$

for any integer n. Let Leb denote the Lebesgue measure of $\mathbb{R}^2$. We have

$$\mathrm{Leb}(\mathcal{X}_n) = \pi\left(R_0^{2 2^{-n}} - R_0^{-2 2^{-n}}\right).$$

Therefore, for large n, we get

$$\mathrm{Leb}(\mathcal{X}_n) = 4\pi 2^{-n} \log R_0,$$

namely the probability of survival up to time n when starting with the Lebesgue measure decays exponentially fast. This suggests looking for an absolutely continuous QSD (see Lemma 8.3). More generally, assume we start with a measure (on $\mathcal{X}_0$) which is absolutely continuous with respect to the Lebesgue measure and with density h. For any integrable function g, we would like to say something for large n about the integral

$$\int_{\mathcal{X}_n} g \circ f^n h \, d\mathrm{Leb}.$$

Instead of using z as the variable, we can try to use $f^n(z)$. The transformation is not bijective, but we can decompose the integral over subsets of $\mathcal{X}_n$ where the map is bijective. We obtain

$$\int_{\mathcal{X}_n} g \circ f^n h \, d\mathrm{Leb} = \int_{\mathcal{X}_0} \Bigg(\sum_{y \in \mathcal{X}_n,\, y^{2^n} = z} \frac{h(y)}{4^n |y|^{2(2^n - 1)}} \Bigg) g(z) \, d\mathrm{Leb}(z).$$

If we define

$$\mathscr{P} h(z) = \sum_{y \in \mathcal{X}_1,\, y^2 = z} \frac{h(y)}{4|y|^2},$$

an easy computation leads to

$$\mathscr{P}^n h(z) = \sum_{y \in \mathcal{X}_n,\, y^{2^n} = z} \frac{h(y)}{4^n |y|^{2(2^n - 1)}},$$

and we have

$$\int_{\mathcal{X}_n} g \circ f^n h \, d\mathrm{Leb} = \int_{\mathcal{X}_0} \mathscr{P}^n h g \, d\mathrm{Leb}.$$

Since the equation $y^2 = z$ has two roots for any $y \in \mathcal{X}_0$, and both roots satisfy $|y|^2 = |z|$, we obtain that the function $h(y) = |y|^{-2}$ (defined on $\mathcal{X}_0$) satisfies

$$\mathscr{P}h = \frac{1}{2}h.$$

Therefore, for any integer n and any integrable function g, we have

$$\int_{\mathcal{X}_n} g \circ f^n h \, d\mathrm{Leb} = 2^{-n} \int_{\mathcal{X}_0} gh \, d\mathrm{Leb},$$

and the measure $h \, d\mathrm{Leb}$ is a QSD with rate of survival $1/2$.

The careful reader is invited to verify that, as expected from Lemma 8.3, for any n (not only asymptotically),

$$\int_{\mathcal{X}_n} h \, d\mathrm{Leb} = 4\pi 2^{-n} \log R_0.$$

We now state some more general hypotheses in view of generalizing the above example. We will denote by $\mathrm{Int}\, F$ the interior of a set F.

H1 The set $\mathcal{X}$ is $\mathbb{R}^d$ for some $d \geq 1$, finite.
H2 The set $\mathcal{X}_0$ is open, connected with compact closure (in particular, it has a positive Lebesgue measure), and $\overline{\mathcal{X}}_0 \subseteq \mathrm{Int}\, f(\mathcal{X}_0)$.
H3 The map f is $C^{1+\alpha}$ in a neighborhood V of $\overline{\mathcal{X}}_0$, and there is a constant $a < 1$ such that

$$\sup_{x \in V} \left\| (Df_x)^{-1} \right\| \leq a.$$

Note that by the Open Mapping Theorem (see Dieudonné 1969), it follows from Hypotheses that $f(\mathrm{Int}\, \mathcal{X}_0)$ is an open set. We define recursively as before for $n \geq 1$

$$\mathcal{X}_n = \mathcal{X}_{n-1} \cap f^{-n}(\mathcal{X}_0).$$

We first prove two simple topological lemmas.

Lemma 8.21 *For any integer $n \geq 1$, we have $f(\mathrm{Int}\, \mathcal{X}_n) = \mathrm{Int}\, \mathcal{X}_{n-1}$ and $f(\overline{\mathcal{X}}_n) = \overline{\mathcal{X}}_{n-1}$. Moreover, the set*

$$\overline{\mathcal{X}}_\infty = \bigcap_n \overline{\mathcal{X}}_n = \lim_{n \to \infty} \overline{\mathcal{X}}_n$$

is nonempty, compact and invariant.

Proof Since the sequence $(\overline{\mathcal{X}}_n)$ is a decreasing sequence of compact sets, none of them being empty (since $f^n(\overline{\mathcal{X}}_n) = \overline{\mathcal{X}}_0$), the fact that $\overline{\mathcal{X}}_\infty$ is nonempty and compact follows (see Dieudonné 1969). □

If r is a positive number and $x \in \mathbb{R}^d$, we denote by $B_r(x)$ the ball of $\mathbb{R}^d$ of radius r and center x. We will use several times the following version of the Inverse Function Theorem.

Theorem 8.22 *There are two positive numbers ϵ_0 and $\epsilon_1 < \epsilon_0$ such that if $y_0 \in \overline{\mathcal{X}}_0$ and $x_1 \in f(\overline{\mathcal{X}}_0) \cap \overline{\mathcal{X}}_0$ satisfy*

$$\left|x_1 - f(y_0)\right| \leq \epsilon_0,$$

then there exists a unique $y_1 \in \overline{\mathcal{X}}_0$ such that

$$f(y_1) = x_1 \quad \textit{and} \quad |y_1 - y_0| \leq \epsilon_1.$$

Moreover, the map $x_1 \to y_1$ defined on the ball $B_{\epsilon_0}(y_0)$ is C^1. Its differential has a norm of at most a.

Proof By Hypotheses **H2** and **H3**, and since the map f is open (see Dieudonné 1969), there is a number $\delta_0 > 0$ such that

$$d\big(\overline{\mathcal{X}}_0, \partial f(\overline{\mathcal{X}}_0)\big) > \delta_0,$$

where $\partial f(\overline{\mathcal{X}}_0)$ denotes the boundary of $f(\overline{\mathcal{X}}_0)$ (recall that $\overline{\mathcal{X}}_0 \subset f(\overline{\mathcal{X}}_0)$). Since $\overline{\mathcal{X}}_0$ is compact, there is a number $\delta_1 > 0$ with $\delta_1 \leq \delta_0$ such that for any $z_0 \in \overline{\mathcal{X}}_0$ we have

$$f\big(B_{\delta_1}(z_0)\big) \subseteq B_{\delta_0}(z_0).$$

Since $\overline{\mathcal{X}}_0$ is compact, we can choose this number δ_1 independent of z_0. By the Inverse Function Theorem (see Dieudonné 1969), and Hypothesis **H3**, for any $y_0 \in \overline{\mathcal{X}}_0$ we can find two numbers $\epsilon_1 < \epsilon_0 < \delta_1$ such that f is a C^1- diffeomorphism from $B_{\epsilon_1}(y_0)$ to its image which contains $B_{\epsilon_0}(x_0)$. Since $\overline{\mathcal{X}}_0$ is compact, we can choose these numbers ϵ_1 and ϵ_0 independent of y_0. The results follows from the fact that by hypothesis $x_1 \in B_{\epsilon_0}(x_0) \subseteq f(B_{\epsilon_1}(y_0))$. □

We now define the Ruelle–Perron–Frobenius linear operator $\mathscr{P}$, mapping functions on $\mathcal{X}_0$ to functions on $f(\mathcal{X}_0)$ by

$$\mathscr{P}g(x) = \sum_{y \in \mathcal{X}_0, f(y)=x} \frac{g(y)}{J(f)(y)},$$

where $J(f)$ denotes the Jacobian of f. As before, this operator is useful due to the following relation whose recursive proof is left to the reader

$$\int_{\mathcal{X}_n} g \circ f^n \, d\text{Leb} = \int_{\mathcal{X}_0} \mathscr{P}^n \mathbf{1} g \, d\text{Leb}. \tag{8.14}$$

We now state some elementary properties of the operator $\mathscr{P}$. We denote as in Sect. 8.2.2 by $\mathcal{C}(F)$ (respectively, $\mathcal{H}_\beta(F)$) the set of continuous (respectively, β-Hölder ($\beta \in]0,1]$)) functions on the bounded set F. These are Banach spaces when equipped with the respective norms

$$\|g\|_{\mathcal{C}(F)} = \sup_{x\in F}\bigl|g(x)\bigr| \quad \text{and} \quad \|g\|_{\mathcal{H}_\beta(F)} = \sup_{x,y\in F, x\neq y} \frac{|g(x)-g(y)|}{|x-y|} + \sup_{x\in F}\bigl|g(x)\bigr|.$$

Lemma 8.23 *The operator $\mathscr{P}$ is continuous on $\mathcal{C}(\overline{\mathcal{X}}_0)$ and also on $\mathcal{H}_\beta(\overline{\mathcal{X}}_0)$ for any $\beta \in]0,\alpha]$. It maps the cone $\mathcal{C}(\overline{\mathcal{X}}_0)^+$ of nonnegative continuous functions into itself.*

The proof is left to the reader.

By Krein's Lemma (see Oikhberg and Troitsky 2005), we conclude that there is a positive measure ζ on $\overline{\mathcal{X}}_0$ and a number $\lambda_0 \geq 0$ such that for any $g \in \mathcal{C}(\overline{\mathcal{X}}_0)$ we have

$$\int_{\overline{\mathcal{X}}_0} \mathscr{P} g \, d\zeta = \lambda_0 \int_{\overline{\mathcal{X}}_0} g \, d\zeta.$$

It is easy to verify that for any $x \in \overline{\mathcal{X}}_0$, we have

$$\mathscr{P}\mathbb{1}_{\overline{\mathcal{X}}_0} \geq a\mathbb{1}_{\overline{\mathcal{X}}_0},$$

and since $\mathbb{1}_{\overline{\mathcal{X}}_0} \in \mathcal{C}(\overline{\mathcal{X}}_0)$, we conclude that

$$a\zeta(\overline{\mathcal{X}}_0) \leq \lambda_0 \zeta(\overline{\mathcal{X}}_0),$$

which implies (since $\zeta(\overline{\mathcal{X}}_0) > 0$, by Krein's Lemma) $\lambda_0 \geq a$. We now define the operator $\tilde{\mathscr{P}}$ by

$$\tilde{\mathscr{P}} = \lambda_0^{-1}\mathscr{P}. \tag{8.15}$$

It is easy to verify that for any integer $p \geq 1$

$$\tilde{\mathscr{P}}^p g(x) = \lambda_0^{-p} \sum_{y\in\mathcal{X}_p, f^p(y)=x} \frac{g(y)}{J(f^p)(y)}. \tag{8.16}$$

The following lemma will be used several times.

Lemma 8.24 *Let x and x' in $\overline{\mathcal{X}}_0$ be such that $|x-x'| \leq \epsilon_0$ (see Lemma 8.22). Then for any integer p, there exists a bijection $F^{(p)}_{x,x'}$ between the two sets*

$$\bigl\{y \in \overline{\mathcal{X}}_p : f^p(y) = x\bigr\} \quad \textit{and} \quad \bigl\{y' \in \overline{\mathcal{X}}_p : f^p\bigl(y'\bigr) = x'\bigr\}$$

such that for any $y \in \overline{\mathcal{X}}_p$ with $f^p(y) = x$ we have for any $0 \leq j \leq p$

$$\bigl|f^j(y) - f^j\bigl(F^{(p)}_{x,x'}(y)\bigr)\bigr| \leq a^{p-j}\bigl|x-x'\bigr|.$$

There is also a constant $C > 0$ such that for any integer p

$$\sup_{\substack{x,x',\\ |x-x'|\le\epsilon_0}} \ \sup_{\substack{y\in\overline{\mathcal{X}}_p,\\ f^p(y)=x}} \left|1 - \frac{J(f^p)(y)}{J(f^p)(F^{(p)}_{x,x'}(y))}\right| \le C|x-x'|^\alpha .$$

Proof Let $y \in \mathcal{X}_p$ be such that $f^p(y) = x$. Using recursively Lemma 8.22, we conclude that there exists a unique $y' \in \mathcal{X}_p$ such that $f^p(y') = x'$, and for any $0 \le j \le p-1$ we have $|f^j(y) - f^j(y')| \le \epsilon_1$. There is indeed a better estimate, namely, from Hypothesis **H3**, we get

$$\left|f^j(y) - f^j(y')\right| \le a^{p-j}\left|x - x'\right|.$$

Therefore, we have constructed a bijection $F^{(p)}_{x,x'}$ between the two sets

$$\left\{y \in \overline{\mathcal{X}}_p : f^p(y) = x\right\} \quad \text{and} \quad \left\{y' \in \overline{\mathcal{X}}_p : f^p(y') = x'\right\}$$

such that for any $y \in \overline{\mathcal{X}}_p$ with $f^p(y) = x$ and for any $0 \le j \le p$ we have

$$\left|f^j(y) - f^j\left(F^{(p)}_{x,x'}(y)\right)\right| \le a^{p-j}\left|x - x'\right|.$$

This finishes the proof of the first part of the lemma. We now observe that there is a constant $D > 0$ such that for any x and x' in $\overline{\mathcal{X}}_0$ we have

$$\frac{J(x)}{J(x')} \le e^{D|x-x'|^\alpha}.$$

This follows from the fact that the function $\log J(x)$ is Hölder of exponent α.

Therefore,

$$\begin{aligned}
\frac{J(f^p)(y)}{J(f^p)(F^{(p)}_{x,x'}(y))} &= \prod_{j=0}^{p-1} \frac{J(f)(f^j(y))}{J(f)(f^j(F^{(p)}_{x,x'}(y)))} \\
&\le e^{D\sum_{j=0}^{p-1}|(f^j(y)-f^j(F^{(p)}_{x,x'}(y))|^\alpha} \le e^{D\sum_{j=0}^{p-1} a^{\alpha(p-j)}|x-x'|^\alpha} \\
&\le e^{D|x-x'|^\alpha/(1-a^\alpha)}.
\end{aligned}$$

□

We will also need the following lemma.

Lemma 8.25 *For any $\epsilon > 0$ there exists an integer $p = p(\epsilon) > 0$ such that for any $y \in \overline{\mathcal{X}}_p$ and any $x' \in \overline{\mathcal{X}}_0$, we can find $y' \in \overline{\mathcal{X}}_p$ such that $f^p(y') = x'$ and $|y - y'| \le \epsilon$.*

Proof Let ϵ_0 be as in Lemma 8.22. Since $\overline{\mathcal{X}}_0$ is compact, it can be covered by a finite number M of open balls of radius $\epsilon_0/3$. Denote by $x_1, \ldots, x_M$ the centers of these balls (which belong to $\overline{\mathcal{X}}_0$). We claim that for any $x \in \overline{\mathcal{X}}_0$ and any $x' \in \overline{\mathcal{X}}_0$, there is a sequence of q indexes $1 \le j_1 < j_2 < \cdots < j_{q-1} < j_q$ with $q \le M$ such that $x \in B_{\epsilon_0}(x_{j_1})$, $x' \in B_{\epsilon_0}(x_{j_q})$, and for any $1 \le \ell < q$

$$B_{\epsilon_0}(x_{j_\ell}) \cap B_{\epsilon_0}(x_{j_{\ell+1}}) \neq \emptyset.$$

This follows at once from the connectedness of $\overline{\mathcal{X}}_0$. Let

$$p = -\frac{\log((M+1)\epsilon^{-1})}{\log a}.$$

We can now apply Lemma 8.24 to construct q points y_ℓ, $1 \le \ell \le q$ in $\overline{\mathcal{X}}_p$, which satisfy

- $f^p(y_\ell) = x_{j_\ell}$ for any $1 \le \ell \le q$,
- $\|f^j(y_\ell) - f^j(y_{\ell+1})\| \le a^{p-j}\epsilon_0$ for any $1 \le \ell \le q-1$ and any $0 \le j \le p$,
- $\|f^j(y) - f^j(y_1)\| \le a^{p-j}\epsilon_0$ for any $1 \le \ell \le q$,
- there exists $y' \in \overline{\mathcal{X}}_p$, $f^p(y') = x'$, and $\|f^j(y_q) - f^j(y')\| \le a^{p-j}\epsilon_0$ for any $1 \le \ell \le q$.

By the triangle inequality, we have

$$\left|y' - y\right| \le |y - y_1| + \left|y_q - y'\right| + \sum_{\ell=1}^{q-1} |y_\ell - y_{\ell+1}| \le (M+1)\epsilon_0 a^p \le \epsilon.$$

□

We now derive a uniform estimate on the functions $\tilde{\mathscr{P}}^p\mathbf{1}(x)$.

Lemma 8.26 *There exists a constant $C > 1$ such that for any x in $\overline{\mathcal{X}}_0$, and any integer p, we have*

$$\frac{1}{C} \le \tilde{\mathscr{P}}^p\mathbf{1}(x) \le C.$$

Proof Assume first x and x' are such that $|x - x'| \le \epsilon_0$ (see Lemma 8.22). By Lemma 8.24, we have

$$\frac{\tilde{\mathscr{P}}^p\mathbf{1}(x)}{\tilde{\mathscr{P}}^p\mathbf{1}(x')} = \frac{\sum_{y\in\mathcal{X}_p, f^p(y)=x} \frac{1}{J(f^p)(y)}}{\sum_{y'\in\mathcal{X}_p, f^p(y')=x'} \frac{1}{J(f^p)(y)}} = \frac{\sum_{y\in\mathcal{X}_p, f^p(y)=x} \frac{1}{J(f^p)(y)}}{\sum_{y\in\mathcal{X}_p, f^p(y)=x} \frac{1}{J(f^p)(F^{(p)}_{x,x'}(y))}}$$

$$\le \sup_{y\in\mathcal{X}_p, f^p(y)=x} \frac{J(f^p)(F^{(p)}_{x,x'}(y))}{J(f^p)(y)} \le C_1.$$

Since $\overline{\mathcal{X}}_0$ is compact and connected, there is an integer M such that for any x and x' in $\overline{\mathcal{X}}_0$, we can find a finite sequence $x_0, \ldots, x_k$ with $k \le M$, $x_0 = x$, $x_k = x'$, and $|x_j - x_{j+1}| \le \epsilon_0$ for $0 \le j \le k-1$ (see Dieudonné 1969).

Therefore, for any x and x' in $\overline{\mathcal{X}}_0$, we have

$$\frac{\tilde{\mathscr{P}}^p\mathbf{1}(x)}{\tilde{\mathscr{P}}^p\mathbf{1}(x')} = \prod_{j=0}^{k-1} \frac{\tilde{\mathscr{P}}^p\mathbf{1}(x_j)}{\tilde{\mathscr{P}}^p\mathbf{1}(x_{j+1})} \le C_1^M.$$

This estimate implies for any $x \in \overline{\mathcal{X}}_0$

$$C_1^{-M} = \frac{C_1^{-M}}{\zeta(\overline{\mathcal{X}}_0)} \int_{\overline{\mathcal{X}}_0} \tilde{\mathscr{P}}^p\mathbf{1}\, d\zeta \le \tilde{\mathscr{P}}^p\mathbf{1}(x) \le \frac{C_1^{1M}}{\zeta(\overline{\mathcal{X}}_0)} \int_{\overline{\mathcal{X}}_0} \tilde{\mathscr{P}}^p\mathbf{1}\, d\zeta \le C_1^M.$$

The lemma is proved with $C = C_1^M$. □

Proposition 8.27 *The operator $\tilde{\mathscr{P}}$ is quasi-compact in $\mathcal{H}_\beta(\overline{\mathcal{X}}_0)$ for any $\beta \in]0, \alpha[$.*

We will denote by D the diameter of $\overline{\mathcal{X}}_0$.

Proof From Hypotheses **H2**, **H3**, the compactness of $\overline{\mathcal{X}}_0$ and Lemma 8.21, it follows that there is a number $\delta > 0$ such that for any $y \in \overline{\mathcal{X}}_1$, the ball $B_\delta(y)$ of radius δ centered at y is contained in $\overline{\mathcal{X}}_0$.

Let η be a nonnegative C^∞ function in $\mathbb{R}^d$ such that $\eta(x) = 0$ if $|x| \ge 2$, and satisfying

$$\int_{\mathbb{R}^d} \eta(z)\, dz = 1.$$

For any integer p, we define the operator Δ_p by

$$\Delta_p g(x) = \frac{1}{2Da^p\lambda_0^p} \sum_{y\in\mathcal{X}_p,\, f^p(y)=x} \frac{1}{J(f^p)(y)} \int \eta\left(\frac{y-z}{2Da^p}\right) g(z)\, dz.$$

We first observe that since the integral of η is one

$$\left|\frac{1}{2Da^p}\int \eta\left(\frac{y-z}{2Da^p}\right) g(z)\, dz - g(y)\right|$$
$$= \left|\frac{1}{2Da^p}\int \eta\left(\frac{s}{2Da^p}\right)\big(g(y+s) - g(y)\big)\, ds\right| \le \big(4Da^p\big)^\beta \|g\|_{\mathcal{H}_\beta(\overline{\mathcal{X}}_0)}.$$

Therefore, using Lemma 8.26,

$$\left|\tilde{\mathscr{P}}^p g(x) - \Delta_p g(x)\right|$$
$$= \frac{1}{\lambda_0^p}\left|\sum_{y\in\mathcal{X}_p,\, f^p(y)=x} \frac{1}{J(f^p)(y)}\left(g(y) - \frac{1}{2Da^p}\int \eta\left(\frac{y-z}{2Da^p}\right) g(z)\, dz\right)\right|$$
$$\le \big(4Da^p\big)^\beta \|g\|_{\mathcal{H}_\beta(\overline{\mathcal{X}}_0)} \tilde{\mathscr{P}}^p\mathbf{1}(x) \le C\big(4Da^p\big)^\beta \|g\|_{\mathcal{H}_\beta(\overline{\mathcal{X}}_0)}.$$

We conclude that if $|x-x'|>a^{p/2}$ we have

$$\big|\big(\tilde{\mathscr{P}}^p g(x)-\Delta_p g(x)\big)-\big(\tilde{\mathscr{P}}^p g(x')-\Delta_p g(x')\big)\big|\le 2C(4D)^\beta a^{\beta p/2}|x-x'|^\beta.$$

Assume now $|x-x'|<a^{p/2}$, and $p>2\log\epsilon_1/\log a$. Then, using Lemmas 8.24 and 8.26, we have

$$\begin{aligned}
\big|\tilde{\mathscr{P}}^p g(x)-\tilde{\mathscr{P}}^p g(x')\big| &= \frac{1}{\lambda_0^p}\Bigg|\sum_{y\in\mathcal{X}_p,\,f^p(y)=x}\frac{g(y)}{J(f^p)(y)}-\sum_{y'\in\mathcal{X}_p,\,f^p(y')=x'}\frac{g(y')}{J(f^p)(y')}\Bigg|\\
&= \frac{1}{\lambda_0^p}\Bigg|\sum_{y\in\mathcal{X}_p,\,f^p(y)=x}\frac{g(y)}{J(f^p)(y)}-\frac{g(F^{(p)}_{x,x'}(y))}{J(f^p)(F^{(p)}_{x,x'}(y))}\Bigg|\\
&\le \frac{1}{\lambda_0^p}\Bigg|\sum_{y\in\mathcal{X}_p,\,f^p(y)=x}\frac{g(y)}{J(f^p)(y)}-\frac{g(F^{(p)}_{x,x'}(y))}{J(f^p)(y)}\Bigg|\\
&\quad+\frac{1}{\lambda_0^p}\Bigg|\sum_{y\in\mathcal{X}_p,\,f^p(y)=x}\frac{g(F^{(p)}_{x,x'}(y))}{J(f^p)(y)}-\frac{g(F^{(p)}_{x,x'}(y))}{J(f^p)(F^{(p)}_{x,x'}(y))}\Bigg|\\
&\le a^{p\beta}|x-x'|^\beta\|g\|_{\mathcal{H}_\beta(\overline{\mathcal{X}}_0)}\tilde{\mathscr{P}}^p\mathbf{1}(x)\\
&\quad+\|g\|_{\mathcal{H}_\beta(\overline{\mathcal{X}}_0)}\sup_{\substack{y\in\mathcal{X}_p,\\ f^p(y)=x}}\Bigg|1-\frac{J(f^p)(y)}{J(f^p)(F^{(p)}_{x,x'}(y))}\Bigg|\tilde{\mathscr{P}}^p\mathbf{1}(x)\\
&\le C\big(a^{p\beta}|x-x'|^\beta+C|x-x'|^\alpha\big)\|g\|_{\mathcal{H}_\beta(\overline{\mathcal{X}}_0)}\\
&\le C\big(a^{p\beta}+Ca^{p(\alpha-\beta)/2}\big)|x-x'|^\beta\|g\|_{\mathcal{H}_\beta(\overline{\mathcal{X}}_0)}.
\end{aligned}$$

Similarly, we have

$$\begin{aligned}
&\big|\Delta_p(x)-\Delta_p(x')\big|\\
&\quad=\frac{1}{\lambda_0^p}\Bigg|\sum_{y\in\mathcal{X}_p,\,f^p(y)=x}\frac{1}{J(f^p)(y)}\frac{1}{2Da^p}\int\eta\Big(\frac{s}{2Da^p}\Big)g(y-s)\,ds\\
&\qquad-\sum_{y'\in\mathcal{X}_p,\,f^p(y')=x'}\frac{1}{J(f^p)(y')}\frac{1}{2Da^p}\int\eta\Big(\frac{s}{2Da^p}\Big)g(y'-s)\,ds\Bigg|\\
&\quad\le\frac{1}{\lambda_0^p}\Bigg|\sum_{\substack{y\in\mathcal{X}_p,\\ f^p(y)=x}}\Big(\frac{1}{J(f^p)(y)}-\frac{1}{J(f^p)(F_{x,x'}(y))}\Big)\frac{1}{2Da^p}
\end{aligned}$$

$$\times \int \eta\left(\frac{s}{2Da^p}\right) g(y-s)\,ds\Bigg| + \frac{1}{\lambda_0^p}\Bigg| \sum_{y\in\mathcal{X}_p,\, f^p(y)=x} \frac{1}{J(f^p)(F_{x,x'}(y))}$$

$$\times \left(\int \eta\left(\frac{s}{2Da^p}\right) g(y-s)\,ds - \int \eta\left(\frac{s}{2Da^p}\right) g\left(y'-s\right) ds\right)\Bigg|$$

$$\le C\left(a^{p\beta} + Ca^{p(\alpha-\beta)/2}\right)\left|x-x'\right|^{\beta} \|g\|_{\mathcal{H}_\beta(\overline{\mathcal{X}}_0)}.$$

This estimate implies

$$\lim_{p\to\infty} \|\tilde{\mathscr{P}}^p - \Delta_p\|_{\mathcal{H}_\beta(\overline{\mathcal{X}}_0)} = 0.$$

It is easy to verify that the operator Δ_p maps continuously $\mathcal{H}_\beta(\overline{\mathcal{X}}_0)$ to $\mathcal{H}_\alpha(\overline{\mathcal{X}}_0)$, and therefore is compact as an operator from the Banach space $\mathcal{H}_\beta(\overline{\mathcal{X}}_0)$ to itself. This finishes the proof of the proposition using Definition 8.27 of quasi-compact operators. □

We can now apply the Uniform Ergodic Theorem 8.16. From the existence of the measure ζ, we conclude that 1 belongs to the spectrum of $\tilde{\mathscr{P}}$ (recall that the spectrum of the adjoint of $\tilde{\mathscr{P}}$ is the complex conjugate of the spectrum).

We will now prove that the only eigenvalue of modulus one is the number 1, and this is an eigenvalue of finite multiplicity.

Lemma 8.28 *Let $g \in \mathcal{H}_\beta(\overline{\mathcal{X}}_0)$ be a nonnegative function such that*

$$\lim_{n\to\infty} \|g\|_{\mathcal{C}(\overline{\mathcal{X}}_n)} > 0.$$

Then there is an integer p such that

$$\inf_x \tilde{\mathscr{P}}^p g(x) > 0.$$

Proof Let $x_0 \in \overline{\mathcal{X}}_\infty$ be such that

$$\lim_{n\to\infty} \|g\|_{\mathcal{C}(\overline{\mathcal{X}}_n)} = g(x_0).$$

Let

$$\epsilon = \left(\frac{g(x_0)}{2\|g\|_{\mathcal{H}_\beta(\overline{\mathcal{X}}_0)}}\right)^{1/\beta}.$$

Using Lemma 8.25 applied to any $x' \in \overline{\mathcal{X}}_0$ and $y = x_0 \in \overline{\mathcal{X}}_{p(\epsilon)}$, we conclude that there exists $y' \in \overline{\mathcal{X}}_{p(\epsilon)}$ with $f^p(y') = x'$ and

$$g\left(y'\right) \ge g(x_0) - \|g\|_{\mathcal{H}_\beta(\overline{\mathcal{X}}_0)}\epsilon^\beta \ge \frac{g(x_0)}{2}.$$

Using formula 8.16 and the nonnegativity of g, we get

$$\tilde{\mathscr{P}}^p g(x') \geq \lambda_0^{-p} A^{-p} \frac{g(x_0)}{2},$$

where

$$A = \sup_{x \in \overline{\mathcal{X}}_0} J(f)(x).$$

□

Lemma 8.29 *The operator* $\tilde{\mathscr{P}}$ *has an eigenvalue* 1 *of multiplicity* 1, *and the eigenvector* φ *can be chosen strictly positive on* $\overline{\mathcal{X}}_0$.

Proof It follows from Lemma 8.26 that the functions g_n defined for any $n \geq 1$ by

$$g_n = \frac{1}{n} \sum_{j=0}^{n-1} \tilde{\mathscr{P}}^j \mathbf{1}$$

satisfies $g_n \geq C^{-1}$. Moreover, by the Uniform Ergodic Theorem 8.16 and Proposition 8.27, it follows that this sequence converges in $\mathcal{H}_\beta(\overline{\mathcal{X}}_0)$ to a function φ satisfying also $\varphi \geq C^{-1}$. Since

$$\tilde{\mathscr{P}} g_n = \frac{n+1}{n} g_{n+1} - \frac{1}{n+1},$$

we conclude by taking the limit that φ is an eigenvector of $\tilde{\mathscr{P}}$ of eigenvalue 1 which is obviously strictly positive. □

Lemma 8.30 *Let* $g \in \mathcal{H}_\beta(\overline{\mathcal{X}}_0)$ *be a nonnegative function. Then there exists a constant* $\gamma \geq 0$ *such that for any* $x \in \overline{\mathcal{X}}_0$ *we have*

$$\lim_{n \to \infty} \left| \left(\tilde{\mathscr{P}}^n g \right)(x) - \gamma \varphi(x) \right| = 0.$$

Proof We assume that the function g is not proportional to φ, otherwise the result is obvious. We define recursively a sequence of nonnegative numbers (γ_n) and a sequence of nonnegative function (g_n) in $\mathcal{H}_\beta(\overline{\mathcal{X}}_0)$ by

$$\gamma_n = \inf_{x \in \overline{\mathcal{X}}_0} \frac{\tilde{\mathscr{P}}^n g(x)}{\varphi(x)}$$

and

$$g_n = \tilde{\mathscr{P}}^n g - \gamma_n \varphi.$$

In particular, $g_n \geq 0$ and

$$\inf_{x \in \overline{\mathcal{X}}_0} g_n(x) = 0.$$

From the relation

$$\tilde{\mathscr{P}}^{n+1}g = \gamma_{n+1}\varphi + g_{n+1} = \tilde{\mathscr{P}}\big(\tilde{\mathscr{P}}^n g\big) = \tilde{\mathscr{P}}(\gamma_n\varphi + g_n) = \gamma_n\varphi + \tilde{\mathscr{P}}(g_n),$$

we derive $\gamma_{n+1} \geq \gamma_n$ and

$$\tilde{\mathscr{P}}(g_n) \geq g_{n+1}. \tag{8.17}$$

Since the operator $\tilde{\mathscr{P}}$ is quasi-compact (see the definition and Proposition 8.27), we have for any n

$$\gamma_n \leq \frac{\|\tilde{\mathscr{P}}^n g\|_{\mathcal{H}_\beta(\overline{\mathcal{X}}_0)}}{\inf_{x\in\overline{\mathcal{X}}_0}\varphi(x)} \leq C\frac{\|g\|_{\mathcal{H}_\beta(\overline{\mathcal{X}}_0)}}{\inf_{x\in\overline{\mathcal{X}}_0}\varphi(x)} < \infty.$$

Therefore, the nondecreasing sequence (γ_n) converges to a nonnegative number γ. Assume that (g_n) does not converge to zero in $\mathcal{C}(\overline{\mathcal{X}}_0)$. We are going to derive a contradiction. Since the sequence (g_n) is bounded in $\mathcal{H}_\beta(\overline{\mathcal{X}}_0)$, it follows from Ascoli's Theorem (see, for example, Dieudonné 1969) that this sequence is compact in $\mathcal{C}(\overline{\mathcal{X}}_0)$. Let g_* denote a nonzero accumulation point, namely

$$g_* = \lim_{j\to\infty} g_{n_j}$$

for some infinite sequence of integers (n_j). If $g_*|_{\overline{\mathcal{X}}_\infty} \geq 0$ is not identically zero, it follows from Lemma 8.28 that there is an integer p such that

$$\tilde{\mathscr{P}}^p g_* > 0.$$

In particular, there is a number $\delta > 0$ such that

$$\tilde{\mathscr{P}}^p g_* = \delta\varphi + u$$

with $u \geq 0$. We have

$$\begin{aligned}\tilde{\mathscr{P}}^{n_j+p}g &= \gamma_{n_j+p}\varphi + g_{n_j+p} = \tilde{\mathscr{P}}^p\big(\tilde{\mathscr{P}}^{n_j}g\big)\\ &= \gamma_{n_j}\varphi + \tilde{\mathscr{P}}^p g_{n_j} \geq \left(\gamma_{n_j} + \inf_x \frac{\tilde{\mathscr{P}}^p g_{n_j}(x)}{\varphi(x)}\right)\varphi.\end{aligned}$$

However,

$$\lim_{j\to\infty}\left(\gamma_{n_j} + \inf_x \frac{\tilde{\mathscr{P}}^p g_{n_j}(x)}{\varphi(x)}\right) \geq \gamma + \delta,$$

and for large j, we have a contradiction to the definition of $\gamma_{n_j+p} \leq \gamma$.

It only remains to consider the case where all accumulation points g_* vanish on $\overline{\mathcal{X}}_\infty$. Let (n_j) be a diverging increasing sequence of integers such that (in $\mathcal{C}(\overline{\mathcal{X}}_0)$)

$$\lim_{j\to\infty} g_{n_j} = g_*,$$

where g_* vanishes on $\overline{\mathcal{X}}_\infty$. Using iteratively (8.17), we get for any $j > j'$

$$g_{n_j} \le \tilde{\mathscr{P}}^{n_j - n_{j'}} g_{n_{j'}},$$

which implies for any $x \in \overline{\mathcal{X}}_0$

$$g_{n_j}(x) \le \sup_{y \in \overline{\mathcal{X}}_{n_j - n_{j'}}} g_{n_{j'}}(y) \tilde{\mathscr{P}}^{n_j - n_{j'}} \mathbf{1}(x) \le C \sup_{y \in \overline{\mathcal{X}}_{n_j - n_{j'}}} g_{n_{j'}}(y), \tag{8.18}$$

where the last inequality follows from Lemma 8.26. We now choose two increasing sequences of integers (j_ℓ) and (j'_ℓ) such that the sequences $(n_{j'_\ell})$ and $(n_{j_\ell} - n_{j'_\ell})$ diverge. Since the sequence of compact sets $(\overline{\mathcal{X}}_n)$ is decreasing, we have (see Dieudonné 1969)

$$\lim_{\ell \to \infty} \sup_{y \in \overline{\mathcal{X}}_{n_{j_\ell} - n_{j'_\ell}}} d(y, \overline{\mathcal{X}}_\infty) = 0.$$

Since the functions (g_{n_j}) are uniformly continuous and g_* vanishes on $\overline{\mathcal{X}}_\infty$, we conclude that

$$\lim_{\ell \to \infty} \sup_{y \in \overline{\mathcal{X}}_{n_{j_\ell} - n_{j'_\ell}}} g_{n_{j'_\ell}}(y) = 0.$$

Using (8.18), we conclude that

$$g_* = \lim_{\ell \to \infty} g_{n_{j_\ell}} = 0,$$

namely, all the accumulation points of the sequence (g_n) are equal to zero, and hence this sequence converges to zero. □

Theorem 8.31 *Under Hypotheses* **H1**, **H2**, **H3**, *for any* $\beta \in]0, \alpha[$, *the spectrum of the operator* $\tilde{\mathscr{P}}$ *in the Banach space* $\mathcal{H}_\beta(\overline{\mathcal{X}}_0)$ *consists of a positive eigenvalue* $\lambda < 1$ *with eigenspace of dimension* 1 *and an eigenvector* φ *which can be chosen strictly positive. The function* φ *and the number* λ *do not depend on* β. *The corresponding linear form which is an eigenvector of the adjoint for the same eigenvalue can be chosen as a probability measure* ζ *independent of* β. *Moreover, there exists* $\rho_\beta \in]0, \lambda[$ *such that the rest of the spectrum of* $\tilde{\mathscr{P}}$ *is contained in the complex disk* $|z| \le \rho_\beta$.

Proof Let φ^* be an eigenvector of $\tilde{\mathscr{P}}$ in $\mathcal{H}_\beta(\overline{\mathcal{X}}_0)$ with eigenvalue λ satisfying $|\lambda| = 1$. We can write

$$\varphi^* = \varphi^*_{R,+} - \varphi^*_{R,-} + i\big(\varphi^*_{I,+} - \varphi^*_{I,-}\big)$$

where $\varphi^*_{R,\pm}$ and $\varphi^*_{I,\pm}$ denote four nonnegative functions in $\mathcal{H}_\beta(\overline{\mathcal{X}}_0)$, satisfying $\varphi^*_{R,+}\varphi^*_{R,-} = 0$ and $\varphi^*_{I,+}\varphi^*_{I,-} = 0$. Applying Lemma 8.30, we conclude that

$$\tilde{\mathscr{P}}^n \varphi^* = \lambda^n \varphi^*$$

converges when n tends to infinity to a multiple of φ. Therefore, $\lambda = 1$ and this eigenvalue is nondegenerate. The rest of the theorem follows by applying the Uniform Ergodic Theorem 8.16 and the definition (8.15). □

We can now state the main result for this section.

Theorem 8.32 *Under Hypotheses* **H1**, **H2**, **H3**, *the map* f *has a unique QSD* ν *on* $\mathcal{X}_0$ *absolutely continuous with respect to the Lebesgue measure. Moreover, if* h *denotes the density of this QSD* ($d\nu = h\,d\text{Leb}$), *there are two constants* $C > 0$ *and* $\rho \in]0, 1[$ *such that for any* $g \in \mathcal{C}(\mathcal{X}^M)$ *we have*

$$\left| \frac{1}{\nu(\mathcal{X}_n)} \int_{\mathcal{X}_n} g \circ f^n \, d\text{Leb} - \int_{\mathcal{X}_0} gh \, d\text{Leb} \right| \leq C \|g\|_{\mathcal{C}(\mathcal{X}_0)} \rho^n .$$

Proof This result follows at once from (8.14) and Theorem 8.31. □

8.3.1 An Application to the Newton Method

As an example, we consider the problem of solving the polynomial equation $z^3 = 1$. The roots are, of course, well known (the three cube roots of unity $1, j, j^2$), but one can investigate on this simple example the behavior of the Newton method which is often used to find roots of more complicated equations. This method consists in iterating a map (see, for example, Henrici 1974) which in this case is given by

$$R(z) = \frac{2z}{3} + \frac{1}{3z^2} .$$

It is easy to verify that the three roots of unity are the only fixed points ($R(z) = z$), and the derivative of R vanishes at these points. It is known that if one starts with an initial condition z_0 near one of the roots, the iteration

$$z_{n+1} = R(z_n),$$

converges very fast (doubly exponentially fast) to the root. However, there are starting points whose orbits never converge. We refer to Eckmann (1983), Hubbard et al. (2001) for more information.

Let Ξ_∞ denote the set of initial conditions whose orbit does not converge to any of the fixed points 1, j or j^2. This set turns out to be in this case the Julia set of the map R (see, for example, Blanchard 1984 and Lyubich 1986 for definitions and results).

The part of this set near the origin is drawn in black in Fig. 8.1.

In order to apply the preceding results about the existence and properties of QSDs, we first conjugate the transformation by a map sending the fixed point 1

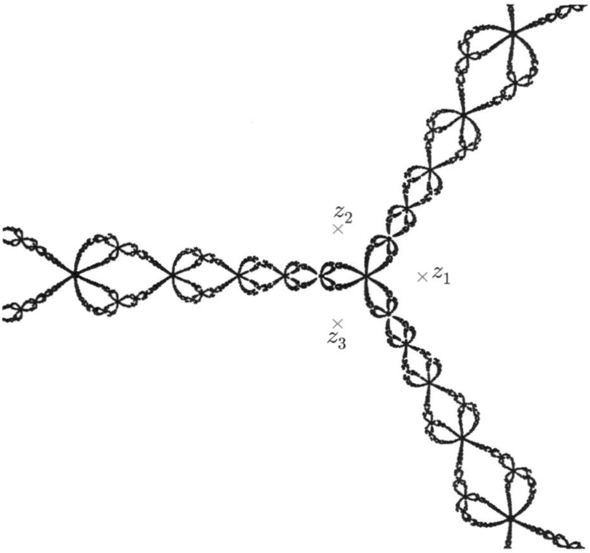

Fig. 8.1 The Julia set for the Newton method of $z^3 = 1$

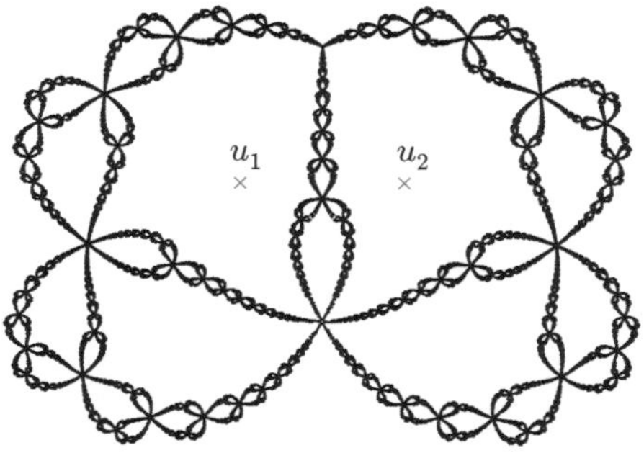

Fig. 8.2 The Julia set for the map f

to infinity. One can use, for example, the (bijective) map

$$H(z) = \frac{1}{z-1}.$$

Instead of using R, it is therefore equivalent to use the map

$$f(z) = H \circ R \circ H^{-1}(z) = \frac{2z^3 + 6z^2 + 3z}{3z + 2}.$$

This map has two fixed points with zero derivative at $1 + j$ and $1 + j^2$ (and a third one at infinity).

The map f is again a rational map. We will denote by J its Julia set which is now a compact set (see Fig. 8.2). We have also conjugated the map by a rotation to obtain nicer pictures.

Recall that a critical point of a rational map is a point where the derivative vanishes. The map R and hence the map f have only three critical points which are all fixed points. Therefore, we conclude by Theorem 8.1 in Blanchard (1984) that the Julia set J is hyperbolic, namely, there is a number $K_0 > 1$ and an integer n_0 such that

$$\inf_{z\in J}\left|f^{n_0\prime}(z)\right| > K_0.$$

By the chain rule, we have for any z

$$f^{n_0\prime}(z) = \prod_{j=0}^{n_0-1} f'\bigl(f^j(z)\bigr).$$

Since $-2/3 \notin J$, f' is continuous (and therefore bounded on J). This implies that $f^{n_0\prime}$ is continuous on a neighborhood of the compact set J. Therefore, there exists a number $\epsilon_0 > 0$ and a number $K > 1$ such that

$$\inf_{z, d(z,J)\le\epsilon_0}\left|f^{n_0\prime}(z)\right| > K,$$

where $d(z, J)$ denotes the distance from z to J, namely,

$$d(z, J) = \inf_{u\in J}|u - z|.$$

Lemma 8.33 *There exists a number* $\epsilon_1 > 0$, $\epsilon_1 < \epsilon_0$, *such that if we define*

$$\mathcal{X}_0 = \bigl\{z : d(z, J) < \epsilon_1\bigr\},$$

then

$$f^{n_0}(\mathcal{X}_0) \supset \mathcal{X}_0.$$

Proof Let $z \in J$. Since $|f^{n_0\prime}(z)| > K$, we can apply the Inverse Function Theorem to f^{n_0} (see Dieudonné 1969). This implies that there exists a number $\tilde{\epsilon}_1 > 0$, which may depend on z such that $f^{n_0\prime}$ is a bijection from $B_{\tilde{\epsilon}_1}(z)$ to its image. Moreover, we can assume $\tilde{\epsilon}_1 > 0$ small enough so that $\epsilon_0 > \tilde{\epsilon}_1 > 0$ and for any $\epsilon \le \tilde{\epsilon}_1$

$$f^{n_0}\bigl(B_\epsilon(z)\bigr) \supset B_{(1+K)\epsilon/2}\bigl(f^{n_0}(z)\bigr).$$

Moreover, it follows from the Inverse Function Theorem that the numbers $\tilde{\epsilon}_1$ are bounded away from zero uniformly in $z \in J$. We can take for ϵ_1 the infimum of these numbers, and the result is proved. □

A set $\mathcal{X}_0$ with $f(\mathcal{X}_0) \supset \mathcal{X}_0$ is shown in Fig. 8.3, together with its first iterate. Figure 8.4 shows the same sets with the Julia set. Figures 8.5 and 8.6 show the equivalent sets in the original coordinates.

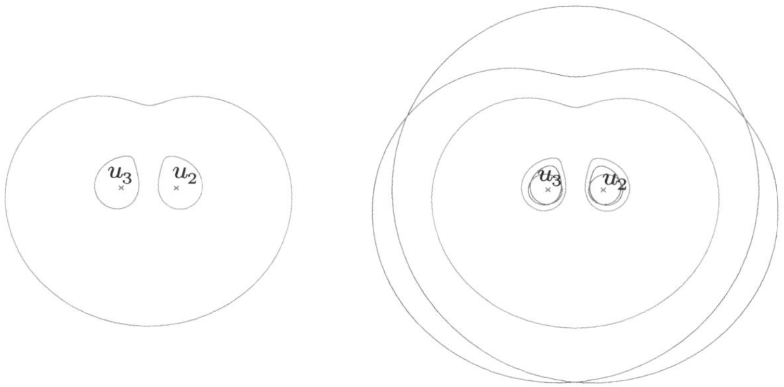

Fig. 8.3 The boundary of a neighborhood of the three critical points of the map f *in green on the left*, together with their image *in magenta on the right*. The connected set $\mathcal{X}_0$ satisfies $f(\mathcal{X}_0) \supset \mathcal{X}_0$

Fig. 8.4 Julia set, boundary of a neighborhood of the three critical points of the map f *in green*, together with their image *in magenta*. The connected set $\mathcal{X}_0$ satisfies $f(\mathcal{X}_0) \supset \mathcal{X}_0$

It is easy to verify that Hypotheses **H1**, **H2** and **H3** of Sect. 8.3 are satisfied. We can therefore apply Theorem 8.32 to conclude that the map f^{n_0} has an absolutely continuous QSD. The density is bounded above and below away from zero.

We can now apply the inverse of the conjugation H to come back to our initial picture.

Going back to the initial picture, we get

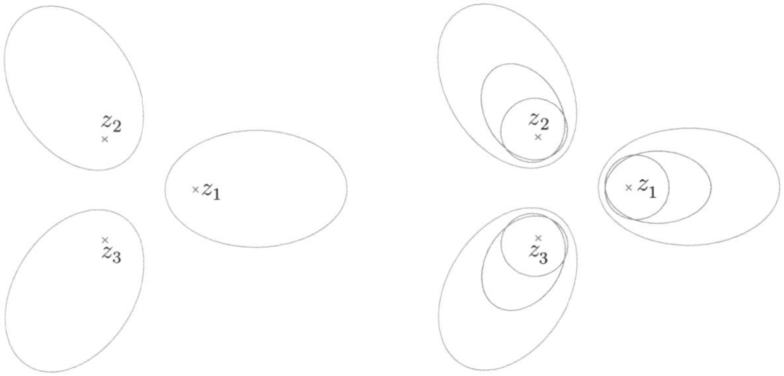

Fig. 8.5 The boundary of a neighborhood of the three critical points of the map R *in green on the left*, together with their image *in magenta on the right*

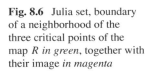

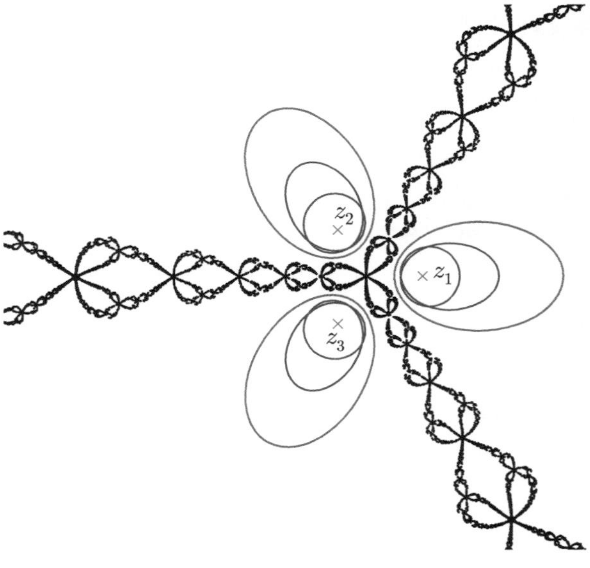

Fig. 8.6 Julia set, boundary of a neighborhood of the three critical points of the map R *in green*, together with their image *in magenta*

8.4 Other Results

8.4.1 Other Proofs

There are several different proofs of the results we have presented in the previous section. In both cases, we have used the method of quasi-compact operators. One can also use projective techniques (based, for example, on the contraction properties of Hilbert projective metric of cones due to Birkhoff). We refer to Pianigiani and Yorke (1979), Collet et al. (1994, 1995b) for results using this technique, and to Demers and Young (2006) for a general review.

8.4.2 Absolutely Continuous QSD

Other examples of QSDs which are absolutely continuous with the one dimensional conformal measure have been constructed in Liverani and Maume (2003). See also Collet and Galves (1995) for a case of small holes. In higher dimension and for hyperbolic systems, one can try to construct QSDs which are absolutely continuous in the unstable direction. We refer to Collet et al. (2000), Chernov et al. (1998, 2000), Chernov and Markarian (1997a, 1997b), Demers (2005a, 2005b), Demers and Liverani (2008), Bruin et al. (2010), Demers et al. (2010) and Demers and Young (2006) for more results and details.

8.4.3 Properties of a QSD

Although a QSD is not invariant, one can wonder about the process obtained by conditioning upon survival. In Collet and Martínez (1999), the following Central Limit Theorem was proved for the Gibbs QSD (8.11) described in Sect. 8.2, Theorem 8.20. The result can be stated as follows (using the same notations).

Theorem 8.34 *Assume the hypothesis of Theorem* 8.20 *hold (in particular, the potential* U *is* α*-Hölder). Let* g *be an* α*-Hölder function such that*

$$\int \varphi^* g\, d\zeta = 0$$

and

$$\lim_{N\to\infty} \frac{1}{N} \frac{\int_{\mathcal{X}_n} (\sum_{n=0}^{N-1} g \circ \sigma^n)^2 \varphi^*\, d\zeta}{\int_{\mathcal{X}_n} \varphi^*\, d\zeta} \neq 0.$$

The limit always exists, and we denote by η *the nonnegative square root of this nonnegative number. Then*

$$\lim_{N\to\infty} \frac{\int_{\mathcal{X}_n} \mathbb{1}_{\sum_{n=0}^{N-1} g\circ\sigma^n \le t\sqrt{N}} \varphi^*\, d\zeta}{\int_{\mathcal{X}_n} \varphi^*\, d\zeta} = \frac{1}{\sqrt{2\pi}} \int_{-\infty}^{t} e^{-u^2/(2\eta)}\, du.$$

We refer to Collet and Martínez (1999) for the proof which uses the properties of the Ruelle–Perron–Frobenius operator given above. The proof of this theorem is similar to the one of Theorem 3.4 in Chap. 3 which is an analogous result for finite Markov chains.

8.4.4 Decay of Probabilities

The result of Lemma 8.3 suggest asking more generally if one can prove at least an asymptotically exponential decay for dynamical systems with holes. This question was investigated in several papers starting with Eckmann and Ruelle (1985), in connection with Lyapunov exponents for an invariant measure on a repeller. We refer to Lopes and Markarian (1996), Vaienti (1987), Froyland and Stancevic (2010), Bahsoun and Bose (2010), Keller and Liverani (2009), Araújo and Pacifico (2006), Bahsoun (2006), Collet et al. (1999), Baladi et al. (1999) and Afraimovich and Bunimovich (2010) for details.

References

Afraimovich, V. S., & Bunimovich, L. A. (2010). Which hole is leaking the most: a topological approach to study open systems. *Nonlinearity*, *23*, 643–656.

Anderson, W. J. (1991). *Continuous-time Markov chains*. New York: Springer.

Araújo, V., & Pacifico, M. J. (2006). Large deviations for non-uniformly expanding maps. *Journal of Statistical Physics*, *125*, 415–457.

Asmussen, S. (1987). *Applied probability and queues*. New York: Wiley.

Athreya, K., & Ney, P. (1972). *Branching processes*. New York: Springer.

Azencott, R. (1974). Behavior of diffusion semigroups at infinity. *Bulletin de la Société Mathématique de France*, *102*, 193–240.

Bahsoun, W. (2006). Rigorous numerical approximation of escape rates. *Nonlinearity*, *19*, 2529–2542.

Bahsoun, W., & Bose, C. (2010). Quasi-invariant measures, escape rates and the effect of the hole. *Discrete and Continuous Dynamical Systems*, *27*, 1107–1121.

Baladi, V., Bonatti, C., & Schmitt, B. (1999). Abnormal escape rates from nonuniformly hyperbolic sets. *Ergodic Theory & Dynamical Systems*, *19*, 1111–1125.

Berezin, F. A., & Shubin, M. A. (1991). *The Schrödinger equation*. Dordrecht: Kluwer Academic.

Billingsley, P. (1968). *Convergence of probability measures*. New York: Wiley.

Blanchard, P. (1984). Complex analytic dynamics on the Riemann sphere. *Bulletin of the American Mathematical Society*, *11*, 85–141.

Blumenthal, R. M., & Getoor, R. K. (1968). *Markov processes and potential theory*. New York: Academic Press.

Bowen, R. (1975). *Lecture notes in mathematics: Vol. 470. Equilibrium states in ergodic theory and Anosov diffeomorphisms*. Berlin: Springer. 2nd revised edition by Chazottes, J.-R.

Brown, L. D. (1968) *IMS lecture notes: Vol. 9. Fundamentals of statistical exponential families*.

Bruin, H., Demers, M., & Melbourne, I. (2010). Existence and convergence properties of physical measures for certain dynamical systems with holes. *Ergodic Theory & Dynamical Systems*, *30*, 687–720.

Cattiaux, P., Collet, P., Lambert, A., Martínez, S., Méléard, S., & San Martín, J. (2009). Quasi-stationary distributions and diffusions models in population dynamics. *Annals of Probability*, *37*, 1926–1969.

Cavender, J. A. (1978). Quasi-stationary distributions of birth and death processes. *Advances in Applied Probability*, *10*, 570–586.

Chernov, N., & Markarian, R. (1997a). Anosov maps with rectangular holes. Nonergodic cases. *Boletim Da Sociedade Brasileira de Matemática*, *28*, 315–342.

Chernov, N., & Markarian, R. (1997b). Ergodic properties of Anosov maps with rectangular holes. *Boletim Da Sociedade Brasileira de Matemática*, *28*, 271–314.

P. Collet et al., *Quasi-Stationary Distributions*, Probability and Its Applications, DOI 10.1007/978-3-642-33131-2, © Springer-Verlag Berlin Heidelberg 2013

Chernov, N., Markarian, R., & Troubetzkoy, S. (1998). Conditionally invariant measures for Anosov maps with small holes. *Ergodic Theory & Dynamical Systems*, *18*, 1049–1073.
Chernov, N., Markarian, R., & Troubetzkoy, S. (2000). Invariant measures for Anosov maps with small holes. *Ergodic Theory & Dynamical Systems*, *20*, 1007–1044.
Chihara, T. S. (1972). Convergent sequences of orthogonal polynomials. *Journal of Mathematical Analysis and Applications*, *38*, 335–347.
Coddington, E. A., & Levinson, N. (1955). *Theory of ordinary differential equations*. New York: McGraw-Hill.
Collet, P., & Galves, A. (1995). Asymptotic distribution of entrance times for expanding maps of the interval. In *Ser. appl. anal.: Vol. 4. Dynamical systems and applications* (pp. 139–152). River Edge: World Scientific.
Collet, P., & Martínez, S. (1999). Diffusion coefficient in transient chaos. *Nonlinearity*, *12*, 445–450.
Collet, P., Martínez, S., & Schmitt, B. (1994). The Yorke–Pianigiani measure and the asymptotic law on the limit cantor set of expanding systems. *Nonlinearity*, *7*, 1437–1443.
Collet, P., Martínez, S., & San Martín, J. (1995a). Asymptotic laws for one dimensional diffusions conditioned to nonabsorption. *Annals of Probability*, *23*, 1300–1314.
Collet, P., Martínez, S., & Schmitt, B. (1995b). The Pianigiani–Yorke measure for topological Markov chains. *Israel Journal of Mathematics*, *97*, 61–71.
Collet, P., Martínez, S., & Schmitt, B. (1999). On the enhancement of diffusion by chaos, escape rates and stochastic instability. *Transactions of the American Mathematical Society*, *351*, 2875–2897.
Collet, P., Maume, V., & Martínez, S. (2000). On the existence of conditionally invariant probability measures in dynamical systems (with V. Maume and S. Martínez). *Nonlinearity*, *13*, 1263–1274. Erratum, *Nonlinearity 17*, 1985–1987 (2004).
Collet, P., López, J., & Martínez, S. (2003). Order relations of measures when avoiding decreasing sets. *Statistics & Probability Letters*, *65*(3), 165–175.
Collet, P., Martínez, S., Méléard, S., & San Martín, J. (2011). Quasi-stationary distributions for structured birth and death processes with mutations. *Probability Theory and Related Fields*, *151*(1–2), 191–231.
Conway, J. B. (1990). *A course in functional analysis* (2nd ed.). New York: Springer.
Darroch, J. N., & Seneta, E. (1965). On quasi-stationary distributions in absorbing discrete-time finite Markov chains. *Journal of Applied Probability*, *2*, 88–100.
Davies, B. (1989). *Heat kernels and spectral theory*. Cambridge: Cambridge University Press.
Demers, M. (2005a). Markov extensions and conditionally invariant measures for certain logistic maps with small holes. *Ergodic Theory & Dynamical Systems*, *25*, 1139–1171.
Demers, M. (2005b). Markov extensions for dynamical systems with holes: an application to expanding maps of the interval. *Israel Journal of Mathematics*, *146*, 189–221.
Demers, M., & Liverani, C. (2008). Stability of statistical properties in two-dimensional piecewise hyperbolic maps. *Transactions of the American Mathematical Society*, *360*, 4777–4814.
Demers, M., & Young, L. S. (2006). Escape rates and conditionally invariant measures. *Nonlinearity*, *19*, 377–397.
Demers, M., Wright, P., & Young, L. S. (2010). Escape rates and physically relevant measures for billiards with small holes. *Communications in Mathematical Physics*, *294*, 353–388.
Denker, M., Grillenberg, C., & Sigmund, K. (1976). *Lecture notes in mathematics: Vol. 527. Ergodic theory of compact spaces*. Berlin: Springer.
Dieudonné, J. (1969). *Foundations of modern analysis*. San Diego: Academic Press.
Dunford, N., & Schwartz, J. (1958). *Linear operators* (Vol. I). New York: Wiley-Interscience.
Eckmann, J.-P. (1983). Savez-vous résoudre $z^3 = 1$? *La Recherche*, *14*, 260–262.
Eckmann, J.-P., & Ruelle, D. (1985). Ergodic theory of chaos and strange attractors. *Reviews of Modern Physics*, *57*, 617–656.
Eveson, S., & Nussbaum, R. (1995). Applications of the Birkhoff–Hopf theorem to the spectral theory of positive linear operators. *Mathematical Proceedings of the Cambridge Philosophical Society*, *117*, 491–512.

Feller, W. (1952). The parabolic differential equation and the associated semi-groups of transformations. *Annals of Mathematics*, *55*, 468–519.
Feller, W. (1971). *An introduction to probability theory and its applications* (2nd ed., Vol. II). New York: Wiley.
Ferrari, P. A., Kesten, H., Martínez, S., & Picco, P. (1995a). Existence of quasi-stationary distributions. A renewal dynamical approach. *Annals of Probability*, *23*, 501–521.
Ferrari, P. A., Martínez, S., & Picco, P. (1995b). Existence of non-trivial quasi-stationary distributions in the birth-death chain. *Advances in Applied Probability*, *24*, 795–813.
Ferrari, P. A., Kesten, H., & Martínez, S. (1996). R-positivity, quasi-stationary distributions and ratio limit theorems for a class of probabilistic automata. *The Annals of Applied Probability*, *6*, 577–616.
Freedman, D. (1971). In *Markov chains*. San Francisco: Holden-Day.
Froyland, G., & Stancevic, O. (2010). Escape rates and Perron–Frobenius operators: open and closed dynamical systems. *Discrete and Continuous Dynamical Systems. Series B*, *14*, 457–472.
Fukushima, M. (1980). *Dirichlet forms and Markov processes*. Amsterdam: Kodansha/North-Holland.
Fukushima, M., Oshima, Y., & Takeda, M. (1994). *Studies in mathematics: Vol. 19. Dirichlet forms and symmetric Markov processes*. Berlin: de Gruyter.
Gikhman, I., & Skorokhod, A. (1996). *Introduction to the theory of random processes*. Mineola: Dover.
Good, P. (1968). The limiting behavior of transient birth and death processes conditioned on survival. *Journal of the Australian Mathematical Society*, *8*, 716–722. (Also see review MR02430879(39#2224)).
Hart, A., Martínez, S., & San Martín, J. (2003). The λ-classification of continuous-time birth-and-death processes. *Advances in Applied Probability*, *35*, 1111–1130.
Henrici, P. (1974). *Applied and computational complex analysis* (Vol. I). New York: Wiley.
Hubbard, J., Schleicher, D., & Sutherland, S. (2001). How to fin all roots of complex polynomials by Newton's method. *Inventiones Mathematicae*, *146*, 1–33.
Ikeda, N., & Watanabe, S. (1988). *Stochastic differential equations and diffusion processes* (2nd ed.). Amsterdam: North-Holland.
Ionescu Tulcea, C., & Marinescu, G. (1950). Théorie ergodique pour des classes d'opérateurs non complètement continus. *Annals of Mathematics (2)*, *52*, 140–147.
Jacka, S. D., & Roberts, G. O. (1995). Weak convergence of conditioned process on a countable state space. *Journal of Applied Probability*, *32*, 902–916.
Kallenberg, O. (1997). *Foundations of modern probability*. New York: Springer.
Karatzas, I., & Shreve, S. E. (1988). *Brownian motion and stochastic calculus*. New York: Springer.
Karlin, S., & McGregor, J. L. (1957a). The classification of birth and death processes. *Transactions of the American Mathematical Society*, *86*, 366–400.
Karlin, S., & McGregor, J. L. (1957b). The differential equations of birth-and-death processes, and the Stieltjes moment problem. *Transactions of the American Mathematical Society*, *85*, 489–546.
Karlin, S., & Taylor, H. M. (1975). *A first course in stochastic processes* (2nd ed.). New York: Academic Press.
Kato, T. (1995). *Perturbation theory for linear operators*. Berlin: Springer. Reprint of 1980 edition.
Keller, G., & Liverani, C. (2009). Rare events, escape rates and quasistationarity: some exact formulas. *Journal of Statistical Physics*, *135*, 519–534.
Kesten, H. (1976). Existence and uniqueness of countable one-dimensional random fields. *Annals of Probability*, *4*, 557–569.
Kesten, H. (1995). A ratio limit for (sub)Markov chains on $\{1, 2, \dots\}$ with bounded jumps. *Advances in Applied Probability*, *27*, 652–691.
Kijima, M., Nair, M. G., Pollett, P., & Van Doorn, E. (1997). Limiting conditional distributions for birth-death processes. *Advances in Applied Probability*, *29*, 185–204.

Kingman, J. F. C. (1963). The exponential decay of Markov transition probabilities. *Proceedings of the London Mathematical Society*, *13*, 337–358.

Knight, F. B. (1969). Brownian local times and Taboo processes. *Transactions of the American Society*, *143*, 173–185.

Kolb, M., & Steinsaltz, D. (2012). Quasilimiting behavior of one-dimensional diffusions with killing. *Annals of Probability*, *40*(1), 162–212.

Kolmogorov, A. (1938). Zur Lösung einer biolischen Aufgabe. *Izvestiya Nauchno-Issledovatelskogo Instituta Matematiki I Mechaniki pri Tomskom Gosudarstvennom Universitete*, *2*, 1–6.

Kreer, M. (1994). Analytic birth-death processes: a Hilbert space approach. *Stochastic Processes and Their Applications*, *49*, 65–74.

Krylov, N. V., & Safonov, M. V. (1981). A certain property of solutions of parabolic equations with measurable coefficients. *Mathematics of the USSR. Izvestija*, *16*, 151–164.

Lamperti, J., & Ney, J. (1968). Conditioned branching process and their limiting diffusions. *Theory of Probability and Its Applications*, *13*, 128–139.

Lindvall, T. (1992). *Lectures on the coupling method*. New York: Wiley.

Liverani, C., & Maume, V. (2003). Lasota–Yorke maps with holes: conditionally invariant probability measures and invariant measures on the survivor set. *Annales de L'Institut Henri Poincaré*, *39*, 385–412.

Lladser, M., & San Martín, J. (2000). Domain of attraction of the quasi-stationary distributions for the Ornstein–Uhlenbeck process. *Journal of Applied Probability*, *37*, 511–520.

Lopes, A., & Markarian, R. (1996). Open billiards: invariant and conditionally invariant probabilities on cantor sets. *SIAM Journal on Applied Mathematics*, *56*, 651–680.

Lund, R., Meyn, S., & Tweedie, R. L. (1996). Computable exponential convergence rates for stochastically ordered Markov processes. *The Annals of Applied Probability*, *6*, 218–237.

Lyubich, M. Y. (1986). The dynamics of rational transforms: the topological picture. *Russian Mathematical Surveys*, *41*(4), 43–117.

Mandl, P. (1961). Spectral theory of semi-groups connected with diffusion processes and its applications. *Czechoslovak Mathematical Journal*, *11*, 558–569.

Martínez, S. (2008). Notes and a remark on quasi-stationary distributions. In *Monogr. semin. mat. García Galdeano: Vol. 34. Pyrenees international workshop on statistics, probability and operations research: SPO 2007* (pp. 61–80). Zaragoza: Prensas Univ. Zaragoza.

Martínez, S., & San Martín, J. (1994). Quasi-stationary distributions for a Brownian motion with drift and associated limit laws. *Journal of Applied Probability*, *31*, 911–920.

Martínez, S., & San Martín, J. (2004). Classification of killed one-dimensional diffusions. *Annals of Probability*, *32*, 530–552.

Martínez, S., & Vares, M. E. (1995). Markov chain associated to the minimal QSD of birth-death chains. *Journal of Applied Probability*, *32*, 25–38.

Martínez, S., & Ycart, B. (2001). Decay rates and cutoff for convergence and hitting times of Markov chains with countably infinite state space. *Advances in Applied Probability*, *33*, 188–205.

Martínez, S., Picco, P., & San Martín, J. (1998). Domain of attraction of quasi-stationary distributions for the Brownian motion with drift. *Advances in Applied Probability*, *30*, 385–408.

Nair, M. G., & Pollett, P. (1993). On the relationship between μ-invariant measures and quasi-stationary distributions for continuous-time Markov chains. *Advances in Applied Probability*, *25*, 82–102.

Nussbaum, R. (1970). The radius of the essential spectrum. *Duke Mathematical Journal*, *37*, 473–478.

Oikhberg, T., & Troitsky, V. (2005). A theorem of Krein revisited. *The Rocky Mountain Journal of Mathematics*, *35*, 195–210.

Pianigiani, G., & Yorke, J. A. (1979). Expanding maps on sets which are almost invariant: decay and chaos. *Transactions of the American Mathematical Society*, *252*, 351–356.

Pinsky, R. (1985). On the convergence of diffusion processes conditioned to remain in bounded region for large time to limiting positive recurrent diffusion processes. *Annals of Probability*, *13*, 363–378.

Pollett, P. (1986). On the equivalence of $\mu-$invariant measures for the minimal process and its q-matrix. *Stochastic Processes and Their Applications*, *22*, 203–221.

Revuz, D., & Yor, M. (1999). *A series of comprehensive studies. Continuous martingales and Brownian motion* (3rd ed.). Berlin: Springer.

Riesz, F., & Sz.-Nagy, B. (1955). *Functional analysis*. New York: Ungar.

Robinson, C. (2008). *Dynamical systems: stability, symbolic dynamics, and chaos*. Boca Raton: CRC Press.

Rogers, L., & Pitman, J. (1981). Markov functions. *Annals of Probability*, *9*, 573–582.

Royer, G. (1999). *Une initiation aux inégalités de Sobolev logarithmiques*. Paris: S.M.F.

Ruelle, D. (1978). *Encyclopedia of mathematics and its applications: Vol. 5. Thermodynamic formalism*. Reading: Addison-Wesley.

Seneta, E. (1981). *Springer series in statistics. Non-negative matrices and Markov chains*. New York: Springer.

Seneta, E., & Vere-Jones, D. (1996). On quasi-stationary distributions in discrete-time Markov chains with denumerable infinity of states. *Journal of Applied Probability*, *3*, 403–434.

Steinsaltz, D., & Evans, S. (2004). Markov mortality models: implications of quasistationarity and varying initial distributions. *Theoretical Population Biology*, *65*(4), 319–337.

Steinsaltz, D., & Evans, S. (2007). Quasistationary distributions for one-dimensional diffusions with killing. *Transactions of the American Mathematical Society*, *359*, 1285–1324.

Tweedie, R. L. (1974). Some properties of the Feller minimal process. *Quarterly Journal of Mathematics*, *25*, 485–495.

Tychonov, A. (1935). Ein Fixpunktsatz. *Mathematische Annalen*, *117*, 767–776.

Vaienti, S. (1987). Lyapunov exponent and bounds for the Hausdorff dimension of Julia sets of polynomial maps. *Nuovo Cimento B*, *99*, 77–91.

Van Doorn, E. (1985). Conditions for exponential ergodicity and bounds for the decay parameter of a birth-death process. *Advances in Applied Probability*, *17*, 514–530.

Van Doorn, E. (1986). On orthogonal polynomials with positive zeros and the associated kernel polynomials. *Journal of Mathematical Analysis and Applications*, *113*, 441–450.

Van Doorn, E. (1991). Quasi-stationary distributions and convergence to quasi-stationary of birth-death processes. *Advances in Applied Probability*, *23*, 683–700.

Van Doorn, E., & Schrijner, P. (1995). Geometric ergodicity and quasi-stationarity in discrete time birth-death processes. *Journal of the Australian Mathematical Society, Series B*, *37*, 121–144.

Vere-Jones, D. (1962). Geometric ergodicity in denumerable Markov chains. *Quarterly Journal of Mathematics Oxford (2)*, *13*, 7–28.

Vere-Jones, D. (1967). Ergodic properties of nonnegative matrices. I. *Pacific Journal of Mathematics*, *22*, 361–386.

Williams, D. (1970). Decomposing the Brownian path. *Bulletin of the American Mathematical Society*, *76*, 871–873.

Yaglom, A. M. (1947). Certain limit theorems of the theory of branching processes. *Doklady Akademii Nauk SSSR*, *56*, 795–798 (in Russian).

Yosida, K. (1980). *GMW: Vol. 123. Functional analysis*. Berlin: Springer.

Yosida, K., & Kakutani, S. (1941). Operator-theoretical treatment of Markoff's process and mean ergodic Theorem. *Annals of Mathematics*, *42*, 168–228.

Index

P. Collet et al., *Quasi-Stationary Distributions*, Probability and Its Applications,
DOI 10.1007/978-3-642-33131-2, © Springer-Verlag Berlin Heidelberg 2013

Table of Notations

a_p^q sequence $(a_p, \dots, a_q)$, 232
$|\mathcal{A}|$ cardinality of the alphabet $\mathcal{A}$, 232
$B_r(x)$ ball of $\mathbb{R}^d$ of radius r and center x, 252
$\mathcal{B}(\mathcal{X})$ Borel σ-field on $\mathcal{X}$, 17
$\mathcal{C}(\mathcal{X})$ set of continuous functions on $\mathcal{X}$, 17
$\mathcal{C}_b(\mathcal{X})$ set of bounded continuous functions on $\mathcal{X}$, 17
$\mathcal{C}^+(\mathcal{X}^{M,+})$ set of continuous nonnegative functions on $\mathcal{X}^{M,+}$, 239
$\mathcal{C}(\mathcal{A}^{\mathbb{Z}})$ set of continuous functions on $\mathcal{A}^{\mathbb{Z}}$, 232
$\mathcal{C}(F)$ set of continuous functions on F, 233
$C_{a_p^q}$ cylinder associated to a_p^q, 232
$D(\mathcal{L})$ domain of $\mathcal{L}$, 200
$\delta(i,j)$ Kronecker's delta, 36
δ_x Dirac measure at the point x, 27
$\partial\mathcal{X}$ set of forbidden states, 18
η exponential rate for T, 141
η_k eigenfunction of $\tilde{\mathcal{L}}$, 202
f transformation, 227
$\mathcal{H}_\alpha(F)$ set of α-Hölder continuous functions on F, 233
$\mathcal{H}_\alpha(\mathcal{A}^{\mathbb{Z}})$ set of α-Hölder functions on $\mathcal{A}^{\mathbb{Z}}$, 233
$\operatorname{Int} F$ interior of a set F, 251
Leb Lebesgue measure, 250
$\mathcal{L}$ infinitesimal generator, 114
$\tilde{\mathcal{L}}$ infinitesimal generator of $(\tilde{P}_t)$, 201
Λ scale function, 114
$\underline{\lambda}(z)$ top of spectrum for $-\mathcal{L}$ on $[z,\infty)$, 162
$\underline{\lambda}$ top of spectrum for $-\mathcal{L}$, 129
$\mathcal{M}(\mathcal{X})$ set of real measurable functions on $\mathcal{X}$, 17
$\mathcal{M}_+(\mathcal{X})$ set of positive real measurable functions on $\mathcal{X}$, 17
μ speed measure, 114
ν quasi-stationary distribution (QSD), 19
$\mathcal{P}_M$ Ruelle–Perron–Frobenius operator associated to M, 234
$\mathcal{P}_{M',M}$ Ruelle–Perron–Frobenius operator for the pair of matrices M, M', 237
$\tilde{\mathcal{P}}$ normalized Ruelle–Perron–Frobenius operator for the transformation f, 253
$\tilde{\mathcal{P}}_{M',M}$ normalized Ruelle–Perron–Frobenius operator for the pair of matrices M, M', 240
P_t semigroup of the process before killing, 24
$P_t^\dagger$ semigroup in the spaces of measures, 24
p transition density of X, 114
p_t^D Dirichlet heat kernel in $\mathbb{R}^+$, 207
$\mathcal{P}(\mathcal{X})$ set of probability measures, 17
φ_λ left-eigenfunction, 115
ψ_λ right-eigenfunction, 115
Q jump rate matrix, 31
r density of X on $[0,\infty)$, 131
$r^{(b)}$ density of X on $[0,b]$ w.r.t. μ, 117
σ shift operator, 233
τ holding time, 39
T killing time, 18
T_a hitting time of a, 113
$\theta(\nu)$ exponential rate of survival of a QSD, 19
θ^* exponential rate of survival, 47
θ_x^* exponential rate of survival starting from x, 20
U potential, 234
$X = (X_t : t \geq 0)$ diffusion process, 113

P. Collet et al., *Quasi-Stationary Distributions*, Probability and Its Applications,
DOI 10.1007/978-3-642-33131-2,

$\mathcal{X}$ state space, 17
$\mathcal{X}^{\mathbf{a}}$ set of allowed states, 18
$\mathcal{X}^{\mathrm{tr}}$ trap in the phase space, 11
$\mathcal{X}^{M}$ set of M compatible sequences, 233
$\mathcal{X}^{M,+}$ set of M compatible unilateral sequences, 234
$\mathcal{X}_n$ set of points whose orbit stays in $\mathcal{X}_0$ up to time n, 11
$x_n = f^n(x_0)$ n-th iterate of x_0, 10
$Y = (Y_t : t \geq 0)$ Markov process, 17
ξ_1 supremum positive polynomials, 72
ζ transverse measure, 235

Citations Index

P. Collet et al., *Quasi-Stationary Distributions*, Probability and Its Applications,
DOI 10.1007/978-3-642-33131-2, © Springer-Verlag Berlin Heidelberg 2013

Printed by Printforce, the Netherlands